Engineering Design Synthesis

Springer
*London
Berlin
Heidelberg
New York
Barcelona
Hong Kong
Milan
Paris
Singapore
Tokyo*

http://www.springer.de/phys/

Amaresh Chakrabarti (Ed)

Engineering Design Synthesis

Understanding, Approaches and Tools

Springer

Amaresh Chakrabarti
Associate Professor
Centre for Product Design and Manufacturing (CPDM)
Indian Institute of Science
Bangalore 560012
Karnataka, India

British Library Cataloguing in Publication Data
Engineering design synthesis: understanding, approaches and tools
 1. Engineering design 2. Engineering design – Data processing 3. Engineering design – Computer programs
 I. Chakrabarti, Amaresh
 620'.0042
 ISBN 1852334924

Library of Congress Cataloging-in-Publication Data
Engineering design synthesis: understanding, approaches and tools / Amaresh Chakrabarti (ed).
   p. cm.
 Includes bibliographical references and index.
 ISBN 1-85233-492-4 (alk. paper)
 1. Engineering design. I. Chakrabarti, Amaresh.
 TA174.E545 2001
 620'.0042 – dc21   2001038410

Apart from any fair dealing for the purposes of research or private study, or criticism or review, as permitted under the Copyright, Designs and Patents Act 1988, this publication may only be reproduced, stored or transmitted, in any form or by any means, with the prior permission in writing of the publishers, or in the case of reprographic reproduction in accordance with the terms of licences issued by the Copyright Licensing Agency. Enquiries concerning reproduction outside those terms should be sent to the publishers.

ISBN 1-85233-492-4 Springer-Verlag London Berlin Heidelberg
a member of BertelsmannSpringer Science+Business Media GmbH
http://www.springer.co.uk

© Springer-Verlag London Limited 2002
Printed in Great Britain

The use of registered names, trademarks, etc. in this publication does not imply, even in the absence of a specific statement, that such names are exempt from the relevant laws and regulations and therefore free for general use.

The publisher makes no representation, express or implied, with regard to the accuracy of the information contained in this book and cannot accept any legal responsibility or liability for any errors or omissions that may be made.

Typesetting: Best-set Typesetter Ltd., Hong Kong
Printed and bound at The Cromwell Press, Trowbridge, Wiltshire, England
69/3830-543210 Printed on acid-free paper SPIN 10835033

Preface

This book is an attempt to bring together some of the most influential pieces of research that collectively underpin today's understanding of what constitutes and contributes to design synthesis, and the approaches and tools for supporting this important activity.

The book has three parts. Part 1 – Understanding – is intended to provide an overview of some of the major findings as to what constitutes design synthesis, and some of its major influencing factors. Part 2 – Approaches – provides descriptions of some of the major prescriptive approaches to design synthesis that together influenced many of the computational tools described in the final part. Part 3 – Tools – is a selection of the diverse range of computational approaches being developed to support synthesis in the major strands of synthesis research – composition, retrieval, adaptation and change.

In addition, the book contains an editorial introduction to the chapters and the broader context of research it represents, and a supplementary bibliography to help locate this broader expanse of work. With the wide variety of methods and tools covered, this book is intended primarily for graduate students and researchers in product design and development; but it will also be beneficial for educators and practitioners of engineering design, for whom it should act as a valuable sourcebook of ideas for teaching or enhancing design creativity.

The general idea of the need to bring together works of research in design synthesis, both manual and computational, had its seeds in the feeling that grew in me in the early 1990s while participating in design conferences. It seemed that conferences that were largely design(er)-centred had a great deal in common, in the goals pursued and means used, with those with a strong computational flavour; yet, there was little information exchange or synergy between the two. A synthesis of ideas developed in these two research communities seemed necessary. This culminated in an earlier attempt at bringing together functional representation and reasoning research in the form of an *Artificial Intelligence in Engineering Design, Analysis, and Manufacturing* special issue in 1996. Taking on a project of this breadth, however, required a closer feasibility study, and working out the modalities. A workshop in design synthesis in Cambridge in 1998, in which about 30 researchers from around the world participated, provided this, and I am grateful to Lucienne Blessing and Tetsuo Tomiyama who helped make that possible.

This book would not be possible without the many spirited discussions with Lucienne Blessing when the idea seemed far too ambitious and unclear to be pursued at all. When the idea eventually became expressible enough, Nicholas Pinfield, the then Engineering Editor of Springer London, gave the much-needed encouragement for the project to take off on a serious note. I am thankful to all the contributors for their enthusiastic response, without which there would not be a credible proposal with which to proceed. In particular, I am thankful to Susan Finger for her excitement at the idea of this book, and her suggestions for its improvement. In the more advanced stages, Oliver Jackson of Springer London has been extremely helpful with editorial support. The Cambridge Engineering Centre, my employer until recently, and John Clarkson, its director, have been generous with the facility; in particular, Andrew Flintham, the Computer Associate of the Centre, has been a great help in sorting out the computational problems faced. I am also grateful to Rob Bracewell for lending a patient ear whenever needed.

I would also like to thank the Centre for Product Design and Manufacturing at the Indian Institute of Science, my present employer, for its logistic support during the copy-editing and proof-reading stages of this book.

Finally, I would like to thank Ken Wallace and Thomas Bligh of Cambridge University for their effort in creating an ambience that fosters discussion, collaboration and integration, with creativity as an emergent, natural consequence. This book is as much a product of the effort of individuals as it is of this collective ambience.

Amaresh Chakrabarti
Bangalore
October 2001

Editor's introduction

Amaresh Chakrabarti

Engineering design, a central part of product development, is distinguished from other areas of human endeavour by its creative aspects, generally termed synthesis, whereby novel products are conceived. Engineering design synthesis is, therefore, a central area of design research. Traditionally, there have been consistent efforts in behavioural sciences to identify what constitutes creativity, and how it manifests itself in various aspects of human endeavour. Systematic research into design synthesis is relatively new. However, in the last few decades, especially with the increasing realisation of the potential of systematic design methods in enhancing design competence, and the advancement of computers as a potential design aid, the area has seen unprecedented growth. Descriptive studies and experiments have been undertaken, often in conjunction with psychologists and sociologists, to understand better the factors that influence this complex aspect of design. Many approaches have been, and are being, developed in order to enable, assist or even automate aspects of design synthesis; some of these approaches are theoretical, others are empirical, some are manual and others are computational.

This book brings together some of the most influential pieces of research undertaken around the world in design synthesis. It is the first, comprehensive attempt of this kind, and covers all three aspects of design synthesis research. Part 1 – Understanding – provides an overview of some of the major findings as to what constitutes synthesis and some of its major influencing factors. Part 2 – Approaches – provides a detailed description of some of the major prescriptive approaches to design synthesis, which together influenced many of the computational tools described in the final part. Part 3 – Tools – provides a selection of the diverse range of computational support techniques for synthesis in its major strands of research. It is to be noted that the parts have some overlap in content: the chapters in Understanding often propose approaches, and the chapters in Approaches and in Tools sometimes have well-developed theories that form part of the corpus of knowledge on which the current understanding of design synthesis is based. However, the part in which a chapter is placed signifies its main emphasis.

The chapters together provide an extensive coverage of the outcomes of design synthesis research in the last four decades: these include cutting-edge findings, as well as established, ready-to-use methods to help designers synthesise better ideas. The chapters are contributed by eminent researchers from four continents. Together, these chapters cover all major generic synthesis approaches, *i.e.*, composition, retrieval and change, and tackle problems faced in a wide variety of engineering domains and in many areas of application, including clocks, sensors and medical devices.

The rest of this article provides a summary of the chapters in the wider context of design synthesis research.

Part 1: Understanding

This part has five chapters. Together, these chapters provide insights into what constitutes and influences synthesis. Although all the chapters in this book are based, implicitly or explicitly, on some definition of synthesis, the first and last in this part are attempts to define and model the synthesis process. The other chapters in this part discuss the function and nature of knowledge necessary for synthesis.

In Chapter 1, Norbert Roozenburg provides an overview of the existing definitions of design synthesis, and their relationships to analysis. Synthesis, he argues, has taken two broad meanings in design research: as a distinct phase in designing, and as a part of the problem-solving process. Taking synthesis as the process of progressing from function to form, he analyses the logic of synthesis, and argues that certain kinds of synthesis cannot be attained by deduction alone and should require innovative abduction. This he terms innoduction, and defines as a reasoning process in which, given the intended function of a product, one must discover not only a form that can fulfil this function, but also the law that ascertains that the function can indeed be fulfilled by that form.

In Chapter 2, Michael French argues that insight into engineering science is the single most important influencing factor for good design synthesis. Drawing numerous examples from the history of designed artefacts, both industrial and household, and from both ancient and recent, he demonstrates that this engineering insight can often be encapsulated into a variety of "design principles". Research into, and use of, these principles should be very useful, he argues, but they are presently largely ignored and hardly researched.

In Chapter 3, Yoram Reich introduces the General Design Theory (GDT) of Hiroyuki Yoshikawa, which is one of the most mathematical of design theories. GDT is an axiomatic theory of design, which tries to establish the nature of knowledge necessary for engineering design in an idealistic sense, and the nature of designing given this knowledge. It also indicates how the nature of designing should change for existing engineering knowledge, which is far from ideal. Reich uses the domain of chairs as a simple example to explain GDT, and how designing is envisaged to proceed according to this theory.

In Chapter 4, Vladimir Hubka and Ernst Eder discuss their theory of technical systems and what it tells about synthesis. The theory of technical systems describes a technical system as one that fulfils a purpose using technical means, and proposes that it can be described at four levels of detail: process, function, organ and assembly. It prescribes synthesis as the process whereby these levels are achieved; in order to achieve these transformations, they suggest the use of various creative and systematic methods, such as brainstorming and morphological charts.

In Chapter 5, Tomiyama, Yoshioka and Tsumaya describe a model of the synthesis process developed by the "Modeling of Synthesis" project in Japan. Design is seen to be synonymous to synthesis; the relationship between the thought processes involved in synthesis and analysis are discussed, and synthesis is modelled in terms of knowledge and actions on knowledge. The theory is verified by developing a "reference model" from protocol data of designing sessions, and comparing the constructs of this model with that "predicted" by the model of synthesis developed.

Part 2: Approaches

This section has four chapters, each providing (the basis for) a prescriptive approach to design synthesis. It is interesting to note that most of these are based on theories of artefacts, although the nature and level of the approaches proposed vary considerably. Whereas the first three chapters provide outcome-based approaches of various degrees of detail, the fourth chapter provides a set of guidelines as to how areas of improvement can be found in a product, and how improvements can be effected. Together, the chapters provide guidelines as to how function and form can be developed.

Chapter 6 is by Claus Thorp Hansen and Mogens Myrup Andreasen and describes the domain theory of artefacts, which has been influenced by the theory of technical systems but which has evolved into one in which an artefact is described at three levels (called domains here): transformation, organ and part. Transformation between these is prescribed to take place using relationships that link functions to means, where each choice of a means leads to uncovering further functions and then to further means and so on, developing into a function–means tree. In this sense, synthesis of form for a given function could be seen, in a normative sense, as one of a bootstrapping process of developing means to fulfil a function and identifying functions required as a result.

In Chapter 7, Gerhard Pahl and Ken Wallace describe the function structures approach popularised by Pahl and Beitz. The function structures approach starts with the overall function necessary to be fulfilled by the intended product, and develops this into an assemblage of simpler subfunctions – a function structure. This is followed by a search for principles (means in Andreasen's terminology) that can fulfil each of these subfunctions, and combining them into concept variants. Using Krumhauer's generally valid functions as functional building blocks, they suggest the use of a morphological matrix to systematise the development of alternative concept variants, which, they argue, should lead to innovative designs.

Chapter 8, by Karlheinz Roth, describes another way of describing the various levels of an artefact description, but promulgates the use of design catalogues, with components made out of existing designs, for use by designers to achieve attainment of these levels. He argues that development and use of design catalogues, where each existing product or its components is described at multiple levels ranging from function through principles to form, should allow designers to reuse existing knowledge in an effective way. He uses several catalogues as examples to illustrate the variety and usefulness of design catalogues in designing.

In Chapter 9, Denis Cavallucci, an expert on the TRIZ approach developed by Genrich Altshuller of the former Soviet Union, introduces its basic components. Altshuller analysed a vast number of Russian patents to identify a set of "laws" that he believed were behind these patents. The laws are divided into three categories: static, cinematic and dynamic. Together, they help identify the areas in which an existing design can be improved and guidelines as to how this improvement can be pursued. Cavallucci also provides a comprehensive list of references on this approach, especially for the English-speaking reader.

Part 3: Tools

This part has ten chapters, which together exemplify all the major directions of research into computational (support to) synthesis of designs. Computational synthesis research has taken two major directions in the past: one is compositional synthesis, in which solutions are developed by combining a set of building blocks, and the other is retrieval of an existing design and its change for various purposes. The change effected may be to adapt the original design for the purpose at hand, or to modify it into other innovative designs.

The first two chapters are on automated compositional synthesis of concepts for fulfilment of a given function.

Chapter 10 is by Karl Ulrich and Warren Seering, and is one of earliest attempts at automated compositional synthesis of concepts. The area of application is sensors. The representational language is bond graphs, the algorithm is search, and the system developed is limited to synthesis of single-input single-output systems. Synthesis is performed at the topological level, and the resulting concepts are intended to be evaluated by the designer.

Chapter 11 is by Amaresh Chakrabarti, Patrick Langdon, Ying-Chieh Liu and Thomas Bligh, and is on the development of FuncSION – a multiple I/O concept synthesis software for mechanical transmissions and devices. The representation is based on systems theory and symbolic geometry, and the algorithm is search. Synthesis is performed at three levels: topology, spatial and generic physical. FuncSION has been tested using case studies, product compendia and patent catalogues. The designs synthesised are intended to be evaluated, modified and explored by the designer.

The next two chapters are examples of development of function into a function structure and support of compositional synthesis.

Chapter 12, by Rob Bracewell, is on the concepts underlying the Schemebuilder software for supporting design of mechatronic systems, involving mechanical, electrical and software elements. The representation is based on function–means trees and bond graphs. Using this system, a designer should be able interactively to develop the function and concept by a progressive proliferation of a function–means tree. The software has been tested using examples from several case studies.

Chapter 13 is by Ralf Lossack, and is for supporting the design of physical systems. The approach – DIICAD Entwurf – is a synthesis of systematic methodologies, and is based on the concept of a "working space" within which the design interacts with its inputs and outputs. Synthesis is done by designers selecting and concatenating means from a database. The software has been tested using several case studies and its use in student projects.

The next two chapters are examples of retrieval of existing designs.

Chapter 14 is by Tamotsu Murakami, and is on retrieval of existing mechanisms to fulfil a given, specified mechanical function. The representation used is based on qualitative configuration space, and the number of designs retrieved is one in each case. This has been tested using several cases, some of which are used as examples in the chapter. Retrieval is based on matching of the characteristics of intended function with that of the stored designs. The resulting designs are intended to be explored by the designer, but that is not currently supported within the framework.

Chapter 15 is by Lena Qian, and is on retrieval of mechanical, structural, hydraulic and software systems. The retrieval is done using analogy at three levels: function, behaviour and structure. The degree of similarity between the target and retrieved domains determines the choice of level used. Retrieved designs may be from a different discipline, and it is the task of the designer to transform the insight gained into an artefact appropriate for the domain in these cases.

The next two chapters are examples of changing retrieved designs for adapting to the current purpose.

Chapter 16 is by Sambasiva Bhatta and Ashok Goel, and is for adaptation using analogy. The current areas of application are electronic and mechanical controllers. The representation is based on logic and systems theory, and the adaptation mechanism is based on the use of design patterns with associated knowledge of what they can change into and how. The approach has been tested using several example cases.

Chapter 17 is by Boi Faltings on the FAMING system for adaptation of mechanisms. The software requires input from the designer for deciding the direction of modification and adapts the initial design using simple rules of replacement and envisionment. The representation is based on qualitative configuration space. This has been tested using several example cases, including those from architecture.

The final two chapters are on change from existing designs for generating innovative designs.

Chapter 18 is by Susan Finger and James Rinderle, and is on software that uses transformational grammar for changing a given intended behaviour or an existing design into new, behaviour-preserving designs. The current application is gear transmissions, and the representation used is bond graphs. This is one of the earliest papers that use grammars for generating designs, and is a precursor to much work on various generative grammars, not covered in this book.

Chapter 19 is by John Koza and is on software that uses genetic programming, which is based on the concept of genetic algorithms but uses programs that evolve in order to transform given designs to generate innovative designs with better performance in terms of the given criteria. The applications are electrical and electronic circuits and chemical reactions. The software has been tested using several case studies and patent catalogues.

Summary

Together, the chapters in this book provide a collection of views on the definition and nature of synthesis and some of its influencing factors, and a collection of approaches to synthesis. Below is a summary of these.

Definition of synthesis

There are five overlapping definitions of synthesis on which the chapters in this book are explicitly or implicitly based. These are:

- synthesis as designing;
- synthesis as problem solving;
- synthesis as design solution generation;
- synthesis as design problem and solution generation;
- synthesis as exploration.

According to the first definition (synthesis as designing), designing and synthesis are synonymous, as is propounded by Tomiyama and Yoshikawa. This appears to be used implicitly by Koza, who uses many cycles of generation and evaluation, operating at many levels of abstraction (topological, parametric, *etc.*) to develop a solution.

The second definition (synthesis as problem solving) means that one is operating at a particular level of abstraction, and uses a process involving both generation and decision (evaluation and selection) in order to develop a design solution at that level. In other words, synthesis is synonymous with problem solving. One example is the work of Finger and Rinderle, who use behaviour preservation as the evaluation process embedded in the algorithm to modify a given original graph representing an initial design to generate variants.

The third definition (synthesis as design solution generation) takes synthesis as a single part of the basic problem-solving process, which requires evaluation and selection in addition to this in order to complete the problem-solving cycle. In this sense, synthesis is synonymous with generation. Roozenburg mentions the ubiquity of this definition in design-process diagrams. This definition can be extended further to encompass generation of any design-related construction (synthesis as design problem and solution generation), if the problem-solving cycle is seen as cycling through at each level of design description through which a design develops: problem statement, requirements, functions, concept, embodiment, *etc.* It is the view taken in many engineering design methodologies not explicitly featured here, and is one way of describing the design process in practice [1].

The fifth view (synthesis as exploration) is different from the fourth in that it requires that synthesis be the process whereby clarity of the state of knowledge is increased. This is the definition implicitly used by the opportunistic strategy promulgated by Michael French, and many approaches described in this book try to support this process. Smithers [2], who takes this view of synthesis, gives a formal definition of exploration: it is the process by which a state of well-structured knowledge results from that of ill-structured knowledge.

Nature of synthesis and influencing factors

In designing, *designers* create an *artefact* by carrying out *activities* in an *environment* (settings, management, tools, *etc.*). Therefore, aspects of the human (designers, team), the artefact, design activities, and environment all affect design and its underlying synthesis process. Issues related largely to human and environmental aspects are not covered here. For human aspects, which include psychological studies of creativity, methods for enhancing idea generation, *etc.*, see among others Adams [3], Sternberg [4] and Frankenberger and Badke-Schaub [5]. For effects of environment, see Ottosson [6].

The chapters in the first two parts of this book cover some important aspects of the artefact, activities and underlying knowledge that make synthesis possible. Whereas Hubka and Eder, Hansen and Andreasen, and Roth highlight the necessity of artefactual knowledge and provide various views on the nature of this knowledge, Tomiyama *et al.* in particular present what they propose are the activities prevalent in design and syn-thesis. All chapters provide a viewpoint on the knowledge needed for synthesis. For instance, GDT (Reich) takes the view that this knowledge must lie in the relationships between entities and the functions that these entities are capable of performing. French claims that insight of engineering science is of essence, while TRIZ (Cavallucci) and other models provide various domain-neutral, procedural guidelines as to how these explorations may be carried out.

Between them, they propose three influences that are crucial for synthesis: (1) knowledge of artefact states; (2) knowledge of possible activities as progress from one state to another; and (3) knowledge of how these activities can be carried out.

Approaches to synthesis

Together, the chapters exemplify two major directions to synthesis: composition from scratch, and building on an existing design. Whereas compositional synthesis is often believed to enable generation of more innovative ideas, retrieval-based approaches are seen to be more efficient [7].

The essence of compositional synthesis is to bring the state of knowledge of the intended function of an artefact sufficiently close to that of the structural world such that a mapping between the two becomes possible. One way of doing this is to restructure the functional description such that each of its parts can be satisfied by composition of fragments of available artefacts. Another way of doing this is by decomposing the functional description using the functional descriptions of the existing artefacts themselves; this makes the generation process capable of being automated, with or without the intention of handing the resulting solutions to designers for exploration. The first two chapters, *i.e.*, Ulrich and Seering and Chakrabarti *et al.*, serve this purpose. The same can also be done by either decomposing the functional description sufficiently and then (composing and) replacing each with artefact fragments, thereby developing a composite artefact that fulfils the overall function. The chapters by Bracewell and by Lossack are intended to support this process.

Pure retrieval is seen as the most efficient way of developing a design, which requires no development at all. However, often the retrieved designs do not adequately fulfil the required functions, and need modification. The two chapters by Murakami and by Qian are primarily focused on retrieval, but both with the intention that the solutions retrieved should be modified, if necessary, by the designer to fulfil the requirements of the domain or the purpose. The issue of adaptation to fulfil the purpose is dealt with in the two chapters by Bhatta and Goel, and by Faltings. Once an initial design is retrieved (and adapted) for a given function, it can be used as a starting point for further modifications for generating other ideas either to produce variants or to optimise the design. Change is the theme of the last part, and is dealt with by the chapters by Finger and Rinderle, and by Koza.

The wider body of literature

Any anthology of this sort has to be indicative only of the body of literature at large, and cannot aspire to be exhaustive on any account. I mention some of the many interesting and useful studies, approaches and tools as pointers for readers who would like to delve into the wider body of literature beyond this book.

A number of researchers have developed theories of design and synthesis. Some notable ones are the knowledge level theory of designing by Tim Smithers [2], the situated model of design by Gero and Kannengiesser [8] and the reflection in action model of designing by Schön [9], and their implications on synthesis.

Many descriptive studies comment on the nature of synthesis in practice. See Fricke [10] and Ehrlenspiel *et al.* [11] for a case study where designers were observed, their attributes and design processes analysed and their solutions evaluated in order to measure success and success-promoting abilities. It was found, for instance, that balanced expansion of solution space and frequent evaluation of solutions are success-promoting factors. For an overview of descriptive studies with implications on synthesis, see Blessing [12].

A number of approaches to synthesis have been developed, for instance the functional reasoning approach developed by Freeman and Newell [13], and the prototyping approach developed by Gero and coworkers [14,15]. For a comprehensive review of synthesis techniques in various domains, see Flemming *et al.* [16].

Computational tools have been developed in a wide variety of domains and applications. For instance, several other researchers use compositional synthesis. Braha [17] uses adaptive search in his approach for finding optimal solutions in car configuration problems, Kota and Chiou [18] use search for mechanisms synthesis, Welch and Dixon [19] concatenate bond graph elements for synthesis of physical systems. Maher [20], Hundal [21], Umeda *et al.* [22], Malmqvist [23], and Alberts and Dikker [24] each developed an integrated framework for supporting synthesis of solutions, with goals broadly similar to Lossack and Bracewell.

Retrieval and repair has been a major theme of synthesis research, especially in case-based design [25]. For examples of (mainly) retrieval-based synthesis see Galletti and Giannotti [26] and McGarva [27], who use trial-and-error-based interactive selection of mechanisms from catalogues. For examples of retrieved designs see Sycara and Navinchandra [28], Madhusudan *et al.* [29], Joskowicz and Addanki [30], and Murthy and Addanki [31]. For mainly associative systems for innovative designs see the reviews by Navinchandra [32,33].

Changing existing designs for generation of new designs has been a continuing theme of synthesis research. Taura and Yoshikawa [34] use a metric space approach with adaptive search for this purpose. Grammar-based approaches use rules from a formal grammar to change designs. For examples of this see Shea and Cagan [35], Schmidt and Cagan [36], Heisserman [37] and Woodbury *et al.* [38], among others.

Most of the above references focus on the synthesis of solutions. However, the quality of the solution developed depends as much on the quality of solution synthesis as it does on the quality of problem finding. A number of interesting researches exist in development of support for identifying and representing requirements and functions, *e.g.*, see Wood and Antonsson [39] and O'Shaughnessy and Sturges [40].

For a more comprehensive coverage of articles related to design synthesis, the reader may find the following, by no means comprehensive, list of journals and conference proceedings useful: Research in Engineering Design (Springer); Design Studies (Elsevier); Journal of Engineering Design (Computational Mechanics); Proceedings of the International Conferences in Engineering Design (WDK); Proceedings of AI in Design Conferences (Kluwer); and Proceedings of the ASME Design Theory and Methodology Conferences (ASME).

References

[1] Blessing LTM. A process based approach to computer supported engineering design. Ph.D. thesis, University of Twente, Enschede, The Netherlands, 1994 [published in Cambridge by Blessing].
[2] Smithers T. Synthesis in designing as exploration. In: Proceedings of the 2000 Tokyo International Symposium on the Modeling of Synthesis, University of Tokyo, Japan, 11–13 December, 2000; 89–100.
[3] Adams JL. Conceptual blockbusting. 3rd ed. Reading (MA): Addison-Wesley, 1992.
[4] Sternberg RJ, editor. Handbook of creativity. New York: Cambridge University Press, 1999.

[5] Frankenberger E, Badke-Schaub P. Integration of group, individual and external influences in the design process. In: Frankenberger E, Badke-Schaub P, Birkhofer H, editors. Designers – the key to successful product development. London: Springer, 1998; 149–64.
[6] Ottosson S. Planetary organisations offer advantages in project work. In: Frankenberger E, Badke-Schaub P, Birkhofer H, editors. Designers – the key to successful product development. London: Springer, 1998; 196–201.
[7] Chakrabarti A. Towards hybrid methods for synthesis. In: International Conference on Engineering Design (ICED01), Design Research – Theories, Methodologies, and Product Modelling, Glasgow, 2001; 379–86.
[8] Gero JS, Kannengiesser U. Towards a situated function–behaviour–structure framework as the basis for a theory of designing. In: Smithers T, editor. Workshop on Development and Application of Design Theories in AI in Design Research, AI in Design'00 (AID00) Conference, Worcester, MA, July, 2000.
[9] Schön D. The reflective practitioner: how professionals think in action. New York: Basic Books, 1983.
[10] Fricke G. Successful individual approaches in engineering design. Res Eng Des 1996;8:151–65.
[11] Ehrlenspiel K, Dylla N, Guenther J. Experimental investigation of individual processes in engineering design. In: ICED '93, The Hague, 1993.
[12] Blessing L. Descriptive studies and design synthesis. In: Chakrabarti A, Blessing L, editors. Proceedings of the 1st Cambridge Workshop on Design Synthesis, Churchill College, Cambridge, 1999.
[13] Freeman P, Newell A. A model for functional reasoning in design. In: Proceedings of 2nd International Joint Conference in Artificial Intelligence, London, 1971; 621–40.
[14] Gero JS. Design prototypes: a knowledge representation schema for design. AI Mag 1990;(Winter): 26–36.
[15] Gero JS, Maher ML, Zhang W. Chunking structural design knowledge as prototypes. In: EDRC-12-25-88, Carnegie Mellon University, 1988.
[16] Flemming U, Adams J, Carlson C, Coyne R, Fenves S, Finger S, et al. Computational models for form–function synthesis in engineering design. In: EDRC 48-25-92, CMU, 1992.
[17] Braha D. Satisfying moments in synthesis. In: Chakrabarti A, Blessing L, editors. Proceedings of the 1st Cambridge Design Synthesis Workshop, Churchill College, Cambridge, UK, 7–8 October, 1999.
[18] Kota S, Chiou S-J. Conceptual design of mechanisms based on computational synthesis and simulation of kinematic building blocks. Res Eng Des 1992;4:75–87.
[19] Welch RV, Dixon JR. Representing function, behavior and structure in conceptual design. In: Proceedings of the ASME Design Theory and Methodology Conference, DE-vol. 42, 1992.
[20] Maher ML. Synthesis and evaluation of preliminary designs. In: Proceedings of the International Conference on the Application of AI in Engineering, Cambridge, UK, July, 1989.
[21] Hundal MS. Use of functional variants in product development. In: ASME Design Theory and Methodology Conference, DE-vol. 31, 1991; 159–64.
[22] Umeda Y, Ishii M, Yoshioka M, Tomiyama T. Supporting conceptual design based on the function–behavior–state modeler. Artif Intell Eng Des Anal Manuf 1996;10(4):275–88.
[23] Malmqvist J. Computational synthesis and simulation of dynamic systems. In: ICED '95, Praha, 1995.
[24] Alberts LK, Dikker F. Integrating standards and synthesis knowledge using the YMIR ontology. In: Gero JS, Sudweeks F, editors. AI in Design '94. Kluwer, 1994; 517–34.
[25] Kolodner JL. Case-based reasoning. San Mateo (CA): Morgan Kaufmann Publishers, 1993.
[26] Galletti CU, Giannotti EI. Interactive computer system for the functional design of mechanisms. Comput Aided Des 1981;12(3):159–63.
[27] McGarva JR. Rapid search and selection of path generating mechanisms from a library. Mech Mach Theory 1994;29(2):223–35.
[28] Sycara K, Navinchandra D. Retrieval strategies in a case-based design system. In: Tong C, Sriram D, editors. Artificial intelligence in engineering design, vol. II. Academic Press, 1992.
[29] Madhusudan TN, Sycara K, Navinchandra D. A case based reasoning approach for synthesis of electro-mechanical devices using bond graphs. EDRC report, Carnegie-Mellon University, USA, 1995.
[30] Joskowicz L, Addanki S. From kinematics to shape: an approach to innovative design. In: Proceedings of AAAI-88, St Paul MN, 1988; 347–52.
[31] Murthy SS, Addanki S. PROMPT: an innovative design tool. In: Proceedings of the National Conference of the American Association for Artificial Intelligence, 1987; 637–42.

[32] Navinchandra D. Innovative design systems, where are we and where do go from here? Part I: design by association. Knowl Eng Rev 1992;7(3):183–213.
[33] Navinchandra D. Innovative design systems, where are we and where do we go from here? Part II: design by exploration. Knowl Eng Rev 1992;7(4):345–62
[34] Taura T, Yoshikawa H. A metric space for intelligent CAD. In: Brown DC, Waldron M, Yoshikawa H, editors. IFIP intelligent computer aided design. Elsevier/North-Holland, 1992; 133–61.
[35] Shea K, Cagan J. Generating structural essays from languages of discrete structures. In: Proceedings of Artificial Intelligence in Design AID 98, Lisbon, July, 1998; 365–84.
[36] Schmidt LC, Cagan J. GGREADA: a graph grammar based machine design algorithm. Res Eng Des 1997;9:195–213.
[37] Heisserman J. Generative geometric design. IEEE Comput Graph Appl 1994;14(2):37–45.
[38] Woodbury R, Datta S, Burrow A. Towards an ontological framework for knowledge-based design systems. In: Proceedings of the 6th AI in Design Conference (AID00), 24–26 June, Worcester, MA, 2000.
[39] Wood KL, Antonsson EK. A first class of computational tools for preliminary engineering design. Technical report, California Institute of Technology, 1992.
[40] O'Shaughnessy K, Sturges RH Jr. A systematic approach to conceptual engineering design. In: ASME Design Theory and Methodology Conference, DE-vol. 42, 1992; 283–90.

Contents

Preface v
Editor's introduction vii
Contents xvii
Contributors xxvii

Part 1: Understanding

1 Defining synthesis: on the senses and the logic of design synthesis
Norbert F.M. Roozenburg 3
  1.1 Senses of synthesis 3
      1.1.1 General meanings of synthesis and analysis 3
      1.1.2 Synthesis and analysis as phases of the design process 4
      1.1.3 Synthesis and analysis as functions of problem solving 6
      1.1.4 Synthesis as assemblage of subsystems 7
      1.1.5 Synthesis as integration of ideas 8
  1.2 The logic of synthesis 9
      1.2.1 Form and function 9
      1.2.2 Reasoning from function to form 10
      1.2.3 The pattern of reasoning of synthesis 12
      1.2.4 Conclusions 16

2 Insight, design principles and systematic invention
Michael J. French 19
  2.1 Introduction 19
  2.2 The opportunistic designer 19
      2.2.1 The opportunistic approach 20
  2.3 Parallels with mathematics 20
      2.3.1 Poincaré's sieve 22
      2.3.2 Visual thought 22
  2.4 Insight 23
  2.5 Developing insight 24
      2.5.1 Sufficient insight 25
  2.6 Design principles 25
      2.6.1 Kinematic design (least constraint) 25
      2.6.2 The small, fast principle 26
      2.6.3 Matching 27
      2.6.4 "Prefer pivots to slides and flexures to either" 27
      2.6.5 "Where possible, transfer complexity to the software" 28

2.7	Systematic synthesis	28
	2.7.1 Clothes-peg example	28
2.8	Insight and systematic invention in power from sea waves	29
	2.8.1 Background	29
	2.8.2 An abstract view	29
	2.8.3 Table of options	30
	2.8.4 Embodiment	31
	2.8.5 The checking of systematic design processes-link-breaking	32
2.9	Summary	32
	Appendix 2A	33

3 Synthesis and theory of knowledge: general design theory as a theory of knowledge, and its implication to design
Yoram Reich 35

3.1	Introduction	35
3.2	The domain of chairs	36
3.3	GDT	38
	3.3.1 Preliminary definitions	38
	3.3.2 GDT's axioms	39
	3.3.3 Ideal knowledge	40
	3.3.3.1 Summary of ideal knowledge	42
	3.3.4 Real knowledge	42
	3.3.4.1 Summary of real knowledge	44
3.4	Contribution of GDT	45
	3.4.1 Representation of design knowledge	45
	3.4.2 Design process	46
3.5	Summary	47

4 Theory of technical systems and engineering design synthesis
Vladimir Hubka and W. Ernst Eder 49

4.1	Introduction	49
4.2	Design science and the theory of TSs	50
4.3	Designing – general	56
	4.3.1 Starting designing – clarifying the problem – design specification	61
	4.3.2 Designing – design procedure – novel products	62
	4.3.3 Designing – design procedure – redesigned products	65

5 A knowledge operation model of synthesis
Tetsuo Tomiyama, Masaharu Yoshioka and Akira Tsumaya 67

5.1	Introduction	67
5.2	Related work	68
5.3	Design process modelling	69
5.4	A formal model of synthesis	69
	5.4.1 Mathematical preparation	71
	5.4.2 Analysis versus synthesis	73
	5.4.3 Multiple model-based reasoning	74
	5.4.4 Function modelling	76
	5.4.5 A reasoning framework of design	77

		5.4.5.1	Knowledge operations in design	77
		5.4.5.2	A hypothetical reasoning framework of design	78
		5.4.5.3	Modelling operations in the object-dependent models	78
		5.4.5.4	Logical reasoning operations in the object independent level workspace	79
		5.4.5.5	Formalising knowledge operations in design	79
5.5	The implementation strategy of the framework			79
	5.5.1	Multiple model-based reasoning system		80
	5.5.2	Thought-process model		80
	5.5.3	Model-based abduction		80
		5.5.3.1	Strategy for modelling of abduction operation	80
		5.5.3.2	Model-based abduction	82
		5.5.3.3	Algorithm of model-based abduction	83
5.6	Verification of the model of synthesis			84
	5.6.1	Selection of data for verification		84
	5.6.2	The reference model		85
	5.6.3	Verification of the knowledge operation model		85
	5.6.4	Vocabulary about design		86
	5.6.5	Verification through implementation of the reasoning framework		87
5.7	Conclusions			88

Part 2: Approaches

6 Two approaches to synthesis based on the domain theory
Claus Thorp Hansen and Mogens Myrup Andreasen 93

6.1	Introduction			93
6.2	The domain theory			94
	6.2.1	Systems theory		94
	6.2.2	The domain theory		95
		6.2.2.1	The transformation domain	95
		6.2.2.2	The organ domain	96
		6.2.2.3	The part domain	97
		6.2.2.4	Visualising the domain theory	98
6.3	The function–means law			99
	6.3.1	The function–means tree (F/M-tree)		99
6.4	Engineering design synthesis			99
	6.4.1	The design object and its synthesis		100
	6.4.2	Process-oriented synthesis		101
		6.4.2.1	Problem analysis	105
	6.4.3	Artefact-oriented synthesis approach		105
		6.4.3.1	Utilising the F/M-tree	105
		6.4.3.2	Developing a product model	106
6.5	Implications and conclusion			107

7 Using the concept of functions to help synthesise solutions
Gerhard Pahl and Ken Wallace 109

7.1	Introduction	109
7.2	Functional interrelationship	110

7.3	Handling the concept of functions in practice	114
7.4	Inappropriate use of the concept of functions	118
7.5	Summary of the approach and its advantages	118

8 Design catalogues and their usage
Karlheinz Roth 121

8.1	Purpose of design catalogues		121
8.2	Types and structure of design catalogues		121
	8.2.1	Object catalogues	121
	8.2.2	Solution catalogues	123
	8.2.3	Operation catalogues	125
8.3	Requirements placed on design catalogues		127
8.4	Desirable forms of design catalogue		127
8.5	Use of design catalogues		128

9 TRIZ, the Altshullerian approach to solving innovation problems
Denis Cavallucci 131

9.1	The genesis of a theory			131
	9.1.1	Introduction		131
	9.1.2	Altshuller: evaluation of a life dedicated to others		132
	9.1.3	Opening to the West gives TRIZ the opportunity to develop		133
9.2	An approach to classifying Altshuller's work			133
	9.2.1	Introduction		133
	9.2.2	Basic notions		134
	9.2.3	Spotting Altshuller's original idea: the laws (or regularities) of developing technical systems		134
		9.2.3.1	The "static" laws	134
		9.2.3.2	The "cinematic" laws	136
		9.2.3.3	The "dynamic" laws	136
		9.2.3.4	Summary of the laws	137
	9.2.4	Case study: improving the performance of an intake manifold [10]		137
		9.2.4.1	Description of the problem	137
		9.2.4.2	Positioning the manifold in relation to the laws of evolution	137
		9.2.4.3	Interpreting the positioning	139
		9.2.4.4	Findings of the study	139
	9.2.5	Tools for breaking down the blockages of psychological inertia		140
	9.2.6	Problem-solving tools		141
	9.2.7	ARIZ, the algorithm for applying TRIZ		141
9.3	TRIZ's contribution to integration in the design process			142
	9.3.1	Using TRIZ in an approach consisting of applying a series of tools		142
	9.3.2	TRIZ as a "meta-method"		143
	9.3.3	TRIZ as a component part of an existing method		143
	9.3.4	The intuitive design model approach to methodological integration		143

9.4 Potential development of the theory in research 144
    9.4.1 Contributions to integrating TRIZ in one or more existing methods 144
    9.4.2 Contributions to the development of TRIZ itself 145
    9.4.3 Contributions to other fields of activity 145
9.5 Orchestrating the work in Altshuller's wake 145
9.6 Conclusions 146
    9.6.1 Industrial integration strategies 146
    9.6.2 Creativity and innovation: the missing (or forgotten) link in the design process 147
    9.6.3 An asset for product design 147

Part 3: Tools

10 Synthesis of schematic descriptions in mechanical design
Karl T. Ulrich and Warren P. Seering 153
10.1 Introduction 153
    10.1.1 What is schematic synthesis? 154
    10.1.2 Schematic synthesis of SISO systems 155
    10.1.3 Importance of schematic synthesis 156
        10.1.3.1 Reducing complexity 157
        10.1.3.2 Decoupling functional and physical issues 157
10.2 Domain description 157
    10.2.1 SISO dynamic systems 157
    10.2.2 Representing schematic descriptions 158
    10.2.3 Classifying the behaviour of a schematic description 158
    10.2.4 Specifying a problem 160
        10.2.4.1 Example specification 161
10.3 Solution technique 162
    10.3.1 Generating candidate descriptions 162
        10.3.1.1 Concept of a power spine 162
        10.3.1.2 Connecting input to output 163
    10.3.2 Classifying behaviour 163
    10.3.3 Modifying candidate schematic descriptions 163
        10.3.3.1 Transform the candidate design to a compact description 163
        10.3.3.2 Based on domain knowledge, generate modifications 165
        10.3.3.3 Reverse compacting transformation 166
10.4 A complete example 167
10.5 Discussion 170
    10.5.1 Importance and utility of technique 171
    10.5.2 Completeness of the technique 171
    10.5.3 Extensibility 172
        10.5.3.1 Extension within dynamic systems domain 172
        10.5.3.2 Extension to other domains 173
    10.5.4 Computer implementation 173
10.6 Related work 174

Appendix 10A 175
    10A.1 Determining the type number from the system equations .. 175
    10A.2 An explanation of isolated groups using bond graphs 176

11 An approach to compositional synthesis of mechanical design concepts using computers
Amaresh Chakrabarti, Patrick Langdon, Yieng-Chieh Liu and Thomas P. Bligh 179
    11.1 Objective 179
    11.2 Research approach 180
    11.3 Synthesis approach: representation, reasoning and example 182
        11.3.1 Develop theory from known design problems and solutions 182
        11.3.2 Generate solutions to known problems and compare with existing designs 186
    11.4 Evaluation of the synthesis approach 188
        11.4.1 MAS project case studies 188
        11.4.2 Hands-on experiments by experienced designers 188
    11.5 Further developments 190
        11.5.1 Resolving the first problem: managing the number of solutions generated 191
            11.5.1.1 Improving efficiency of the synthesis procedure ... 191
            11.5.1.2 Using additional constraints 192
            11.5.1.3 Grouping solutions using similarity 192
        11.5.2 Resolving the second problem: strategies for aiding visualisation 193
            11.5.2.1 Embodiment at the generic physical level 193
            11.5.2.2 Three-dimensional representation of the solution space 194
    11.6 Conclusions and further work 194

12 Synthesis based on function–means trees: Schemebuilder
Rob Bracewell 199
    12.1 Background 199
    12.2 Key concepts of Schemebuilder 200
        12.2.1 Hierarchical schematic diagrams 200
        12.2.2 Scheme generation by combination of alternative subsolutions 200
        12.2.3 Function–means trees 201
        12.2.4 Artificial intelligence (AI) support for design context decomposition and recombination 201
        12.2.5 Computer support for simulation and evaluation of schemes 204
        12.2.6 Bond-graph-based functional synthesis 205
    12.3 Design synthesis example: telechiric hand 205
    12.4 Implementation of function–means-based synthesis 209

13 Design processes and context for the support of design synthesis
Ralf-Stefan Lossack 213
    13.1 Introduction and overview of the design process 213

13.2	Solution patterns	217
	13.2.1 Artefact and process knowledge	217
13.3	Design working space	220
13.4	The DIICAD Entwurf design system	224
13.5	Conclusion	224
13.6	Future work	225

14 Retrieval using configuration spaces
Tamotsu Murakami 229

14.1	Introduction	229
14.2	Mechanism library	229
	14.2.1 Mechanism and configuration space	230
	14.2.2 Kinematic behaviour and configuration space	231
	14.2.3 Additional behavioural information description	232
14.3	Required behaviour as retrieval key	233
	14.3.1 Required behaviour description	233
	14.3.1.1 Timing charts of input/output motions	233
	14.3.1.2 Types of input/output motion	234
	14.3.1.3 Motion speed dependence	234
	14.3.2 Required locus pattern generation	234
14.4	Locus pattern and configuration space matching	234
	14.4.1 Locus along region boundary	235
	14.4.1.1 Motion by object contact	235
	14.4.1.2 Compliance	236
	14.4.2 Locus along range limit	236
	14.4.3 Locus through free region	237
	14.4.4 Generation of entire locus from segments	237
	14.4.5 Check additional conditions on motion	238
14.5	Implementation and execution examples	238
	14.5.1 Mechanism library	238
	14.5.2 Specifying required behaviour	239
	14.5.3 Example 1: mechanism for shutter release	239
	14.5.4 Example 2: mechanism in sewing machine	241
14.6	Conclusions and discussions	242

15 Creative design by analogy
Lena Qian 245

15.1	Introduction	245
15.2	Knowledge representation for design retrieval based on analogy	246
	15.2.1 Structure	246
	15.2.1.1 Primitive element and structural element	248
	15.2.1.2 Attribute	248
	15.2.1.3 Relationship	248
	15.2.1.4 Operation and process	249
	15.2.1.5 Static and dynamic structure	249
	15.2.2 Behaviour	250
	15.2.3 Function	253
	15.2.4 Qualitative causal knowledge	256
	15.2.5 Design prototype	257
15.3	An ABD model	258

	15.3.1 Design retrieval process	259
	15.3.2 Analogy elaboration process	260
	15.3.3 Mapping and transference	260
	15.3.4 Analogy evaluation process	263
15.4	Design support system using analogy	263
15.5	An example of designing a new door by behaviour analogy	264
15.6	Conclusion	267

16 Design patterns and creative design
Sambasiva R. Bhatta and Ashok K. Goel 271
16.1 Background, motivations and goals 271
16.2 MBA 272
     16.2.1 SBF models of devices 273
     16.2.2 Design patterns 275
16.3 Acquisition of GTMs 276
16.4 Analogical transfer based on GTMs 277
16.5 Evaluation 282
16.6 Related research 282
16.7 Conclusions 283

17 FAMING: supporting innovative design using adaptation – a description of the approach, implementation, illustrative example and evaluation
Boi Faltings 285
17.1 Introduction 285
     17.1.1 Model-based design 286
     17.1.2 Prototype-based design 287
     17.1.3 Case-based design 287
     17.1.4 Annotating cases with functional models 287
     17.1.5 Case adaptation using SBF models 288
     17.1.6 Innovation in case-based design 288
     17.1.7 FAMING: an interactive design tool 289
17.2 Qualitative SBF models used in FAMING 289
     17.2.1 Structure: metric diagram 290
     17.2.2 Qualitative behaviour 290
          17.2.2.1 Qualitative motions 290
          17.2.2.2 External influences 291
          17.2.2.3 Place vocabulary 291
          17.2.2.4 Behaviour = envisionments of kinematic states ... 292
     17.2.3 A language for specifying function 292
          17.2.3.1 Quantitative constraints on behaviour ... 294
17.3 Inverting the FBS model 294
     17.3.1 Matching behaviour to functional specification ... 294
     17.3.2 S–B inversion 295
17.4 Case adaptation 295
     17.4.1 Case combination 296
     17.4.2 Modification operators 297
     17.4.3 Discovering and satisfying compositional constraints ... 298
17.5 Conclusions 299

18 Transforming behavioural and physical representations of mechanical designs
Susan Finger and James R. Rinderle 303
18.1 Introduction 303
18.2 Related work 304
    18.2.1 A brief introduction to bond graphs 305
    18.2.2 Representation of function and behaviour 305
    18.2.3 Grammars for representation of geometry 306
    18.2.4 Configuration design 307
18.3 Representation of behaviour of specifications and components 307
    18.3.1 Representation of design specifications 307
    18.3.2 Representation of behavioural requirements of mechanical systems 308
    18.3.3 Representation of behavioural characteristics of components 309
    18.3.4 Representation of designs 309
18.4 Transformation of specifications into physical descriptions 310
    18.4.1 Behaviour-preserving transformations 310
    18.4.2 Component-directed transformations 311
18.5 The shaft matrix 312
18.6 Conclusions 316

19 Automatic synthesis of both the topology and numerical parameters for complex structures using genetic programming
John R. Koza 319
19.1 Introduction 319
19.2 Genetic programming 320
19.3 Automatic synthesis of analog electrical circuits 322
    19.3.1 Lowpass filter circuit 324
        19.3.1.1 Preparatory steps for lowpass filter circuit 324
        19.3.1.2 Results for lowpass filter circuit 326
    19.3.2 Squaring computational circuit 329
        19.3.2.1 Preparatory steps for squaring computational circuit 329
        19.3.2.2 Results for squaring computational circuit 329
19.4 Automatic synthesis of controllers 331
19.5 Other examples 333
19.6 Conclusions 335

Index 339

Contributors

Mogens Myrup Andreasen
Section of Engineering Design and Product Development
Department of Mechanical Engineering
Technical University of Denmark
Lyngby
Denmark

Sambasiva R. Bhatta
Verizon Communications
White Plains
USA

Thomas P. Bligh
Engineering Design Centre
Department of Engineering
Cambridge University
UK

Rob Bracewell
Engineering Design Centre
Department of Engineering
Cambridge University
UK

Denis Cavallucci
Ecole Nationale Supérieure des Arts et Industries de Strasbourg
France

Amaresh Chakrabarti
Centre for Product Design and Manufacturing
Indian Institute of Science
India

W. Ernst Eder (Professor Emeritus)
Department of Mechanical Engineering
Royal Military College of Canada
Canada

Boi Faltings
Artificial Intelligence Laboratory (LIA)
Swiss Federal Institute of Technology (EPFL)
Lausanne
Switzerland

Susan Finger
Department of Civil and Environmental Engineering
Carnegie Mellon University
Pittsburgh
USA

Michael J. French (Professor Emeritus)
Engineering Design Centre
Lancaster University
UK

Ashok K. Goel
College of Computing
Georgia Institute of Technology
Atlanta
USA

Claus Thorp Hansen
Section of Engineering Design and Product Development
Department of Mechanical Engineering
Technical University of Denmark
Lyngby
Denmark

Vladimir Hubka (Professor Emeritus)
Swiss Federal Institute of Technology
Zurich
Switzerland

John R. Koza (Consulting Professor)
Department of Electrical Engineering
School of Engineering
Stanford University
USA

Patrick Langdon
Engineering Design Centre
Department of Engineering
Cambridge University
UK

Yieng-Chieh Liu
Engineering Design Centre
Department of Engineering
Cambridge University
UK

Ralf-Stefan Lossack
Institute of Applied Computer Science in
  Mechanical Engineering
University of Karlsruhe
Germany

Tamotsu Murakami
Department of Engineering Synthesis
The University of Tokyo
Japan

Gerhard Pahl (Professor Emeritus)
Fachbereich 16 Maschinenbau
Technische Hochschule Darmstadt
Germany

Lena Qian
Canon Information Systems Research
  Australia
North Ryde
Australia

Yoram Reich
Department of Solid Mechanics, Materials
  and Systems
Faculty of Engineering
Tel Aviv University
Israel

James R. Rinderle
Department of Mechanical and Industrial
  Engineering
University of Massachusetts
Amherst
USA

Norbert F.M. Roozenburg
Faculty of Design, Engineering and
  Production
Technical University of Delft
The Netherlands

Karlheinz Roth (Professor Emeritus)
Institute for Engineering Design, Machine
  and High-Precision Elements
Braunschweig Technical University
Germany

Warren P. Seering
Department of Mechanical Engineering
Massachusetts Institute of Technology
USA

Tetsuo Tomiyama
Research into Artifacts, Center for
  Engineering (RACE)
The University of Tokyo
Japan

Akira Tsumaya
Collaborative Research Center for Advanced
  Science and Technology
Osaka University
Japan

Karl T. Ulrich
Operations and Information Management
  Department
Wharton School
University of Pennsylvania
USA

Ken Wallace
Engineering Design Centre
Department of Engineering
Cambridge University
UK

Masaharu Yoshioka
Rsearch Center for Information Resources
National Institute of Informatics
Tokyo
Japan

Part 1

Understanding

1. Defining synthesis: on the senses and the logic of design synthesis
2. Insight, design principles and systematic invention
3. Synthesis and theory of knowledge: General Design Theory as a theory of knowledge, and its implication to design
4. Theory of Technical Systems and engineering design synthesis
5. A knowledge operation model of synthesis

Defining synthesis: on the senses and the logic of design synthesis

Norbert F.M. Roozenburg

Abstract This chapter comprises two parts. In the first part I shall discuss different meanings of the words "synthesis" and "analysis" and look at some different views on synthesis and analysis in design. Synthesis, in its general sense, is the combining or mixing of ideas or things into new ideas and things. In design, functional and physical representations of subsystems, as well as the viewpoints and interests of stakeholders are synthesised. In design theory and design methodology "synthesis" is looked at as a phase of the design process, as well as a function of problem solving. According to the first view, exhaustive problem analysis must precede solution synthesis. According to the second view, synthesis is part of a general problem-solving cycle that occurs in all phases of the design process. The second part focuses on the pattern of reasoning of synthesis. "Synthesis" can be looked at as reasoning from statements about the functions or behaviour of a new product towards statements about its form or structure. Many authors take abduction as the characteristic pattern of reasoning for this crucial step in design. I shall demonstrate that two fundamentally different forms of abduction can be distinguished: explanatory abduction and innovative abduction. What is usually understood by "abduction" is *explanatory* abduction, but synthesis in the sense of reasoning from function to form follows the pattern of *innovative* abduction, or "innoduction". This means that the form does not indisputably follow from the functions to be fulfilled and that, in principle, there are always many good solutions. The reasoning from function to form is a creative process, which can be encouraged methodically, but cannot be logically guaranteed.

1.1 Senses of synthesis

1.1.1 General meanings of synthesis and analysis

In the early 1960s new systematic design methods began to emerge in different professional fields of design. Ever since, the concept of synthesis has figured prominently in the literature on design theory and design methodology. Synthesis, in the loose sense of generating tentative solutions to design problems, is commonly considered the hallmark of design. It is seen as the creative phase, the idea getting, and the climactical part of problem solving: "for without ideas there is nothing to analyse or to choose between" [1]. So we often speak of design as concerned with "synthesis", whereas science is concerned with "analysis". No doubt this is true to some extent. Yet, "if properly understood, analysis and synthesis, are not two different methods but only two necessary parts of the same method. Each is the relative and correlative of the other" [2]. In design, as well as in science, both synthesis and analysis have a role to play. Therefore, I shall dwell in this chapter on the concept of analysis too.

Etymologically, the word "synthesis" comes from the Greek *sunthesis* (collection) and *suntithenai* (to put together). So, generally speaking, "synthesis" means the combining, assembling, mixing or compounding of anything – whether ideas or beliefs, substances or parts, activities or processes – into a new whole. Simple concepts are said to be synthesised into complex ones, species into genera, individual propositions into systems, functions into function structures, components (electrical, mechanical, hydraulic, pneumatic, *etc.*) into technical systems, preferences into value functions, *etc.*

Synthesis is usually treated as the opposite of analysis.

"Analysis" derives from the Greek *analusis* and literally means to unloose, to dissolve or resolve into elements. Thus the word "analysis" refers to the resolution of an object into its constituent or original elements: a separate examination of the component parts of a subject or a thing, for example the words that compose a sentence, the tones of a tune, the simple statements that make up an argument or the chemical constituents of a compound substance. More generally, "analysis" means the examination of something in detail in order to understand it better or draw conclusions from it.

Whereas the word "synthesis" is also used for the complex, unified whole resulting from synthetic activity, "analysis" often refers to the result of the analysis process: a statement giving details of all constituent elements of something and how they relate to one another.

In the philosophy of science the terms "synthesis" and "analysis" are used in other senses too. In this discipline "analysis" also means resolving knowledge into its original principles, or resolving a problem into its first elements. To analyse in this sense is to generalise, and it has been identified with the process of induction in logic. Similarly, "synthesis" may refer to the process of reasoning from the general to the particular, from first principles to a conclusion, from a cause to an effect, *i.e.*, deductive reasoning. For example, in Newton's Method of Analysis and Synthesis, "analysis" is the process of inductive reasoning from observations and experiments to general principles that serve as explanations, whereas "synthesis" is the deduction of consequences from these principles [3]. This use of the words "analysis" and "synthesis" is rather confusing. With regard to design, the word "synthesis" usually refers to the creative act or process of generating ideas for design proposals, which is far from a deductive process. Moreover "analysis" can refer to problem analysis, as well as to the analysis of the behaviour of a product or a product design, and the latter largely rests on deductive reasoning.

1.1.2 Synthesis and analysis as phases of the design process

In writings on design theory and methodology the processes of "synthesis" and "analysis" are understood in two different ways. In one view synthesis and analysis are phases of the design process; in another view synthesis and analysis are functions that should be fulfilled by any problem-solving process.

Influential early examples of the former view are Archer's *Systematic method for designers* [4], and Jones' *Method of systematic design* [5]. They described synthesis and analysis as phases or stages of a largely linear process. Here, "synthesis" is the stage of finding possible solutions for subproblems and building up complete designs from these, whereas "analysis" includes the identification of subproblems and the preparation of performance specifications. In Jones' model the analysis and synthesis phases are followed by "evaluation". Archer favoured six stages: "synthesis" and

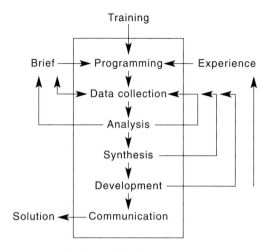

Fig. 1.1 Archer's model of the product design process.

"analysis" formed the creative heart of his model, to which he added "programming" and "data collection" at the beginning, and "development" and "communication" at the end (see Fig. 1.1).

Archer and Jones saw their phases as overlapping and with feedback to early phases. Yet their methods expressed the belief that concentration on possible solutions should not begin until the design problem had been thoroughly studied, and a good working formulation of the problem set down. The designer was to start by exhaustively listing relevant factors and setting performance limits on relevant functional variables. Only then was the designer to start generating partial solutions and "synthesising" them into an overall design proposal. It was hoped that this procedure would almost automatically generate new and more innovative solutions. Rigorously separating synthesis from analysis would help free the designer of any preconceptions and to overcome early fixation on existing solutions.

In the early 1970s, design methodologists began to criticise this "linear" view of the design process. They argued that designers not only do, but also inevitably must prestructure their problems in order to solve them [6]. They suggested that the "analysis–synthesis" model of design was derived from a fallacious view of the role of inductive logic in science and that designers *have* to rely on their prior knowledge of solution types. Therefore, inspired by Popper's view on the method of science, Hillier *et al.* [6] and Darke [7] proposed models of design that emphasised the crucial role of conjectures. In these models the designer must first generate a solution conjecture, which is then subjected to analysis and evaluation, rather than analysis preceding synthesis or conjecture.

Notwithstanding this, and a great deal of other criticism on the "classical" analysis–synthesis–evaluation model [8,9], the idea of analysis and synthesis as more or less clearly delimited phases of the design process firmly persists, and still many design projects are structured and reported in terms of these phases.

1.1.3 Synthesis and analysis as functions of problem solving

Systems engineering, emerging in the late 1950s, provided another view of synthesis and analysis. In his classical account of systems engineering methodology, Hall [10] defined synthesis and analysis as two of seven interrelated functions, comprising the logic of systems methodology. Hall distinguished the logic of systems engineering from the time dimension formed by the successive phases in the generic system's life cycle (see Table 1.1). For Hall, these functions were not linearly related: they had no necessary time sequence and were not "steps". They stand for classes of activity and correspond to a generic model of problem solving applicable to any field. The logic of systems methodology is exercised in every phase; it can be thought of as the finer structure of the systems engineering methodology.

Hall describes synthesis as collecting, searching for, or inventing a set of ideas, alternatives, or options. The general objective of synthesis is to compile an extensive (ideally an exhaustive) list of hypothetical systems, each worked out in enough detail to be evaluated relative to the system objectives. For Hall, synthesis ranges from reproportioning parts and conceiving of new configurations to inventions, with techniques for synthesising ranging from highly logical to purely psychological.

Table 1.1 Hall's morphology of the systems engineering process.

Phases of the coarse structure / Time ↓ \ Elements of the fine structure logic →	1 Problem definition (problem finding and context analysis)	2 Value system design (develop objectives and criterion)	3 Systems synthesis (collect and invent alternatives)	4 Systems analysis (deduce consequences of alternatives)	5 Optimisation of each alternative (iteration of steps 1–4 plus modelling)	6 Decision making (application of value system)	7 Planning for action (to implement next phase)
1 Program planning	A_{11}	A_{12}				A_{16}	A_{17}
2 Project planning (and preliminary design)	A_{21}						
3 System development (implement project plan)							A_{37}
4 Production (or construction)				A_{44}			
5 Distribution (and phase in)							
6 Operations (or consumption)	A_{61}						A_{67}
7 Retirement and phase out	A_{71}	A_{72}				A_{76}	A_{77}

Hall is at variance with Archer and Jones, who associate analysis with the analysis of problems. Hall terms (systems) analysis the deduction of those sets of consequences (of alternative decisions and actions) that were specified as relevant in the value system. These deductions may relate to quality, market, reliability, cost, effectiveness, quality of life, *etc*. The term "analysis" is still in use in both senses, which might cause misunderstanding and controversy. For instance, debates on the question as to whether synthesis must be preceded or succeeded by analysis can easily run astray if it is not recognised that "analysis" in Archer and Jones' models means "problem analysis", whereas for Hillier *et al.* and Dark "analysis" is the deduction of the consequences of a conjecture.

Many models and theories of systematic design in (mechanical) engineering design, architectural design and industrial design have adopted the systems engineering view of synthesis and analysis [11]. For example, in his book on engineering design, Asimow [12] distinguished between the design process and the morphology of design. For the design process Asimow favoured the same stages as Jones: an analysis of the situation, in which the problem is embedded, followed by a synthesis of possible solutions and an evaluation of the solutions. But he considered these stages as parts of an iterative process (the horizontal structure), which can be found, sometimes in full, sometimes only in part, in every phase of the morphology (the vertical structure) of design. This idea clearly resembles Hall's systems engineering model. For architectural design, a similar model was developed by Markus and by Maver [13]; see Fig. 1.2. They suggested that we need to go through the decision sequence of analysis, synthesis, appraisal and decision at increasingly detailed levels of the design process. For product development, Archer took on this view in his writings on the structure of industrial innovation [14].

Like analysis and synthesis as phases, the understanding of analysis and synthesis as functions is also in current use. For instance, the guideline VDI 2221 [15], the work of the "Copenhagen School" [16] and our own analyses of the structure of product development [17] clearly draw upon this view.

1.1.4 Synthesis as assemblage of subsystems

Not surprisingly, in the design methods literature the term "synthesis" is also used in the literal sense of putting together, combining or assembling parts into a new whole. "Synthesis" then refers to particular ways and specific methods for generating schemes, principal solutions, concepts and layouts, *i.e.*, by combining or assembling functional and physical representations of available or conceivable parts and components. Under the heading "compositional synthesis", several chapters of this book are devoted to this approach to synthesis.

Zwicky's morphological method is a classical example of this approach. By systematically combining all possible realisations of essential "elements" of a problem, this method aims at finding all conceivable solutions for a problem. Another typical example is function analysis. German engineering designers, notably Koller [18], Pahl and Beitz [19] and Roth [20], have paid much attention to this method. They regarded the establishment of function structures as an important first step towards the thinking up of new working principles. The underlying idea is that function structures can be built up from a limited number of "elementary" or "general" functions and provide entries to catalogues that hold collections of physical phenomena and technical principles. In this manner, engineering design might be systematised and partially automated, especially in the phase of conceptual design.

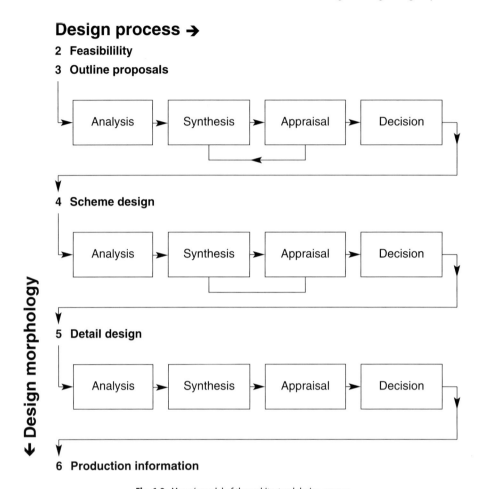

Fig. 1.2 Maver's model of the architectural design process.

Methods like these have much potential, though not for all types of product. The "assembly approach" to synthesis applies in so far as the spatial relationships of the parts of a product are relatively unimportant in comparison with their patterns of interconnections. For "flow systems", such as machines, power tools, electronic circuits, chemical installations, *etc.*, this might be the case. But this is not so for "associative" systems, like chairs, containers, or bridges, in which parts form organic wholes, in the sense that a change in any one subsystem impairs the overall function of the whole. Therefore, notwithstanding successful implementations and applications in some domains, the "assembly approach" to synthesis has its limitations for design in general [21].

1.1.5 Synthesis as integration of ideas

The last sense of "synthesis" to be mentioned is synthesis as integration. When designing a product many factors must be considered. To the design engineer a product is a technical–physical system that has to function efficiently and reliably. An industrial designer considers the product to be an object that functions in a psychological sense and embodies cultural values. Production engineers have to manufacture it, often in large numbers, preferably quickly, cheaply, accurately and with the lowest possible number of faults. A marketer considers it a commodity with added value. Consumers look upon a product as something to be bought and used. Entrepreneurs invest in new products and count on an attractive return. People that are not directly involved may see in the above the other side of the coin: the undesirable and, often, even harmful side effects of production and use. These different points of view – and there a more – must be "synthesised" into one single design for the new product. Here, "synthesis" does not merely connote the combining of separate ideas. The design of a product is more than a collection of solutions to subproblems. Design aims towards an *integrated* solution: a higher conception that goes beyond its stepping stones or constituents; a solution that makes subproblems disappear, so to speak. This sense of synthesis emanates from Hegel's philosophy. Hegel called "synthesis" the new idea that resolves the conflict between the initial proposition (thesis) and its negation (antithesis). Synthesis is the higher conception that involves but transcends both the initial conviction and its opposite.

1.2 The logic of synthesis

1.2.1 Form and function

Synthesis must provide for tentative solutions for design problems. Hence synthesis can be thought of as reasoning from statements on functions (or intended behaviour) to a description of the form (or structure) of the designed object. Taken as a form of reasoning, synthesis belongs to the category of plausible reasoning. Plausible inferences are of a number of different kinds, sometimes lumped together under the ambiguous label of "induction", such as: analogy, first- and second-degree induction, statistical generalisation and specification, reduction or presumption of fact, and abduction. The question I shall address in this section is: which type, or pattern, of plausible inference is typical for the transition from function to form? Is it "abduction" as many authors contend, or is it another pattern or reasoning? This matter is important because both simulating the design process by computers for scientific reasons and building "designing machines" for practical purposes necessitates the modelling of design reasoning in some sort of logical formalism. Moreover, insight into the typical mode(s) of reasoning in design contributes to our understanding of design.

The answer to the above question demands an analysis of the structure of design problems. As a physical system a product is defined by its form (or structure). The form of a product includes the geometrical form as well as the physico-chemical form, *i.e.*, the shape and dimensions of the whole and the parts plus the materials the parts are made off. The description of the geometrical and physico-chemical form makes up the design of the new product.

Because of its form, a product has certain properties, *e.g.*, weight, strength, hardness, and colour. Although we usually describe properties categorically, we actually claim that some corresponding hypothetical statements are true. For example, if we categorically state that "the stiffness of this construction *is* such and such", we claim that the following hypothetical statement on that construction is true: "*if* this construction is loaded in the manner *y*, *then* it will be deformed in the manner *z*". Each property tells us something about the reaction the product will show if we bring it into a certain environment and use it in a certain way. Properties describe the behaviour to be expected under certain conditions.

We can distinguish between intensive and extensive properties. Intensive properties depend on the physico-chemical form only, such as specific gravity. Extensive properties, or "thing properties", are determined by intensive properties plus the geometrical form, *e.g.*, the weight of an object. In designing, we are especially interested in the extensive properties, as these most directly determine the functioning of a product.

Products are designed, made and used for their functions. The function of a product is the intended and deliberately caused ability of the product to bring about change in something that is part of its immediate environment, including ourselves. Functions of products can be described in different ways, such as in normal language, in mathematical formulas, or in the manner of a "black box". Whatever the representation, statements about functions are statements about the intended behaviour of the product. Like statements about properties, these statements have a hypothetical form, For example: in grinding coffee, we aim for a final state S_2 that yields ground coffee with a certain size of granule. We assume that there is a beginning state S_1, which is characterised by beans of a certain kind, the availability of energy in a certain form, no extreme temperatures, *etc*. If we put this intended behaviour into propositional form, we get the hypothetical $S_1 \rightarrow S_2$ (read: if S_1 then S_2, or S_1 implies S_2). Unlike properties, statements about functions are normative. A product either has certain properties or does not have them, irrespective of the purpose of a user. Functions, however, are imposed on products.

1.2.2 Reasoning from function to form

Now, one can think of all sorts of functions and try to design a new product for them, but will the product really behave as intended? Of course, this depends on its form, for the form determines the extensive properties. But the product must also be used in a certain manner. We only notice some of the properties of a product directly. Most properties only become "visible" when we do something with it. Properties are characterised by hypothetical statements. Even if a hypothetical statement is true, its consequence only becomes real when we actualise its antecedent. To do so, we actually have to bring the product into a certain environment. For example: iron has the property that it rusts, but only in contact with water; similarly, a ballpoint pen can write, but not on a vertical surface. A product having the required properties, therefore, functions in the intended manner only if it is used in the environment and in the way that the designer has thought up and prescribed. The use of a product is not a given for the designer, like the function, but is thought up – together with the form of the product – and thus comprises an essential part of the solution to the design problem.

The foregoing is summarised in Fig. 1.3, which shows how the functioning, or actual behaviour of a product, depends both on its form or structure, as well as on the mode and conditions of its use. The arrows in Fig. 1.3 represent causal relations. Given the form and the use of a product, then, by deduction or experiment, one can predict its properties and functional behaviour. The designer, however, must reason against the direction of the arrows. Given a desired function, the

Defining Synthesis: On the Senses and the Logic of Design Synthesis

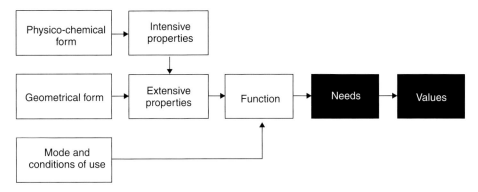

Fig. 1.3 The functioning of a product.

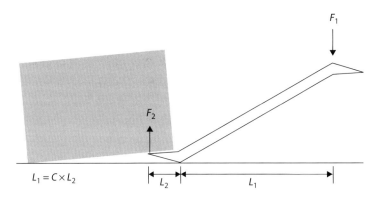

Fig. 1.4 The functional behaviour of a crowbar.

designer must think up the form and its use. The reasoning from form to function is usually called "analysis", whereas the reasoning from function to form is called "synthesis" [22]. Notwithstanding the importance of analysis, in design the essential mode of reasoning is synthesis, for without an idea of form and use there is nothing to analyse.

To clarify Fig. 1.3 I shall give a simple example (see Fig. 1.4). The form of a crowbar can be described in categorical statements: this crowbar *is* made of steel, *has* a length of 850 mm, *has* a diameter of 30 mm, *etc*. The function of a crowbar can be phrased as "amplifying force": *if* a force F_1 is exerted on a crowbar (the antecedent), *then* there results a force F_2 (the consequence). This is a hypothetical statement. If the design of the crowbar is given, *i.e.*, if all dimensions are known and we know what material it is made of, and how and what it is used for, then we can predict, by means of the formulas of mechanics: how large F_2 will be as a result of F_1, how much the crowbar will bend, whether it will resist the stresses involved, *etc*. This reasoning from form to function – or from structure to behaviour – is based on deduction. The conclusion follows necessarily from the premises and, in principle, there is only one answer: the product has or does not have the required properties, it will or will not show the intended behaviour. With a certain form and a certain value of F_1, there is only one corresponding value of F_2.

Nevertheless, a crowbar can fulfil other functions too. If we use it in another environment or use it in another way, other laws of nature operate and other transformations will take place. Each transformation corresponds to a function that the product can fulfil, e.g., that of a hammer. However, the formulas of mechanics do not permit one to deduce conclusively the form of a crowbar from the specification of its function. The stiffness required and strength of the device can be achieved by many different materials, shapes and dimensions. Only by making an arbitrary choice for all degrees of freedom minus one can the remaining form factor be determined unambiguously. The indeterminateness of the transition from function to form goes even further, for a lever is not the only conceivable technical principle for amplifying forces. This can also be accomplished with the help of pulleys and cables, gear wheels and hydraulics, and new principles can be invented.

Reasoning from function to form – synthesis – is a creative process. This does not mean that scientific and technical knowledge do not play a part. Rather, causal models indicate the direction in which the main choices can be made (choice of physical effects and technical principle, choice of material, choice of geometric shapes, choice of one or more key dimensions). Yet this never leads to an unambiguous answer, because the number of solutions is, in principle, innumerable.

Now, which type, or pattern, of non-deductive inference is typical for the transition from function to form?

1.2.3 The pattern of reasoning of synthesis

We saw that, from a description of the form and the use of a product, one can infer by deduction those properties, attributes or performance characteristics that are not already explicitly contained in the description itself. In propositional logic, the pattern of this inference is:

premise	$p \rightarrow q$	if x is of aluminium **then** x does not corrode
premise	p	x is of aluminium
conclusion	q	x does not corrode,

where p stands for a statement describing something of the form of a product; q for a derived (predicted) property and $p \rightarrow q$ for the rule (a rule of thumb, an empirical generalisation, a validated law) upon which the deduction "rests". Deduction plays an important role in design. Before actually realising a design, or making important decisions as to the continuation of an unfinished design process, designers have to determine to what extent the designed artefact possesses the desired performance characteristics, and this is done either by deduction or by experiment. Yet, in design, one does not usually start with a description of the product, but with some ideas about its intended functions. One then tries to arrive at a description of the form of the product in terms of geometrical and material attributes. Under this interpretation q becomes a performance requirement, from which p, a statement on the design, is inferred.

premise	$p \rightarrow q$	if x is of aluminium **then** x does not corrode
premise	q	(it is required that) x is of aluminium
conclusion	p	x is (to be made) of aluminium.

This pattern of reasoning is commonly called "abduction", after the philosopher Ch.S. Peirce [23]. It resembles deduction, but it is actually very different. In abduction the conclusion does not follow necessarily from the premises, as it reasons from consequence to antecedent. One cannot run a deductive system backwards, so to speak.

Now, many authors take the view that the abductive pattern of reasoning exhibited in the above example is characteristic for the reasoning from function to form [24,25]. Some years ago I demonstrated that this view is not correct [26]. Peirce subsumed two different patterns under the name "abduction", which have been called "explanatory" or "non-creative" abduction and "innovative" or "creative" abduction [27,28]. The above example illustrates explanatory or non-creative abduction. However, as I shall go on to explain, the reasoning towards new solution principles for design problems does not follow this pattern, but the pattern of innovative abduction.

Let us take as an example the designing of a kettle. The reason for using kettles is: "boiling water". This we call the function, or purpose, of the kettle. The function of a product always involves change in some object, which is part of the immediate environment of that product. In this case the object is the water; boiling water is the process of transforming water of say 20 to 100°C:

function: $W_{20} \Rightarrow W_{100}$.

How does a kettle boil water? How does it work? Answer: the bottom of the kettle is heated and transports the heat to the water inside. This is called the "mode of action" of the kettle. The mode of action is the (functional) behaviour of the product itself, in response to influences exerted upon it from its environment. The mode of action is also a process, in this case, say, a kettle of 20°C transforms into a kettle of 300°C:

mode of action: $K_{20} \Rightarrow K_{300}$.

Of course, someone has to fill the kettle with water, and place it on a burner. The actual functioning of any product involves, at least once, but often continuously, the action of a user. By that action the product is "connected" to its immediate environment, including the object it has to change. We call this user-action the "actualisation" of the product. Again, this is a process: the kettle-in-rest transforms into a kettle-in-action:

actualisation: $K_{rest} \Rightarrow K_{action}$.

We have now identified three real processes. They are connected by two statements, both having the logical form of the material implication:

actualisation → mode of action

and

mode of action → function.

The actualisation implies the mode of action and the mode of action implies the function (that a particular kettle actually fulfils).

But, how can that be? Why does the water not flow away? Why does the kettle not melt or dissolve? The answer is: because a designer has given the kettle and its parts a proper geometrical form (shape and dimensions), as well as an appropriate physico-chemical form (the chosen materials). The form of a product is no process, but a state, defined by the conjunction of several categorical statements (normally represented in technical drawings), such as: (f_1) the diameter of this kettle is d; (f_2) its form is a hemisphere; (f_3) it is made of stainless steel, *etc.*:

form: $f_1 \wedge f_2 \wedge f_3 \wedge \ldots$

Thus, the form of the kettle and the way it is used (actualised) causes it to behave in a certain way (the mode of action), and, therefore, with this behaviour, it can fulfil its function. Again, this can be represented as a material implication:

((form ∧ actualisation) → mode of action) → function

or, because $((a \rightarrow b) \wedge (b \rightarrow c)) \rightarrow (a \rightarrow c)$

(form ∧ actualisation) → function.

Usually, the function of a new artefact is given to a designer by his client or marketing department. What has to be designed? The usual view is: (the description of) the form of the artefact, including its shape, dimensions, and materials. Indeed, a design problem is solved if the designer has come to a finish with these decisions in drawings and other documents. But, as I said before, a description of the form of an artefact does not suffice to predict its behaviour and function fulfilment; the mode of action depends upon both the form of the artefact *and* its use. Working back from function to form, the designer cannot escape from developing ideas about the actualisation as well. In "routine" design, these ideas may be more or less self-evident, suggested by precedents for example; in "innovative" design they are certainly not. Here, the way the artefact is supposed to be used has to be conceived, from scratch so to speak, and in a close interplay with its form. So, the solution to a design problem includes a description of the artefact's form *and* a description of its actualisation; the latter has also to be designed. Therefore, strictly speaking, the kernel of the design problem is the reasoning from function to the form of a product *and* its actualisation.

Now, the only premise of this reasoning process is a statement q about the function; the new artefact has to transform some object from one state into another:

$q : O_{S_1} \Rightarrow O_{S_2}$.

At the end of the process the designer claims the following: make a thing with form F, do A with it (actualise it as has been prescribed), and you shall see that $O_{S_1} \Rightarrow O_{S_2}$

(it will show the intended mode of action and thereby fulfil its function). Thus the conclusion of the designer's reasoning process is a compound statement, having the form of the material implication:

if form is *F* **and** actualisation is *A* **then** $O_{S_1} \Rightarrow O_{S_2}$.

This statement seems to be given as a second premise, but, at the start of the design process, F and A are still variables to be defined. Therefore, at that time, the conclusion is not known. The act of synthesising the product *is* to determine these variables, such that the conclusion *becomes* true. In other words, the designer must conceive of the form and actualisation of the artefact and, simultaneously, "construct" a material implication that expresses how this form and actualisation "causes" the product to function as intended. This material implication represents the functioning of the artefact and will be true only for those design proposals that can count, so to speak, on the collaboration of the laws of nature. But, designers cannot change laws of nature, can they? Indeed not; nevertheless, the material implication cannot be considered as a given premise. By their design decisions designers "force" nature to cooperate in a certain way. So designers *make* certain laws of nature come into action. In this sense designers "construct" the functional behaviour of the new product, as well as its form, although this behaviour is fully governed by the laws of nature.

To clarify the foregoing we go once more back to the crowbar example (see Fig. 1.4). The functional behaviour of levers is governed by the law

$$F_1 L_1 = F_2 L_2.$$

Someone having designed a particular crowbar claims: make an object with the described form from the described material, and you will see that, if you use it properly, you can lift heavy loads with it. More formally stated, and restricting ourselves to the mechanical behaviour:

$$(((L_1 = CL_2) \wedge F_1) \rightarrow (F_2 = CF_1)) \rightarrow (\text{load}_{S_1} \Rightarrow \text{load}_{S_2})$$

form ∧ actualisation → mode of action → function.

If a designer has decided upon the form and actualisation of a crowbar, its performance capabilities can be calculated from the law of levers. Note that this law refers to characteristics of the form of the crowbar (L_1 and L_2). That is, knowing that this law applies entails the "insight" that something with the form of a lever has the ability to lift loads. Now, let us imagine a designer, who has never seen a crowbar before and who wants to design a functionally equivalent device. This designer does not know which laws of nature apply to the problem, unless the main characteristics of a solution principle are presupposed, and *vice versa*! The ideas for the new product and its behaviour occur simultaneously, and many different solutions are possible.

What does the foregoing entail for the pattern of reasoning from function to form? We saw that the solution for a design problem comprises two parts:

p_F : a description of the form of the artifact

p_A : a description of its actualisation.

So the solution for a design problem can be written as

$$p = p_F \wedge p_A.$$

Hence, the pattern of reasoning from function to form and actualisation is

premise q
_____
conclusion $p \to q$
conclusion p.

The scheme shows clearly that the pattern of reasoning from function to form is different from explanatory abduction; there is only one premise and the rule $p \to q$ has become part of the conclusion. In order to distinguish this pattern from explanatory abduction, it has been called "innovative" and "creative" abduction. For clarity, we prefer a name without the term "abduction"; hence we have introduced the term "innoduction" [17, p. 78].

In explanatory abduction $p \to q$ is taken as a premise. This allows us to reason from an effect q to be explained, to a possible cause p. But as the conclusion p is already contained in the premise $p \to q$, explanatory abduction cannot produce any *new* concepts or ideas; they are already assumed to be known as part of the premises of the argument. Explanatory abduction cannot account for the reasoning from function to form, because producing a tentative solution that is new in at least some respect is precisely what is expected from the reasoning process from function to form. For if a design proposal is identical to the design of an existing product it has not been designed but copied.

If reasoning is to produce a new concept or idea, then the rule must be part of the conclusion and inferred together with the antecedent [28, p. 93]. So the only logical operation that can introduce any new ideas is innovative abduction or "innoduction". Therefore, innoduction is the pattern of reasoning from function to form.

1.2.4 Conclusions

"Synthesis", in the general sense, is the combining or mixing of ideas or things into new ideas and things. In this sense, functional and physical representations of subsystems, viewpoints, and interests of stakeholders are synthesised in design. In design theory and design methodology "synthesis" is looked at as a phase of the design process, as well as a function of problem solving. According to the first view, exhaustive problem analysis must precede solution synthesis. Indeed, designers must study and understand requirements, produce tentative solutions, and predict and evaluate their merit. The idea, however, that these activities occur as identifiable separate phases in that order, seems very questionable. According to the second view, synthesis is one of six interrelated problem-solving functions: problem definition, selecting objectives, systems synthesis, systems analysis, selecting the best system, and communicating results. In this view, synthesis, as well as the other "functions", occurs iteratively in all phases of the design process and on different levels of abstraction (synthesis of functions, of solution principles, of concepts, of layouts, *etc.*).

The pattern of reasoning for synthesis is "innovative abduction" or "innoduction". Because innoduction is inherently non-deductive it cannot be grasped in an algorithm. Reasoning from function to form is an open process, which can provide many good solutions. Synthesis is a creative process, which can be encouraged methodically, but cannot be logically guaranteed.

Innoduction is the key mode of reasoning in all forms of design. Yet innoduction is not unique to design. The pattern applies to the reasoning from ends to means in general. In both research and design, as well as in daily life, continually abductive steps are taken in the search for new concepts, hypotheses and practical ideas. And in both science and technology the different modes of reasoning – deduction, induction, (non-creative) abduction and innoduction – have to work together, to support each other. However, the parallels in the logic employed shall not close our eyes to the methodological differences between empirical research and the design of artefacts [29].

What are the implications of this analysis for the modelling design by computational systems? I can only give a tentative answer. I assume that only those processes that can be formalised in logic can be modelled in a computer system. Furthermore, I assume that all models implemented in present-day computers apply *deductive* logic, even those models that "mimic" non-deductive inferences. (Explanatory abduction is mimicked in computers by assuming a so-called "closed world". Under the closed world assumption it is true that "if $A \rightarrow B$, then $B \rightarrow A$". But that means that abductive inferences are dealt with *as if* they are deductive ones.) Now, most efforts to model abduction in the context of design automation concern *explanatory* abduction. Modelling this pattern of reasoning is certainly not irrelevant for design automation, but it seems to "beg the question" of modelling synthesis in the sense of "inventing" a *new* form or structure for a product, a form hitherto not known to the designing system. This requires the modelling of innoduction. But how is that possible if innoduction cannot be reduced to deduction? Therefore, whatever flight engineering sciences and design methodology might take, it seems likely that, for synthesis, creativity and intuition will continue to play an indispensable and irreplaceable roles.

References

[1] Hall AD. A methodology for systems engineering, 6th ed. Princeton (NJ): Van Nostrand, 1968; 109.
[2] Webster's revised unabridged dictionary. 1913 [www version].
[3] Losee J. A historical introduction to the philosophy of science. Oxford: Oxford University Press, 1993; 85.
[4] Archer LB. Systematic method for designers. London: the Design Council, 1965. In: Cross N, editor. Developments in design methodology. Chichester: Wiley, 1984; 57–82.
[5] Jones JC. A method of systematic design. In: Jones JC, Thornley D, editors. Conference on design methods. Oxford: Pergamon, 1963; 53–75.
[6] Hillier B, Musgrove J, O'Sullivan P. Knowledge and design. In: Mitchell WJ, editor. Environmental design: research and practice. Los Angeles: University of California, 1972. [Reprinted in: Cross N, editor. Developments in design methodology. Chichester: Wiley, 1984; 245–64.]
[7] Darke J. The primary generator and the design process. Des Stud 1979;1(1):36–45.
[8] Rittel H, Webber M. Dilemmas in a general theory of planning. DMG-DRS J 1974;8(1):31–9. [Reprinted in: Cross N, editor. Developments in design methodology. Chichester: Wiley, 1984; 135–45.]
[9] Checkland PB. Systems thinking, systems practice. Chichester: Wiley, 1981.
[10] Hall AD. A methodology for systems engineering, 6th ed. Princeton (NJ): Van Nostrand, 1968; 109.
[11] Roozenburg N, Cross N. Models of the design process: integrating across the disciplines. Des Stud 1991;12(4):215–20.
[12] Asimow M. Introduction to design. Englewood Cliffs (NJ): Prentice-Hall, 1962.
[13] Maver TW. Appraisal in the building design process. In: Moore GT, editor. Emerging methods in environmental design and planning. Cambridge: MIT Press, 1970; 195–203.

[14] Archer LB. Technological innovation: a methodology. Frimley: Inforlink, 1971.
[15] VDI guidelines 2221, systematic approach to the design of technical systems and products. VDI-Verlag, 1987.
[16] Andreasen MM, Hein L. Integrated product development. IFS Publications (UK) Springer, 1987.
[17] Roozenburg NFM, Eekels J. Product design: fundamentals and methods. Chichester: Wiley, 1995.
[18] Koller R. Konstruktionslehre für den Maschinenbau; Grundlagen des methodischen Konstruierens, 2nd ed. Berlin: Springer, 1985.
[19] Pahl G, Beitz W. Konstruktionslehre; Handbuch für Studium und Praxis. 2nd ed. Berlin: Springer, 1986; 115-25.
[20] Roth K. Konstruieren mit Konstruktionskatalogen; Systematisierung und Zweckmässige Aufbereitung technischer Sachverhalte für das methodische Konstruieren. Berlin: Springer, 1982.
[21] Gosling W. The relevance of system engineering. In: Jones JC, Thornley DG, editor. Conference on design methods. Oxford: Pergamon, 1963.
[22] Klirr J, Valach M. Cybernetic modeling. London: Iliffe Books, 1967; 29.
[23] Buchler J, editor. The philosophy of Peirce: selected writings. New York: AMS Press, 1978; 150-56.
[24] March LJ. The logic of design and the question of value. In: March LJ, editor. The architecture of form. Cambridge University Press, 1976. [Reprinted in Cross N, editor. Developments in design methodology. Chichester: Wiley, 1984; 265-76.]
[25] Coyne RD, Rosenman MA, Radford AD, Balachandran M, Gero JS. Knowledge-based design systems. Reading (MA): Addison-Wesley, 1990.
[26] Roozenburg N. On the pattern of reasoning in innovative design. Des Stud 1993;14(1):6-18.
[27] Habermas J. Erkenntnis und Interesse. Frankfurt am Main: Suhrkamp, 1968; 147. [English translation: Knowledge and human interests. 2nd ed. London: Heinemann, 1978.]
[28] Schurz G. The significance of abductive reasoning in epistomology and philosophy of science. In: Schramm A, editor. Philosophie in Österreich. Wien, Hölder-Pichler-Tempsky, 1996; 91-109 [in German].
[29] Eekels J, Roozenburg NFM. A methodological comparison of the structures of scientific research and engineering design: their similarities and differences. Des Stud 1991;12(4):197-203.

Insight, design principles and systematic invention 2

Michael J. French

Abstract Design synthesis depends critically on the nature of the problem and the engineering science involved. Progress depends on the related tasks of recognising the key issues and developing insight into the problem, both of which demand work on the engineering science, although quite rough calculations will often clarify important issues. Engineering designers frequently approach problems in just such an *ad hoc* way, concentrating on the key issues. It is important to gain insight rapidly, and plenty of simple calculations help. Examples are given. Design principles are a powerful aid, but most are not widely recognised. Five sample principles of great value are given, with examples of their use. Often, arguments based on engineering science can be strung together to approximate to systematic design. An example on a wave energy collector is given.

2.1 Introduction

Most worthwhile advances in practical design are made by small, or sometimes large, inventive steps. Inventive steps are, so far, exclusively the province of the human mind, but systematic approaches can be of great help in stimulating them. Perhaps the most important precursor to invention is the development of insight into the problem.

The most valuable systematic aids are not general in nature, but are based upon engineering science (as is obvious in the case of insight). In a few areas these aids are so powerful as to leave little to human ingenuity (for instance, in the design of thermodynamic processes, of which the Linde–Frankl column for the liquefaction of air is a classical example [1]). In rare cases, they virtually deliver the invention (optimum gear tooth form – see Appendix 2A). In general, however, they are merely aids and leave judgement and the real work to the designer. Among the most valuable aids are *design principles*, which encapsulate the experience and insight of designers and often enable minor inventive steps to be made quickly [2]. Design platforms of various kinds may also ease the load and increase the throughput of the designer, but they are the subject of much of the rest of this book, *e.g.* see Chapter 12.

2.2 The opportunistic designer

Most real design is done by experienced designers who work in an ostensibly much less systematic way than academics usually advocate. They have a good grasp of the problem, which is generally in an area with which they are more or less familiar. They know where there is scope for progress and often have several ideas they are waiting for an opportunity to try out. They may not formally construct a table of functions and possible means, but they have its contents in their heads and ideas where there is scope for improvement (and they are probably right most of the time).

It would often be a good thing if they were to use a little system, but they are right not to waste time on elaborate procedures that do not involve the specific engineering science, which is where the nitty-gritty is to be found and progress is to be made. The problem analysis, the conceptual, embodiment and detail stages, and a simple table of functions and alternative means, as suggested by the writer in 1971 [3], are enough of a systematic framework for most purposes; they can help to suggest what to look at next and to avoid the overlooking of possibilities, which is a recurrent risk without such aids (and even with them).

An excellent video demonstration is available of the approach of a very distinguished engineering designer, A.E. Moulton, inventor of the Hydrolastic and Hydrogas suspensions and a groundbreaking line of bicycles, holders of various records. Though not representative in some ways, particularly in the high standard and the highly developed personal approach of the designer, it nevertheless displays much in common with the designer in industry rather than the researcher in academia [4].

2.2.1 The opportunistic approach

Although I advocate these props, I know full well that the opportunistic designer, as he may be called [5], will leave them behind when the bit is between his teeth and he is away at a gallop, concatenating ideas and spotting possibilities at every step. He already works through the stages and he is inwardly conscious of the functions and the options for means (but he dismisses silly ones out of hand, and just occasionally one that is not silly along with them). Usually though, he will not write all this out. He knows which choices are easily and safely made and which offer real alternatives. He also knows which are crucial and which are trivial, and above all, if means $r2$ is chosen for function r and means $s1$ for function s, what consequences that will have for the embodiment of each. Nevertheless, simple charts are worth the small effort they demand, and he would do well to use them. However, what can help the opportunistic designer most, I believe, are non-general, task-specific aids of the type described in this chapter.

The term "opportunistic" often has pejorative implications; it has none here. We are concerned with the inventors of our material civilisation, people of immense (and generally inadequately rewarded) worth. Were it not so clumsy, I would be tempted to call them "catch-as-catch-can" designers, for that term is very descriptive of the opportunistic design process, searching around for something to get hold of, a purchase, something you can confidently start from.

I have been an opportunistic designer, with a great engine manufacturer (Napier). The sequence, problem analysis, conceptual design, embodiment and detailing, given in Ref. [3], is based on the practice I observed and followed there. Except for the last, those titles were not used, but the documentation of each stage was distinctive.

2.3 Parallels with mathematics

There is a close parallel between mathematical and engineering invention, and some observations by mathematicians tie in with the views put forward here. In the early years of the last century, there was much consideration of invention in mathematics [6] and some of the ideas it produced are applicable to engineering design. Prominent in the debate was the great polymath Poincaré, who maintained that mathematical invention hinged on the ability of original mathematical thinkers to recognise among all the vast numbers of possible combinations of entities a few that might lead to valuable results [7]. He described this ability as an aesthetic one, because of the joy accompanying the recognition of a potentially fruitful combination. I maintain that the case with engineering design is closely parallel. Who has not had the experience of spotting some happy mutual support between two elements, and itched to get to the board or the work station to see how it all pans out? This is the p-aesthetic (for Poincaré) as I call it, the delighted recognition by the inventor of a fruitful conjunction.

One can experience the p-aesthetic from another's work, as I did recently with the model aero-engine shown in Fig. 2.1. This is a small single-cylinder four-stroke internal combustion (IC) engine designed for model aircraft. The piston, connecting rod and crankshaft are conventional enough, except that the crankshaft is vertical and the piston moves horizontally. The crankshaft drives the horizontal propeller shaft through a 2:1 reduction bevel gear. For matching most IC engines to most airscrews, reduction gearing of about 2:1 is desirable anyway. So far this is a workable but in no way remarkable design, with the one clear advantage over the usual vertical cylinder arrangement of a low frontal area, highly desirable in aircraft, small or large [8].

The masterstroke is that the propeller shaft is also the cylinder and incorporates the valve (hence the name "rotating cylinder valve engine"). The tubular extension of the cylinder that forms the middle length of the propeller shaft is a rotary valve; a radial port in it comes opposite inlet and outlet ports in the casing in the course of the single rotation that it makes during four strokes of the piston, *i.e.*, one cycle of the engine. At the beginning of the induction stroke, it is opposite the inlet port and the outward stroke of the piston draws in the charge of air and fuel. During the exhaust stroke, it is opposite the exhaust port and the products of combustion escape. This simple rotary valve arrangement replaces the usual relatively complicated camshaft and poppet valves. This novel engine has only four moving parts, yet has

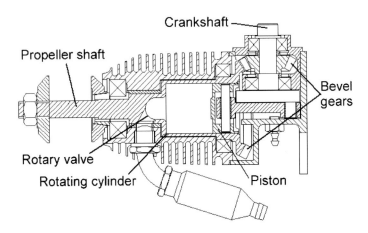

Fig. 2.1 Rotating-cylinder engine.

the advantages of a four-stroke engine, together with low frontal area and reduction gearing.

2.3.1 Poincaré's sieve

Poincaré discussed the concept of a "mechanical sieve" that would sort through all possible combinations of a large number of mathematical entities and sieve out those with potential. He dismissed the possibility of such a thing, for reasons that apply with comparable force to engineering design. The numbers of elements are large, the possible interactions between them are countless, as are the ways they can be varied within themselves to suit one another – a mechanical sieve is unimaginable. The situation has not yet been altered by the advent of computers, which, with all their huge capacity for repetitive operations, still cannot begin to exercise the high-level discrimination required.

A postulate of the time was the division of the inventive process into four steps: preparation, incubation, illumination and verification [6]. In the first, the inventor studies intensely the area into which he proposes to inquire. In the next, he looks at other things, to allow his subconscious to turn over the material with which it is saturated, leading at last to illumination, appearing in the form of the seminal conjunction of elements. There follows verification, in which the new idea is tested and its consequences pursued.

Though I have doubts about incubation as a necessary step, these ideas describe well enough the common experience of designers.

2.3.2 Visual thought

The other subject of this section, visual thought, occupied the minds of mathematicians and others at about the same time. It has been maintained that language is essential to thought above a low level. It is clear, however, that some of the deepest thinking of which humans have proved capable is not done verbally, and even that such thinking is difficult to express in language at all.

Einstein wrote [6]

> ...this combinatory play seems to be the essential feature in productive thought – before there is any connection with logical construction in words or other kinds of signs which can be communicated to others. The above-mentioned elements[1] are, in my case, of visual, and some of muscular, type.

Einstein's account supports the view of Poincaré about the combinatory aspect. It ditches the dictum about language being essential to higher thought and it supports the experience of engineers. Incidentally, it also destroys any argument that his thinking *was* based on a language, the language of mathematics.

We know that the portion of the brain concerned with the processing of visual thinking is much larger than that concerned with language. The parts of the brain devoted to visual (and related) information processing are large and work at a high level.

---

[1] The elements are those that are subject to the combinatory play. I take the last sentence to mean that Einstein was an arm-waving physicist.

The writer believes that the motion of many higher animals, *e.g.*, a gibbon swinging from arm to arm and bough to bough through trees, involves something very close to rapid simulation of the intended trajectory, just in advance of execution, to determine the route and the necessary motions to achieve it, which are then refined in accordance with the feedback from vision and the proprioceptive system. It is convenient to use the term visual thinking, but closely linked with the visual part there are the motor and proprioceptive (self-sensing or feedback) systems; in effect, the elements needed for a dynamic simulation. We might almost say that the gibbon *imagines* its future path, and the writer suggests that human imagination may have evolved in the first place from the use of the visual–motor–proprioceptive system in a simulative role [9].

That engineering design is largely based on visual thinking is beyond dispute. Even if readers do not themselves wave their arms, they surely recognise the role that arm waving plays.

As an example of visual thinking in science, Maxwell on Faraday is fascinating [10]. He admires the way in which Faraday was able to arrive at results with almost no mathematics, by means of powerful visual concepts and imagery. Maxwell himself gives three proofs of one of his own propositions, one of which is wonderfully elegant but demands remarkable visual imagery.

2.4 Insight

Inventive steps, the key to design, are often made possible by insights, crucial advances in understanding. One of the most important of inventions is a clear and well-recorded example. It is Watt's invention of the separate condenser, which vastly increased the scope of the steam engine by reducing its consumption of coal dramatically, and so gave a great impetus to the industrial revolution. For 50 years, at the end of the out stroke the Newcomen engines condensed the steam in the cylinder by spraying cold water into it. This left the cylinder walls cool, so that on the next out stroke they had to be warmed up by fresh steam condensing on them. Almost two-thirds of the steam was used in this way, and so did no work on the piston. Watt studied the Newcomen engine and developed this crucial insight. Based on it, the actual invention of the separate condenser seems simple, almost obvious.

Instead of condensing in the cylinder, you open a valve into an adjacent vessel into which cold water is sprayed, a spray condenser in fact. The cylinder remains hot and the condenser remains cold, and the coal consumption falls to not much over one-third, improving the economics marvellously and making many new niches for the steam engine. But in the absence of much design repertoire, with no parallels or precedents to go on, the inventive step was by no means so obvious as it seems today.

Watt first took an interest in the steam engine in 1760, when a model one used in lectures at the College of Glasgow was sent to him for repair, in his capacity of instrument maker. It was not until 1765 that he made his brilliant invention, after, it has been said, "months of torturing thought" [11], while walking on the Green of Glasgow. It is certainly a case that fits Poincaré's stages.

Of no historical importance, but a remarkable example of insight, was Watt's dismissal of the steam turbine. His partner Boulton had expressed concern at the competition that the steam turbine might offer their engine. Watt wrote to him saying that the turbine would not be practical unless the blades moved at 1,000 ft s^{-1}. He did not discount the possibility in the long run, as the letter shows. It says, "... it will cost much time..." [11]. It took well over 100 years, and his insight was perfect, except that 500 ft s^{-1} would have done. It was based on all the pioneering engineering science he had applied to steam, at a time when engineering science scarcely existed.

A good example of insight that is less ancient, but almost as historic, is to be found in a short paper by Robert Watson-Watt to the Committee for the Scientific Survey of Air Defence of the UK, written in 1935. In a passage of less than 200 words it establishes the feasibility of detecting approaching bombers by land-based radar, using simple arithmetic that would scarcely justify a slide rule, let alone a pocket calculator [3].

The great Spanish civil engineer, Torroja, wrote that a structural designer should grasp the functioning of the structure he was working on as clearly as he understood the fall of a stone to the earth or the release of an arrow from a bow. Though it is to be aimed at, such insight is rare, and we must generally be content with rather less. Moreover, we start with very imperfect insight, and develop it as we go along.

In mechanical and civil engineering at least, visual thought is essential in arriving at the level of insight of which Torroja wrote.

2.5 Developing insight

It is desirable that the designer in a new field should strive to develop insight as rapidly as possible. To achieve this he should look numerately at all aspects, making many rough calculations and checking one against another. Insight-developing studies (IDSs), undertaken primarily to give the designer insight, are a powerful aid.

IDSs should be as simple as possible. Take the extended case given below, where insight was needed into whether it might be practical in a sea wave energy collector (WEC) to use as a source of reaction a mass moving relative to the working surface acted on by the waves. A value of the amplitude of this alternating force is 2 MN, and at the optimum half of this will be balanced by radiation, leaving a net force of 1 MN to be reacted. If we try a mass of 1,000 tonnes, then to react 1 MN it must have an acceleration of 1 m s^{-2}. As the typical angular frequency of waves is about 0.7 rad s^{-1}, to achieve the required acceleration requires the mass to move with an amplitude of about 2 m, which is quite modest. We might consider a smaller mass with bigger amplitude, and so on. It is not long, however, before a more thorough treatment is needed, but it is still quite transparent and comprehensible.

An **abstract view** is often a great help as a prompter of insights. Consider a bolt, which can be regarded in the abstract as a very stiff short piece of string. Ideally, it should not be used to carry shear or bending, only pure tension. This view is useful in the design of cylinder heads and big ends. Similarly, it is often useful to recognise that the ideal joint between structural entities is no joint at all, just the uninterrupted material (again, as in big ends).

A beautiful but rather arcane example is the absorption refrigerator, in which three levels of abstraction proved helpful to the author [3]. The highest of the three is to regard the refrigerator as a reversed distillation column, which immediately makes it possible to derive ideas for improving performance by insights derived directly from distillation engineering.

One small example of abstraction is contained in the advice that functions should be expressed as a verb and a noun, "store liquid" or "guide striker", avoiding forms like "liquid tank" or "striker slide", which assume something in advance about the nature of the means.

However, it is not usually possible to get far without fleshing out the abstract view with some concrete embodiment, in which the abstract view is transformed into a convincing step forward.

2.5.1 Sufficient insight

The extended example given below (Section 2.8) will provide some idea how thorough your grasp should be. To decide when your insight into a problem is sufficiently developed, ask yourself if you could answer *any* question you might be asked on it, with rough values for any quantities involved. This is a weaker form of a harder test. Can you hold all aspects of the design in your mental field at once? Ideally, the designer should be able to bring all aspects of the system being designed "into field" at once. At such a level of understanding, it is possible instantly to perceive the relation of any two parts, without intermediate stages. Torroja's dictum could be rendered as "the designer should be able to hold the whole structure in his mental field at once", which is not difficult for the fall of a stone or the release of an arrow. I cannot quite manage it for the wave-energy converter described below, but that is a hard case.

Insights generally clarify areas of the problem, making them fit into less of the mental field and so extending the fraction that can be grasped all at once. Especially in the case of redesign, it fits well to quote selectively.

> To grasp the ... scheme of things entire ... and then
> Re-mould it nearer to the Heart's Desire!

(with apologies to Fitzgerald and Omar Khayyam).

2.6 Design principles

Design principles are a very valuable, very neglected, little known and mostly unresearched aid to designers. Most designers know a few, and use others without consciously realising it. I originally omitted one important principle from the list [1], though I had used it often. Five examples will be given.

2.6.1 Kinematic design (least constraint)

This principle has been known since the mid-19th century and has an interesting history, having been prominent in instrument design and later in the gas turbine. It may be summarised as "when locating or guiding one body relative to another, use the minimum constraints that will do", or less formally but vividly, "let it all float about, and sort itself out". It is widely understood and generally observed in metrology and instrument design, an early example being the mounting of theodolites on their bases. The three feet on the theodolite are each terminated in balls. One ball was located in a conical hole, removing three degrees of freedom, one in a V-section groove, removing two degrees of freedom, and one rested on a flat, removing one degree of freedom. With six degrees of freedom removed in all, the theodolite was fully constrained but not over-constrained (least constraint, statically determinate). Before the introduction of the new arrangement, theodolites had commonly been over-constrained so that when levelling screws were adjusted, internal forces were generated which caused "stiction" and irregular movement and strained the instrument, leading to inconsistencies in measurement. Modern coordinate measuring machines (CMMs) are textbook examples of least constraint.

The idea of kinematic design is wider, including the introduction of additional degrees of freedom to balance out forces. Helicopter rotor blades have each two hinges at the root, one about a roughly vertical axis (the drag hinge) and one about a roughly horizontal axis (the flap hinge), to balance out horizontal and vertical forces respectively.

In epicyclic gears, if either the sun pinion or the annulus (or for that matter, the planet carrier) is free to float, then it will move so as to balance out the loads on the planet gears. In simple epicyclic gears there is a large torque in the planet carrier, which must be made very stiff or else the wind-up in it will concentrate the loads on the gear teeth towards one end. An alternative is to allow the planet a rotational freedom about a radial axis so that the teeth self-align. The author was able to save much weight in the Napier (later Rolls-Royce) Gazelle helicopter gas turbine by mounting the planet gears on diaphragms to give them such a freedom, and using a light, relatively flexible, planet carrier. The same result was achieved by Hicks in his Compact Orbital gearing using flexible pins [3].

2.6.2 The small, fast principle

The author has only recently listed this principle, after using it dozens of times. It is dramatically illustrated by the comparison of the solenoid and the electric motor, where the former is much larger than the latter for the same task, because the motor uses its working space, the air gap, repeatedly instead of only once in an operation. A given cylinder and piston will give more power if it operates at higher revolutions per minute, and so on. Limits generally exist to what can be done – for instance, the compactness of engines is limited by the piston speed, and the compactness of electrical machines by the rotor surface speed at which inertial loadings become prohibitive. The economical compactness of hydraulic machinery is limited by the falling-off of performance, again due to inertia, but here that of the working fluid. Sometimes this principle manifests itself as a preference for higher frequencies. In wind turbines, the desire to eliminate step-up gearing between the turbine shaft and the generator, itself an example of the principle, has led to another manifestation, that of increasing the frequency of use of the air gap by using very large numbers of poles. The generator still has to be large in diameter, but, because of the high frequency of use, the quantity of material in the magnetic and electrical circuits is reduced.

The wind turbine itself exemplifies the small, fast principle. It has narrow blades moving at high speed (around ten times the wind speed at the tips) and using little material to capture energy from a lot of air. This high speed also reduces the cost of the step-up gear, if there is one, or the generator, if there is not. The cost of gearing or electrical machines is almost directly proportional to its torque capacity (at the slow end, in the case of gearing).

The small, fast principle is very obvious, but in some cases it may be overlooked.

Insight, Design Principles and Systematic Invention

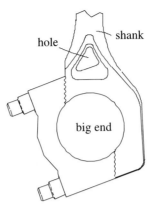

Fig. 2.2 WASA diesel big end.

2.6.3 Matching

The need for matching is well recognised in most important areas (particularly in the relationship of driver and driven), but it is often overlooked in less conspicuous ones (*e.g.*, the matching of the stiffness pattern of a big end to the demands of the oil film in its bearing [12]). Figure 2.2 shows a big end of unconventional form. The split shank matches the deflection of the bearing ring to the needs of the hydrodynamic film, so increasing the load capacity by about 40%. The single shank of the conventional form of connecting rod makes a hard spot in the big end, like a brick under a mattress, but the form of Fig. 2.2 is more like a yielding hammock, strung across the bifurcation of the shank. An interesting point is that the massive big end looks too rigid at first sight to need its deformation to be taken into account, but the evaluation of a simple dimensionless parameter soon shows that it is not.

2.6.4 "Prefer pivots to slides and flexures to either"

The preference for pivots rather than slides is well established in some fields. Pivots have many advantages over slides, as was recognised by Watt, which is why he used straight-line motions (and why they are used today, *e.g.*, in some car suspensions). They are cheaper and easier to make, they do not have exposed working surfaces that must be protected, they usually involve less friction and they are easier to seal.

Where a flexible element (or elements) can be used instead of a pivot or a slide, it has the advantages of not requiring lubrication, freedom from stiction (valuable in instruments such as CMMs) and freedom from wear. A dramatic example is given by modern helicopter rotor heads, which have replaced the drag and flap hinges, and even the pitch bearings, by a flexible link that bends in the flap and drag senses and is twisted by the controls to vary pitch. The resulting simplification of a very complex arrangement is spectacular. Taking another example used above, the two designs of epicyclic gear with self-aligning planets mentioned both use flexures. In this application, it is difficult to ensure that a pivot will not have too much residual friction to work properly.

2.6.5 "Where possible, transfer complexity to the software"

This principle is central in mechatronics, where good design is often characterised by simple but highly refined mechanical parts, with all the complexity put into the software that can be. As an example, from CMMs again, instead of making the axes very precisely orthogonal, which is very expensive, it is cheaper to make them less accurately orthogonal and use the software to correct for the errors. It is interesting to note the conflict between this practice and that advocated by Suh [13] in his second general principle, that the best solution, other things being equal, is the one that demands the least information. Clearly, the practice in CMMs requires far more information, but it is a better solution.

2.7 Systematic synthesis

Engineering science is too extensively analytical, and too little concerned with synthesis. *Systematic synthesis* is occasionally possible, particularly for components or aspects of design. Sometimes, as has been noted, the engineering science does nearly all of the work. Well-known examples come from structures, such as discs and frames [5,14,15], and applications of thermodynamics, as in the refining of steam power cycles and the separation of gases [1]. Occasionally, it is very difficult to see how to make a start at all, as in the case of an extensible sheet [16] for tank walls in liquid natural-gas tankers. Some cases, like the combined wind turbine and heat pump [5], are straightforward once the right starting point is found.

2.7.1 Clothes-peg example

In some simple products, where engineering science is not a major consideration, it is possible to proceed on similar lines with great advantage. In Ref. [21], I looked at the design of a clothes peg. An important consideration is the matching of the hand to the peg, where there is a conflict between providing enough grip on the clothes and the easy "half-a-hand" operation that is desirable, where the hand applying the peg must also support an item of washing that may be quite heavy. It is shown how this problem is avoided in the traditional peg made from a single piece of wood, where the line and the item wrapped round it are forced into the jaws of the peg until tight. The all-wood peg, however, is deficient in its ability to store strain energy. An alternative design was proposed that, like the all-wooden peg, has less energy demand and ample strain energy storage capacity. It is also a monolithic piece of plastic and would be cheap to produce. Unfortunately, it is difficult to use because of the combination of movements required.

This example shows some of the features of systematic synthesis: the need to develop insights, to describe properties like "half-hand operation", "shear strength" (to prevent pegs holding heavy items pulling together) and "corner work", and to recognise the scope for design principles like matching.

Sometimes, however, by tackling the aspects of a problem in the right order, an approach to true systematic synthesis can be achieved. This is probably possible only in special cases, like the example that follows (Section 2.8). Unfortunately, the mechanics of sea waves are not familiar to most engineers, but the lines of the argument should still be clear.

2.8 Insight and systematic invention in power from sea waves

2.8.1 Background

The collection of energy from sea waves, desirable in the search for renewable energy, has proved to be remarkable as a design problem because of the enormous number of distinct solutions possible.

Work began seriously in the UK in the late 1970s, mostly on large devices like the well-known Salter duck [17], spurred partly by the preference of the then future customer, the Central Electricity Generating Board, for large units. The Norwegians Budal and Falnes, however, recognised the desirability of relatively small units, kept small by the use of resonance to give large movements and hence more power per unit area of the working surface (an example of the small fast design principle). A small unit can collect the energy incident in a wave front more than twice its own width, whereas the capture of big devices is limited to about their own width.

There is a close parallel with a radio aerial, where, by virtue of resonance, a narrow antenna draws in energy from around it. Small wave-energy collectors, say less than 30 m in maximum extent, are called "*point absorbers*", because in many ways the waves "see" them as a point, as an electrical engineer might say.

2.8.2 An abstract view

Let us take the abstract view recommended above. If a body floats freely in the sea, the waves exert a *wave force* on it under which it moves around. To draw power from the body we must apply a reaction to it that opposes the wave force, and it has already been stated in Section 2.5 that a reaction of the order of 1 MN is appropriate for the small units that Budal and Falnes advocated. The provision of this reaction is an essential function in wave power, and it turns out to be crucial.

We can now form a very abstract picture of our WEC. We have a floating body with a *working surface* on which the waves exert a *wave force* that is opposed by a *reaction*. The instantaneous power output is the product of the net wave force and the velocity of the working surface (the qualification "net" is because the working surface will itself radiate waves, so reducing the power and also the reaction required). The WEC is resonant, *i.e.*, it is a mass–spring system, in order to increase the amplitude and hence the velocity of the working surface, typically to about 2.5 times the particle velocity in the waves. In the kernel table of functions and means [3], these two insights lead to two functions, one of which, the spring function, essential for resonance, would not have been included at all without the resonance insight. The other, to provide a reaction, might also have been overlooked.

The remaining key choice is that of the motion of the working surface – up and down, as in a bobbing buoy, which is called "heave", or to and fro in the direction of the waves, which is called "surge". In a small body, no other motion is strongly coupled to the sea.

To see this, imagine a vertical square working surface F that is oscillating about a horizontal axis through its centre C (Fig. 2.3). If F turns clockwise, its top half forces water to the right, but the bottom half draws a roughly equal amount from the right, so the net effect at a distance is very small. An alternating rotation about C (pitching) thus propagates only small waves. Now, one way of viewing the action of a good

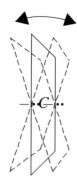

Fig. 2.3 Coupling in pitch.

WEC is that its motion generates a pattern of waves that cancels ("interferes destructively") with the incident train of waves, so that the amplitude, and hence the energy, of those waves is decreased. The difference is the energy captured (barring certain losses). Simple experiments with a square of plywood in a bath of water will confirm the poor coupling in pitch, and similar simple arguments dispose of the other motions, leaving only heave, surge and a combination of the two.

2.8.3 Table of options

Having identified a resonant point absorber as the most promising option, because it is likely to be the least expensive, we can then draw up a simple table, Table 2.1, of the major remaining options.

Function 1, spring. In the first line, we need a spring to achieve resonance in what has to be a mass–spring system. Gravity acting on a weight, as in a pendulum, yields the same characteristics, and is more attractive than a gas spring with its leaky seals and expensive, well-finished bore. Note this and pass on to the second function, "provide a reaction".

Function 2, provide a reaction. A small device could be anchored stiffly to the seabed to provide a reaction, but the wave forces in storms would be very high, demanding a very strong and, therefore, very expensive structure. In the case of a long device, the wave forces on it will vary in phase along the length (phase diversity), and so will largely balance each other, so that there is no need for another source of reaction. This option has been omitted because it does not exist for point

Table 2.1 Simple table of options for point absorber.

Line	Function	Means or configuration 1	Means or configuration 2	Means or configuration 3
1	spring	gravity spring	gas spring	
2	provide reaction	sea bed	amplitude diversity	reaction mass
3	motion	heave	surge	combined (heave and surge)

absorbers. It is also possible to use a difference in amplitude, the fact that wave amplitude falls with depth, "amplitude diversity", but this has the same disadvantage as a poor work ratio in thermodynamics, leading to poor net power because the negative work takes away from the positive.

This leaves "react against a mass", which does not at first sound promising. However, a simple IDS, the first step of which has already been given above, serves to examine how practical the idea is, and it turns out to be surprisingly so.

Function 3, motion. Then go to line 3, and consider the motion. Surge makes it possible to capture the power in a wave front of length (wavelength/π), but heave can collect only half as much, On the other hand, heave works equally well with waves from any direction. But there is another difference, which is decisive. In surge, the reaction mass moves horizontally, but in heave it must move vertically, and that involves large fruitless cycling of energy, and hence losses.

2.8.4 Embodiment

After all this abstract stuff, look at the concrete side, shown in Fig. 2.4. There is a paddle facing the waves, with a handle below it, well ballasted at the bottom. The paddle is the working surface and the ballast is the mass reacted against. This paddle with a weighted handle provides the working surface, the resonant system and the means of reaction, all in one piece; an elegant economy of means. In use it pitches vigorously in the way indicated by the arrow. To force it to move it in the desired way, an alternating moment of the right size and phase must be applied. Our preferred way is by a sliding mass (Fig. 2.4), the movement of which is resisted by hydraulic rams. These rams work predominantly as pumps, and drive a hydraulic motor that drives a generator. By switching the rams between pumping, idling and sometimes doing work, the best motion for power capture can be maintained, even in irregular seas. The mechanics are explained in Ref. [18].

There remain some further important aspects, such as the problem of control to maintain large motions of the working surface (quasi-resonance), but basically the conceptual design is complete. It is very simple, just a hull containing a sliding mass controlled by hydraulic rams. It has no external moving parts and it is anchored by a compliant mooring, the softness of which protects it from big waves. Its output per tonne is high (>400 W t^{-1}) and it appeared as a most promising device in the 1992 Wave Energy Review [19].

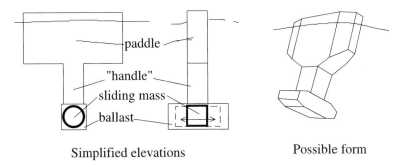

Fig. 2.4 P S Frog.

The whole concept stems from a few crucial insights, particularly those of Budal and Falnes, and our own into the practicability of using a reaction mass and the simple weighted paddle form that embodies it. An IDS led to the rejection of amplitude diversity as a means of reaction. As regards design principles, the Norwegian approach conforms to the small fast principle and their ingenious mooring arrangement was an example of least constraint. At Lancaster we would rather have a pivot than a slide for the reaction mass, and new developments may make that change possible, with several advantages.

Notice how a kernel functions–means table, with just three functions, is enough, but it requires preparation by the designer and the development of crucial insights before such a table can be drawn up. Moreover, there remain further possibilities to explore.

2.8.5 The checking of systematic design processes–link-breaking

It is important to test procedures such as that just described, because there are often links that can be broken, and their breaking may lead to other good designs. In the case of P S Frog, the link that rejects the sea bed as a source of reaction can be broken. To act as a reaction source the mooring must be stiff, preferably able to develop a reaction of the order of 1 MN for a movement of a fraction of a metre, not like an ordinary mooring, which is soft. With a stiff mooring, the device would be subject to huge forces in storms and it and its anchorage would have to be expensively strong. However, it might be possible to devise a mooring that went soft in storms or a device that reduced its exposure to the wave forces to survive, say, by retracting into the seabed (this is not likely to be practical, but there are other possibilities). Thus breaking a link in a logical chain may lead to a radically new design. This is related to ideas about the resolution of paradoxes, *e.g.*, a mooring that is both stiff and soft. Salter's duck WEC [19] has a spine that is rigid but flexes under high enough loads.

2.9 Summary

- Design proceeds by small, and occasionally large, inventive steps that usually derive from insights and are usually based in engineering science. To design requires digging into the engineering science and developing insights. Some ways of developing insights have been suggested.
- Insights can often be linked together in "logical" chains, providing systematic design. These chains are special to the design in question and constitute *ad hoc* inventing engines.
- Some links in these chains may be breakable, often by an inventive step, and may become branch points leading to novel designs.
- Design principles, such as have been described, are a great help to the designer.

Acknowledgements

With grateful acknowledgements for Fig. 2.1, the rotating cylinder valve engine, to Keith Lawes (the designer) and RCV engines, and for Fig. 2.2, the WASA diesel engine big end, to Wartsila Diesel.

Appendix 2A

As an example of a case where the aid leaves little to the human inventor, the author asked the question, "Can we do much better than an involute in the way of gear-tooth forms?". I derived an equation for the local relative radius of curvature R at a contact for a pair of gears of general form, in terms of the distance x from pitch point and the sine of the pressure angle y, which is as follows:

$$R = \{y + x(2p+1-k) + x^2[p^2 + (1-k)p - k]/y\}/(1-k)$$

where k is the gear ratio (<1) and p is dy/dx. R is a measure of the local load capacity of the teeth, so that the overall capacity of the gearing is proportional to C, where C is $(1 + k)$ times the integral of C with respect to x along the path of contact. Only one term in R offers scope for a *large* increase in C, and that is the term in p^2, which on the face of it allows an indefinite increase. As p is increased, the radius of curvature increases as the square and the length of the path of contact decreases, but only as the inverse of p. The net effect is an increase in the term in R in proportion to p. This increase is limited in practice, because as the banana-shaped area of contact increases it eventually runs off the tooth top and bottom. Nevertheless, by introducing a very rapid rate of change of the pressure angle, gearing can be designed with about twice the load capacity of well-designed involutes [20].

Such gearing was invented by Wildhaber around 1923 and it has about twice the load capacity of spur gearing. It was reinvented by Novikov in about 1955 and is generally called "Wildhaber–Novikov" gearing. In spite of being about twice as strong as well-designed involute gearing and not unduly difficult to make, it has failed in the market, partly, no doubt, because a twofold improvement is not enough to justify changing from a well-tried solution. In areas such as helicopter reduction gears, where weight is at a premium, however, it ought to win hands down.

One reason it failed to do so may have been the failure to apply the design principle of least constraint.

All gearing, with the exception of involute, is sensitive to centre distance, but it will run of its own accord at the right centre distance given the chance by a freedom to float perpendicularly to the mean pressure line. In the case of Wildhaber–Novikov gearing this was demonstrated by D.C. Johnson in 1959. Unless least constraint is applied in this way, any advantage in weight will be lost in stiffening the structure. A similar case, given above, is the Gazelle engine gearbox example. Here, the gearing was involute, and the extra freedom provided was a rotation to allow the gears to self-align, saving the great weight of a stiff spider.

References

[1] Ruhemann M. The separation of gases. Oxford University Press, 1949.
[2] French MJ. An annotated list of design principles. Proc Inst Mech Eng 1994;208:229–34.
[3] French MJ. Conceptual design for engineers. 3rd ed. Springer, 1998.
[4] The innovative engineer. Smallpeice Trust, 74 Upper Holly Walk, Leamington Spa, CV32 4JL [in VHS only, with script].
[5] French MJ. The opportunistic route and the role of design principles. Res Eng Des 1992;4:185–90.
[6] Hadamard J. The psychology of invention in the mathematical field. Dover, 1944.

[7] Poincaré H. Science and method. London: Nelson, 1924.
[8] RCV engines, on http://www.rcvengines.com.
[9] Motluk A. Read my mind. New Sci 2001;169(2275):22.
[10] Maxwell JC. In: Niven WD, editor. Scientific papers, vol. 2. Dover, 1965; 171–8.
[11] Dickinson HW, Jenkins R. James Watt and the steam engine. London: Camelot, 1989.
[12] Rosgren C-E. Diesel engine design aspects for heavy fuel operation. Proc Inst Mech Eng 1985; 199:251–3.
[13] Suh NP. The principles of design. Oxford University Press, 1990.
[14] Hemp WS. Optimum structures. Clarendon, 1973.
[15] French MJ. A measure of utility of parts of plane frames. Proc Inst Mech Eng C; 1999;21:623–8.
[16] French MJ. Systematic design of an extensible sheet for LNG carrier tanks. In: Proceedings 6th International Offshore and Polar Engineering Conference, vol. 4, 1996; 529–32.
[17] Salter SH. Power conversion systems for ducks. IEE Conference on Future Energy Systems, Institution of Electrical Engineers, London, 1979.
[18] French MJ. Tadpole: a design problem in the mechanics of the use of sea wave energy. Proc Inst Mech Eng 1996;210:273–7.
[19] Thorpe T. Wave energy review. ETSU Report R26, Dec 1992.
[20] French MJ. Gear conformity and load capacity. Proc Inst Mech Eng 1966;180:1–4.
[21] French, MJ. Invention and Evolution. 2nd ed. Cambridge University Press, 1994.

Synthesis and theory of knowledge: general design theory as a theory of knowledge, and its implication to design

Yoram Reich

Abstract Design knowledge is the driver of successful product synthesis. General design theory (GDT) models design knowledge as a topology and proves theorems about design. We review the theory, its limitations, and discuss how GDT, in spite of its unrealistic assumptions, can still provide insight about building computer-aided design systems.

3.1 Introduction

Knowledge is an elusive and subjective concept. Nevertheless, there are two common views about the nature of knowledge: structural view-knowledge as the content of a representation, and functional view-knowledge as the capability to solve problems [1]. No single view is inclusive; the structural view alone cannot create designs, and the functional view cannot exist in isolation from a representation. Knowledge is better understood as lying between these extreme views.

Synthesis is the creation of artefacts for satisfying given specifications. Clearly, synthesis is an activity; nevertheless, according to the above definition of knowledge, synthesis knowledge could be understood as varying between a similar range: from knowledge about synthesis to the encompassing capability of enacting synthesis.

Among the existing mathematically based design theories, general design theory (GDT) [2-4] is the most familiar. GDT had a comprehensive goal to explain human design and guide the development of computer-aided design (CAD) tools. To this end, GDT focused on the structure of design knowledge. The assumptions made in the theory are so powerful mathematically that they allow one to derive strong implications about design processes: if design knowledge structure is a topology based on all possible artefacts, then design terminates successfully once the specifications, represented by elements of the topology, are provided. Furthermore, GDT assumptions equate design with synthesis; GDT guarantees that when synthesis is complete, it has generated a valid design and not just a design awaiting validation.

Unfortunately, perfect topological structures of knowledge are unrealistic, and since the results about design are not graceful with respect to the topological assumption; any imperfection in the knowledge structure dissolves the remarkable promise. The extended version of GDT tries to remedy this situation by modifying the assumptions, but it still suffers from similar limitations. Therefore, it is fair to question the contribution of GDT to real design beyond its "historical" landmark contribution to design theory.

This chapter reviews GDT and its implications for design in an informal manner; a detailed formal and critical review appeared elsewhere [5]. First, it discusses how GDT models are used in design (or synthesis) with the help of a simple example. Second, it addresses some guidelines for building CAD systems driven by GDT. Third, it briefly mentions recent development related to mathematically based design theory that are more general than GDT, thus placing it in a larger perspective [6].

3.2 The domain of chairs

Figure 3.1 depicts eight chairs, referred to as the chairs domain, that are used to explain the concepts discussed in this chapter. Each chair in the figure is denoted with a letter. The chairs provide some *functionality*, which is summarised in Table 3.1. Each row describes a different function of a chair. The "+" in Table 3.1 denotes that a chair provides the corresponding function, and a "−" denotes its lack thereof.

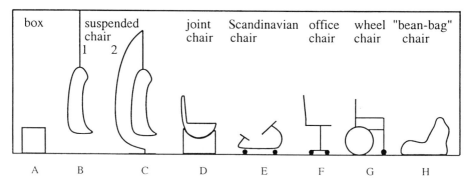

Fig. 3.1 The chairs domain.

Table 3.1 Functional properties of chairs.

Function		Chair							
		A	B	C	D	E	F	G	H
1	*seats* – prevents a downward movement of the body	+	+	+	+	+	+	+	+
2	*supports back* – supports an upright posture	+	+	+	+	+	+	+	+
3	*revolves* – revolves around a vertical axis	+	+	+	+	+	+	+	+
4	*movable* – can be easily moved	+	−	−	−	+	+	+	+
5	*constrains back* – constrains backward movement of back	−	+	+	−	−	+	+	−
6	*easy to manufacture* – has a simple design with standard components	+	−	−	−	−	+	−	+
7	*aesthetic*	−	+	+	+	+	−	−	−

Table 3.2 Observable properties of chairs.

Structure		Chair							
		A	B	C	D	E	F	G	H
1	*has a seat*	+	+	+	+	−	+	+	−
2	*has a back support*	−	+	+	+	−	+	+	−
3	*has legs*	−	−	+	−	−	+	+	−
4	*has wheels*	−	−	−	−	+	+	+	−
5	*has a vertical rotational dof*	−	+	+	+	−	+	−	−
6	*is lightweight*	+	+	−	−	+	+	+	+
7	*has a hanger*	−	+	+	−	−	−	−	−
8	*has a brake*	−	−	−	−	−	−	+	−

In addition to providing functions, each chair has properties that can be observed and which, therefore, describe the *attributes* or the *structure* of the artefact; some of these are summarised in Table 3.2.

Naturally, there are functions that are directly derived from the structure of a chair. For example, a chair that *has wheels* is *movable* or a chair that *has a vertical rotational degree of freedom* (*dof*) can *revolve*. Note that this structure–function relation may be imprecise; *e.g.*, chair C with a *rotational dof* does not allow for 360° rotation. Other functions may be more complex and could not be inferred from one observable property. To illustrate this, a chair can *support back* although it does not have a *back support*. For instance, chair A provides back support owing to its location near a wall; its function is context dependent. Also, chair E provides back support owing to its structure although it does not have a *physical* back support. Some functions may qualify other functions. For example, the function *constrains back* qualifies the function *support back*. This function is quite complex to assess. Chairs F and G constrain back movement owing to their structure, whereas chairs B and C constrain it owing to static considerations; chairs D and H do not constrain back movement, whereas chairs A and E do not even have a physical support.

The previous examples concentrated on inferring potential functionality from artefact structure. This is useful in *analysis*. In contrast, designing is mainly concerned with *synthesis*: the generation of artefact structure that will satisfy a desired function. For example, the specification of a chair that will be *movable* and *constrain back* leads to two potential designs: F and G. These designs can be generated in two ways. The first way starts with {A, E, F, G, H} as the *movable* designs and refines them with the *constrain back* property. The second way starts with {B, C, F, G} as the *constrain back* designs and refines them with the *movable* property. The most concise description of the solution is the chairs that have *physical back support* and *wheels*. Another description, which does not seem relevant but is nonetheless correct, is the chairs that *have legs* and are *lightweight*.

Note that the refinement process was made easier by the use of the eight representative chairs as mediators between the specification and the design description. In the absence of these chairs, the process might have been more difficult.

3.3 GDT

3.3.1 Preliminary definitions

A topological structure of objects provides an interesting perspective of viewing design. Topology can be viewed as a generalisation of the concept of continuity [7]. There are several important properties of continuity that are of interest in various design tasks, such as synthesis, analysis, or redesign, such as *continuity* and *convergence*. *Continuity* is a process-oriented concept. It guarantees that a small change in the design description will result in a small change in the artefact functionality and vice versa. Therefore, if the current candidate's functionality differs slightly from the required function, a small modification to the structure may be sufficient. Convergence is also a process-oriented concept; it provides a different perspective of continuity. Convergence guarantees that a sequence of incremental refinement changes will cause only small incremental changes to functionality.

> **Definition 3.1.** An *entity* is a real object that existed, exists presently, or that will exist in the future. The set of all objects is called the *entity set*.

Any chair that existed since the invention of the chair and that will exist is an entity. For the purpose of simplification, assume that Fig. 3.1 contains all these entities and is the entity set.

> **Definition 3.2.** When an entity is subjected to a situation, it displays a behaviour that is called a *functional property*. The collection of functions observed in different situations is the *functional description* of the entity.

The properties listed in Table 3.1 are all functional properties of chairs. Table 3.1 specifies the functional behaviour manifested by each chair.

> **Definition 3.3.** The representation of an object is called *concept of entity* or *entity*.

The representation of a chair using the function and structure properties from Tables 3.1 and 3.2 is an entity.

> **Definition 3.4.** A classification over the entity set is a division of the entities into several classes. Each class is called an *abstract concept*. The set of all abstract concepts is denoted by $\mathcal{T}$.

Chairs A and B can form a class and the remaining chairs can form another class. A more meaningful classification can be obtained by classifying the chairs based on their properties. For example, the property *has legs* divides the set of chairs into two classes: chairs with legs {C, F, G} and chairs without legs {A, B, D, E, H}.

> **Definition 3.5.** The set of all functions, called the *function space*, is the set of *all* the classes from all the classifications of the functions. It is denoted by $\mathcal{T}_1'$. The set of all artefact descriptions, called the *attribute space*, is the set of *all* classes of all the classifications of attributes. It is denoted by $\mathcal{T}_0$.

In principle, a topology over a set with n entities can contain 2^n classifications. Therefore, in principle, the function (or attribute) space for the chairs domain can contain 2^8 functions (artefact descriptions), each function (description) being a different classification over the set of chairs. In our example we limit the discussion to several functions (artefact descriptions). For example, the functions *support back*, *movable*, and *aesthetic* define a classification that singles out chair E from the whole set of chairs. The number of functions, however, is not accurate, since some of the potential classifications (*e.g.*, *easy to manufacture* and *aesthetic*, or *is movable*, *aesthetic* and *constrain back*) do not contain any chairs. Similarly, the attributes *has legs* and *is lightweight* designate the artefact descriptions of chairs F and G.

3.3.2. GDT's axioms

GDT's axioms convey the assumptions of the theory about the nature of design knowledge.

> **Axiom 3.1 (Recognition).** Any entity can be *recognised* or *described* by its attributes and/or other abstract concepts.

Each of the chairs in Fig. 3.1 can be easily singled out from the set of chairs by using one or more of its artefact description attributes. For example, chair A can be recognised as the only chair that *has no vertical rotational dof* and *no wheels*.

> **Axiom 3.2 (Correspondence).** The entity set and its representation have *one-to-one* correspondence.

If each of the chairs in Fig. 3.1 is perceived as a real object (*i.e.*, entity) and its description given in Tables 3.1 and 3.2 as the concept of entity, the axiom says that there is a one-to-one mapping between them. The discussion on Axiom 3.1 applies to Axiom 3.2, therefore, to the representation of entities. An appropriate representation may require the use of an infinitely long property-value list. Of course, this is practically impossible. Nevertheless, the theory assumes the availability of resources that overcome this difficulty.

> **Axiom 3.3 (Operation).** The set of all abstract concepts is a *topology* of the entity set.

A topology $(S, \mathcal{T})$, sometimes denoted only as $\mathcal{T}$, is a mathematical entity consisting of a set S and the set $\mathcal{T}$ of subsets of S that satisfies the following properties:

1. $\phi \in \mathcal{T}$ and $S \in \mathcal{T}$;
2. for every $s_1, s_2 \in \mathcal{T}$, $s_1 \cap s_2 \in \mathcal{T}$; and
3. for $s_i \in \mathcal{T}$, $i \in \Lambda$, Λ a countable set, $\cup_i s_i \in \mathcal{T}$.

The fact that the set of all abstract concepts is a topology influences both its structure (through properties 1, 2, and 3 of topology) and the possible operations on it (properties 2 and 3). Axioms 3.1 and 3.3 demand that all entities be treated equally.

The simplest topology over the set S of chairs is $\mathcal{T} = \{\phi, S\}$. Another obvious topology is the power set of S (having $2^8 = 256$ elements), also called the discrete topology. Neither of these topologies are too interesting. Another topology $\mathcal{T}$ can be constructed such that $\{\phi, \{A, H\}, \{B, C, D\}, \{E, F, G\}, S\} \subseteq \mathcal{T}$. The inclusion of these subsets of S may be caused by the need to differentiate between the entities in these different classes. To complete the topology, $\mathcal{T}$ must satisfy the three properties listed above. Therefore:

$$\mathcal{T} = \{\phi, \{A,H\}, \{B,C,D\}, \{E,F,G\}, \{A,B,C,D,H\}, \{B,C,D,E,F,G\}, \{A,E,F,G,H\}, S\}.$$

3.3.3 Ideal knowledge

Definition 3.6. *Ideal knowledge* is the one that knows *all* the entities and can describe each of them by abstract concepts without ambiguity.

In the chairs domain, this assumes that one can recognise that two chairs are different through observing their attributes, which is clearly correct for the chair domain. To facilitate the recognition without ambiguity of each chair, $\mathcal{T}_0$ must include each chair. The second and third properties of topology immediately imply $\mathcal{T}_0 = 2^S$. To simplify, we created the abstract concepts (shown in Fig. 3.2) from properties describing chairs in Table 3.2. We will use these concepts as the "topologies" over the set of chairs even though they do not satisfy the properties of topologies. Each line in the figure circles the chairs that satisfy the property number that is written along the line. A number in parentheses denotes that the circled chairs have "−" as their value for this property. We see that each chair can be recognised through its attribute description, since in Fig. 3.2b each chair is enclosed by lines. Figure 3.2a shows two pairs of chairs, $\{B, C\}$ and $\{A, H\}$, each of which contains chairs that cannot be differentiated based on their functionality, since they are not separated by a line. This can be remedied by adding the function *allows for easy floor cleaning*, whose values will be "+" for all the movable chairs and chair B and "−" for chair C, and the function *shapeless*, whose value is "+" for chair H and "−" for the remaining chairs. The inability to differentiate between entities based on function indicates that, given a specification, there may be several candidate designs that satisfy it.

Definition 3.7. The design specification, $T_s \in \mathcal{T}$, designates the function of the required entity by using abstract concepts.

It is natural to describe the specification of an object by the intersection of abstract concepts, since the specification describes functions that the desired chair must fulfil. A specification that a chair must *revolve* and be *movable* is described easily (using Table 3.1) by $\{A, B, C, D, E, F, G, H\} \cap \{A, E, F, G, H\} = \{A, E, F, G, H\}$. A solution to this design problem is any $s \in \{A, E, F, G, H\}$. If the specification were to insist on an *easily manufactured* design $\{A, F, H\}$, the result would be $\{A, E, F, G, H\} \cap \{A, F, H\} = \{A, F, H\}$. We see how the second property of topology is used to identify the set of candidate designs for each specification.

Definition 3.8. A design solution is an entity s that is *included* in its specification and contains its necessary manufacturing information.

Theorem 3.1. The entity concept in the attribute space $\mathcal{T}_0$ is a design solution that is represented by the intersection of classes that belong to $\mathcal{T}_0$. Each of the attributes can be perceived as manufacturing information.

General Design Theory as A Theory of Knowledge, and Its Implication to Design 41

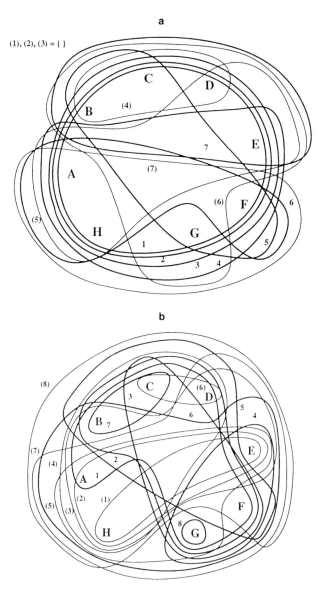

Fig. 3.2 a Function and **b** attribute "topologies".

The attribute space is the one that is created by the observable properties that are needed to manufacture an entity. Therefore, an entity in the attribute space is a design solution. Definition 3.9 formalises the notion of design solution described in Section 3.1, and formalises an intuitive definition of design. Theorem 3.2 establishes the nature of design in the state of ideal knowledge.

Definition 3.9. Design process is the designation of a domain in the attribute space T_0 that corresponds to a domain specifying the specification in T.

Design is a mapping between the function space and the attribute space. This means that design is an algorithmic process, which includes the catalogue design as a special case. Recalling that the function and attribute spaces are topologies over the same set of entities, this mapping is implicitly built into these topological structures.

Theorem 3.2. In the ideal knowledge, design is completed immediately when the specification is described.

Let a specification be to design a chair that *constrains back*, is *movable*, and is *easy to manufacture*. Specification 1 (a chair that *constrains the back*) is {B, C, F, G}. The addition of specification 2 (a *movable* chair), {A, E, F, G, H}, generates the specification {F, G}. The further addition of specification 3 (*easy to manufacture*), {A, F, H}, leads to the combined specification {F}. At this stage all the specification properties have been satisfied and the resulting design solution is {F} (provided that we use conjunctions of properties).

3.3.3.1 Summary of ideal knowledge

The state of ideal knowledge is characterised by the ability to separate between any two entities. This separation may require the use of infinitely long descriptions of entities. The separation between entities and the requirement that the knowledge structure be a topology guarantee that design would terminate with a solution after a specification is given. The operation on large descriptions of entities by means of set operations that manipulate the topology (*i.e.*, properties 2 and 3 of topology), or by other means for handling extensional descriptions of entities, may require infinite memory capacity and processing speed.

GDT-IDEAL, denoting the state of ideal knowledge, restricts the nature of knowledge to have two perfect properties and through them guarantees the termination of design. Design is limited to the selection from a catalogue or the utilisation of an algorithmic mapping between functions to attributes. Real design never has these properties and, therefore, its termination cannot be guaranteed. Nevertheless, several additional design strategies, and assumptions, can guarantee design in a less ideal state. These issues are discussed next.

3.3.4 Real knowledge

Real knowledge cannot be structured as a perfect topology and cannot be manipulated by infinite resources. These discrepancies from the state of ideal knowledge require two important modifications to the theory. First, only finite descriptions of entities can be manipulated. Second, instead of assuring the successful termination of design, the design terminates with candidates that could have undesired behaviour. To illustrate the need for these modifications, consider designing a chair that will be *aesthetic* and *movable*, and will *seat* and *constrain back*. In the present chairs domain this is impossible. The intersection of the classes corresponding to these desired properties does not contain any chair, although the existence of such a chair is guaranteed by GDT-IDEAL if the domain had a topological structure. The design can be accomplished by the addition of a brake to chair D to make it *constrain back*. As we said before, if we elaborated the domain with additional chairs and properties, then this particular problem would have been solved. However, in general, we cannot locate and fix such imperfections *a priori*. In the real knowledge case, when such a void is found, it could be fixed by adding entities, or properties (*e.g.*, abstract concepts). These additions can be accommodated by GDT, but their *findings* are creative acts external to GDT.

Definition 3.10. A *physical law* is a description about the relationship between object properties and its environment. An *attribute* is a physical quantity that is identifiable using a set containing a *finite* number of physical laws. When entities are classified based on physical manifestations of physical laws, the resulting classification is called *concept of physical law*. This classification is a topology.

If each physical law corresponds to one and only one observable property, the topology of physical law is equal to that of the attribute space, otherwise it may be slightly different. Insisting on a finite number of laws has an important implication that is formalised as:

Hypothesis 3.1. The real knowledge is the set of entity concepts that are made compact by coverings selected from the physical law topology.

Any finite space (*e.g.*, the chairs domain) or a space with a finite number of open sets is compact. Formally, insisting on a compact space is a very restrictive hypothesis; it demands the existence of a finite subcover for each cover of the space. This hypothesis immediately allows one to derive most of GDT-REAL's theorems. Conceptually, this hypothesis says that the description of entities is restricted to a finite number of properties.

Definition 3.11. *Feasible design specification* is a non-empty intersection of a finite number of classes from $\mathcal{T}$.

The following definitions and theorems deal with the important concept of *models*. Intuitively, models are some abstraction of entities that focus on few properties. In many cases, models are then subjected to some analysis based on a theory to find behaviours that are attributed to the original entity. For example, modelling the chairs as two-dimensional entities and the application of statics laws lead to estimating the behaviour of chairs A, D, E, and H as *constraining back* and chairs B, C, F, and G as *unconstraining back*.

Definition 3.12. A metamodel M_Λ is an intersection of a finite number of classes from $\mathcal{T}_0$. The metamodel set $\mathcal{M}$ is the set of all metamodels.

This immediately makes $\mathcal{M}$ a topology and, furthermore, $M \subset \mathcal{T}_0$.

Theorem 3.3. In real knowledge, the *necessary* condition for designing is that $\mathcal{M}$ is stronger than the topology of the function space ($M \supset \mathcal{T}_1$).

Since $\mathcal{T}_0 \supset M$, then if $M \supset \mathcal{T}_1$ we obtain $\mathcal{T}_0 \supset M \supset \mathcal{T}_1$. This implies $\mathcal{T}_0 \supset \mathcal{T}_1$, which is the prerequisite for designability, since any specification contains several concepts of entity or design solutions. Conceptually, Theorem 3.3 says that a condition for designing is that, for any complex functional requirement, there will be a model that will allow differentiating the designs that potentially satisfy this requirement from other designs.

Consider designing an *immovable* chair. According to Table 3.1 or Fig. 3.2a, which describe the function topology, there are three chairs that satisfy this requirement: B, C, and D. From Fig. 3.2b we see that the attribute topology is finer at this region than the function topology because we have two separate subsets of {B, C, D}: {B, C}, the chairs that have a *hanger*; and {C, D}, the chairs that are *not lightweight*. A solution exists because we can create a model based on kinematics that will determine that {B, C} cannot be moved away because they are hung. Or, under a different interpretation of the term "immovable", we could use a mechanics model that will determine that {C, D} cannot be pushed by a human and thus are immovable. The fact that we had these models at our disposal and that they were finer than the function topology allowed us to do the mapping between the function and the attribute topologies and focus on the candidates that fit our interpretation of the requirements.

> **Theorem 3.4.** If we choose function elements as the metamodel, design specification is described by the topology of the metamodel, and there exists a design solution that is an element of this metamodel.

Theorem 3.4 does not guarantee arriving at a solution when the specification is given, but guarantees finding an approximate design solution which is an element of $\mathcal{M}$. This corresponds to real design, where products are described by a finite number of properties, thus implicitly representing an infinite number of possible designs.

> **Theorem 3.5.** In the real knowledge, the design solution has unexpected functions.

Since specifications are defined by a finite set of functional and other properties, they cannot fully determine all the attributes or functional behaviours of the set of candidate designs. Therefore, designs will have behaviours that were not dealt with in the design process and thus may be unexpected.

3.3.4.1 Summary of real knowledge

The state of real knowledge is an adaptation of the concept of ideal knowledge to the real world. Entities are described by a finite number of attributes that can be observed or measured by instruments. Therefore, these descriptions can be manipulated in finite time and require finite storage capacity. The main assumption about the topological nature of knowledge, the extensional representation, and need to know all possible designs remain intact. Therefore, GDT-REAL does not and cannot change the incompleteness of design knowledge.

Design in the state of real knowledge starts with specifications formulated by concepts from $\mathcal{M}$. Design requires the ability to intersect models of the designed artefact continually until they evolve into a set of candidates that satisfy the specification. Specifications that cannot be described by $\mathcal{M}$, *e.g.*, related to aesthetics or ethical issues, cannot be handled.

3.4 Contribution of GDT

Since GDT employs unrealistic assumptions about the structure of design knowledge, GDT predictions about design do not hold for real design processes. Nevertheless, GDT offers guidelines for building design support systems. We focus on the guidelines related to the representation of knowledge and design processes.

3.4.1 Representation of design knowledge

Two potential artefact representations exist for representing objects: extensional and intensional.

In the extensional representation, an attribute is expressed as the set of objects having this attribute. In this representation, all objects have the same status; no predefined hierarchy is imposed upon them. For example:

has a seat = $\{A, B, C, D, F, G\}$, or
has a seat (A), *has a seat* (B), K, *has a seat* (G)

are two extensional representations of the attribute *has a seat*. This attribute could also be described as a relation between *seat* and *chair* as two objects. For example, suppose $a, b, \ldots, g$ are seats, then the relation could be

$seat(a)$, $has(a, A)$, $seat(b)$, $has(b, B)$, K, $seat(g)$, $has(g, G)$.

Graphically, Fig. 3.2 represents the chairs domain extensionally.

In the intensional representation the objects are described by the set of attributes characterising them or the components from which they are built. For example, a chair might be described as

chair(seat, back support, legs, wheels, vertical rotational dof,
   weight, hanger, brake).

Tables 3.1 and 3.2 can be thought of as intensional descriptions of chairs.

Each of the two representations has advantages and disadvantages that can be summarised as follows [3]. (1) Extensional descriptions can be easily modified by the addition of new entities, properties, or their new categorisations; thus, they can support innovations, although they do not introduce them themselves. (2) They support recognising the similarity among almost identical objects; thus, they can support designing from precedences. (3) Extensional representations may be hard to understand, since they are not concise. In contrast, intensional descriptions of objects (1) are hard to modify; (2) do not support inferring the similarity between objects; but (3) their descriptions are more concise.

The first two properties are *conceptual* and fundamental to supporting (incremental) design. The third property describes the ability to understand the *representations* of objects, but not necessarily the objects themselves; it impacts on the computational efficiency of operations. Thus, property (3) deals with implementation issues that are important for CAD systems but less relevant to GDT as a theoretical entity. It is no surprise, therefore, that GDT axioms are built upon extensional representations of entities.

The conceptual superiority of extensional over intensional representations and the theoretical predictions of GDT suggest the use of extensional representations in CAD systems. Nevertheless, most CAD systems, including commercial computer graphics software, favour intensional representations owing to their ease of computer implementation, relative comprehensiveness of knowledge, and their computational efficiency in answering some types of query if their expressiveness is limited to first-order logic.

For the purpose of implementing computationally efficient CAD system, extensional descriptions work well when the description is not too detailed; their ability to support the recognition of similarity between objects and their incremental nature make them useful for exploration, which is important in the conceptual or preliminary design stages. In the detail design phase, objects are detailed with many additional attributes, thereby preferring an intensional representation that is more concise.

It is possible to break away from the extensional representation and allow intensional representations to play a major role in design knowledge representation. Casting design knowledge as *closure spaces* [6] permits modelling real design knowledge and processes and yet retains mathematical clarity. Closure spaces are not used to prove theorems about designability, but to explain design phenomena. They can also be shown to include GDT as a special case.

3.4.2 Design process

GDT implicitly assumes that designing is a mapping from the function to the attribute topology. GDT-REAL presents the concept of models as mediating from the function to the attribute topologies. Models that are sufficiently detailed (to satisfy Theorem 3.3) must be used for supporting the incremental modification of the design towards a solution. Theorem 3.4 guarantees arriving at an approximate solution when using this design process with topological knowledge structure. It is hypothesised that such a process at the early stages of design, when operating on design knowledge in forms closer to topology and when managing extensional information, coupled with intensional information in the later stages may lead to improved CAD systems.

Figure 3.3 illustrates the concept of design or synthesis as a converging process using a detailed version of the chairs domain. An arrow emanating from the metamodel denotes a particular model mediating between two consecutive design stages. The function addressed in the transition between design and the physical phenomena used in creating the model are written on the arrow. The first two transitions can be represented extensionally by explicitly intersecting sets. The third stage involves adding a brake to chair E. This involves copying E into a new entity, E′, and refining it extensionally by adding a brake, for example, in either of the following ways:

has a brake(E′)

or

brake(b), *has*(b, E′).

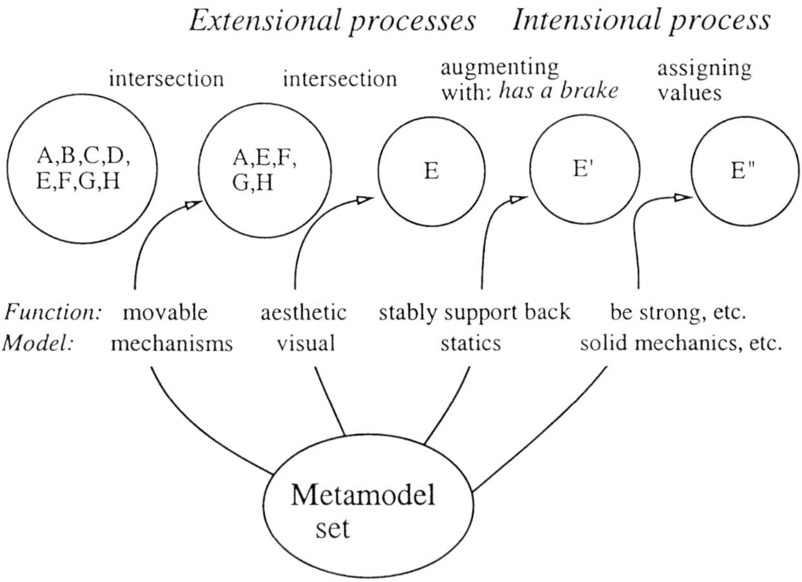

```
                Extensional processes    Intensional process
         intersection   intersection   augmenting        assigning
                                       with: has a brake  values
       ⎛ A,B,C,D, ⎞   ⎛ A,E,F, ⎞     ⎛   ⎞            ⎛   ⎞       ⎛    ⎞
       ⎝ E,F,G,H  ⎠ → ⎝  G,H   ⎠ →   ⎝ E ⎠     →      ⎝ E'⎠  →    ⎝ E" ⎠

      Function:   movable        aesthetic    stably support back   be strong, etc.
      Model:      mechanisms     visual       statics               solid mechanics, etc.

                              ⎛ Metamodel ⎞
                              ⎝    set    ⎠
```

Fig. 3.3 Design as a converging process.

Subsequent refinements that involve sizing parts of the chair may be better dealt with intensionally. This will involve creating the intensional representation of entities and the assignment of values to these representations. For example, the seat may be described as:

seat(*thickness, firmness, width, length,* …)

and the "slots" in this "scheme" or "frame" could be filled by the actual values. After the detailing of the intensional descriptions, the process terminates with a feasible solution.

Knowledge structures need not be created automatically; they can emerge from design given appropriate facilities, such as the support of a "flat" space of objects (*i.e.*, all objects have equal status) and facilities for creating complex categorisations and embedding modelling tools for analysing objects [8].

3.5 Summary

Sceptics may claim that GDT presents a far too ideal model of design; thus, it is unclear what is its contribution to building CAD. Nevertheless, we think that the guidelines mentioned are tangible. The support for this claim remains as a future study through the building of CAD systems that embed these guidelines and their testing in real design tasks. It is hoped that this review presented GDT's concepts in a clear way that may lead researchers to benefit from the insight GDT provides and perhaps undertake this challenge.

References

[1] Reich Y. Measuring the value of knowledge. Int J Hum Comput Stud 1995;42(1):3–30.
[2] Yoshikawa H. General design theory and a CAD system. In: Sata T, Warman E, editors. Man–Machine Communication in CAD/CAM, Proceedings of The IFIP WG5.2–5.3 Working Conference 1980 (Tokyo). North-Holland, Amsterdam, 1981;35–57.
[3] Tomiyama T, Yoshikawa H. Extended general design theory. Technical report CS-R8604, Centre for Mathematics and Computer Science, Amsterdam, 1986.
[4] Tomiyama T. From general design theory to knowledge intensive engineering. Artif Intell Eng Des Anal Manuf 1994;8(4):319–33.
[5] Reich Y. A critical review of general design theory. Res Eng Des 1995;7(1):1–18.
[6] Braha D, Reich Y. Topological structures for modeling design processes. In: Proceedings of the 13th International Conference on Engineering Design (ICED'01), Glasgow, 2001.
[7] Sutherland WA. Introduction to metric and topological spaces. Oxford (UK): Oxford University Press, 1975.
[8] Subrahmanian E, Konda SL, Levy SN, Reich Y, Westerberg AW, Monarch IA. Equations aren't enough: informal modeling in design. Artif Intell Eng Des Anal Manuf 1993;7(4):257–74.

Theory of technical systems and engineering design synthesis

4

Vladimir Hubka and W. Ernst Eder

Abstract "Engineering" or designing a technical system means anticipating its usage, construction, *etc*. A "technical system" is a designed artefact with a substantial technical content. The scope and contents of design science are outlined, showing the role and context of the theory of technical systems. Designing is discussed as a mixture of systematic and intuitive processes. The systematic design processes are based on the theory of technical systems, suitably adapted to the design situation for the particular design problem. This systematic process is described in relation to the theory and to some other known methods. Its major application is for conceptualising products at various abstract levels of modelling, and allowing a wide search for alternatives at each level. The procedure then reaches into the layout and detailing stages. The systematic process is not only applicable to novel design problems (although it is set up for that task), but can also be applied to redesign problems.

4.1 Introduction

"Engineering" a system (a product and/or a process) involves anticipating the possible usage and construction of that system, before any real material is committed. It starts in the phase of product planning, when alternative markets and products are considered, and decisions made about the product range to be offered by an enterprise. This phase is vital for the health of the enterprise. Any resulting ideas for a product (system) must usually be translated into *instructions* for implementing the system, *e.g.*, manufacturing, assembling, packaging, *etc*. This is the scope of *designing*, which involves synthesis and, at times, also creativity.

The requirements for the product should *ab initio* be established to a sufficient degree that designing can proceed in a goal-directed progression. Such a progression still means that intuitive, opportunistic, serendipitous, idiosyncratic and other less systematic ways of thinking and acting are necessary; they are the natural behaviours of humans. Yet only a planned, conscious and systematic (but iterative and recursive) designing procedure can ensure that an optimal solution of the presented problem can be approached in an effective process.

Such systematic procedures should also help to open up further solution possibilities, which can then be used to trigger the opportunistic, *etc*., activities. Results obtained from any opportunistic or intuitive steps should be brought into the systematic schemes, and verified. A constant interplay between systematic and unsystematic actions can result in and lead to a better search of the solution field, and probably to a better design solution. The initial requirements (the design specification) should also be continually revised and updated as designing progresses, of course in consultation with the customers and their representatives.

Many products must function, be capable of performing a duty, and thus also have a technical/engineering content – they are *technical systems* (TSs). In addition to their external appearance and their interactions with humans, they must also be internally configured to perform their intended tasks. Engineering designing (as a constituent of product development, and as a necessary adjunct to industrial design for many types of product) has, therefore, a more restricted, but also more difficult set of tasks. These tasks and activities aim to create viable products of appropriate performance, technical capabilities, economic potential, attractiveness for the customers, *etc*, summarised as appropriate "quality". They should also take into account the considerations of life cycle engineering [1], the associated energy usage, secondary by-products, toxicity, costs, hazards analysis, and many other special disciplines.

We therefore understand and define "engineering designing" as a process of designing (engineering) a *TS*. In this process, certain expressed needs, requirements, preliminary ideas for possible technical (engineering) products, or problems (input information) are transformed into a full description of a manufacturable TS and/or implementable usage process (output information). We are dealing with a transformation of information. This process tries to *anticipate* the future system, and to predict and synthesise its structures and behaviour before it exists in reality. In this way, designing is a major early stage in realising a product, from which the enterprise (industry or any other organisation) intends to obtain a livelihood. The *aim* (*goal*) of designing is to obtain a full description of an optimal TS, as instructions for manufacturing (usually as drawings), within an anticipated time and cost (of designing, and of implementing).

4.2 Design science and the theory of TSs

A dictionary [2] shows for

> **science**: *[from Latin "scientia" – having knowledge]*; 1. a branch of study concerned with observation and classification of facts and esp. with the establishment of verifiable general laws; 2. accumulated systematized knowledge esp. when it relates to the physical world.

It should, therefore, be possible to formulate a science about engineering design.

As Klaus [3] stated in cybernetics, relationships exist between the subject under consideration (its nature as a product or process), the theory, and method. The theory (as expressed in mental, graphical and physical models, verbal explanations, and, where possible, symbolic/mathematical expressions) should describe both the behaviour of the subject (with adequate and sufficient precision), and the methods utilised (for using and/or operating the subject, and for designing it).

The knowledge for engineering design, the "subject" under consideration, is selected, collected, cross-referenced, ordered and categorised within design science [4] (supported by Refs [5–14]). This knowledge is divided into two basic classes:

1. knowledge about TS (products and their usage processes), as the operands of design processes (DesPs), and their development during the DesP;
2. knowledge about DesPs, as elements of the transformation of information from needs to full descriptions of proposed TSs.

Knowledge about each of these two subjects can be classified into:

3. theoretical/descriptive knowledge (theory), and
4. knowledge and advice derived from and/or prepared for practical activities.

Note that we make a distinction between theoretical/descriptive knowledge and descriptive/narrative/observational/protocol studies – the latter aims to elucidate a theory.

This scheme results in a four-quadrant model of knowledge for engineering design, see Fig. 4.1, where the horizontal axis represents items 1 and 2, and the vertical axis shows 3 and 4. Relationships to other forms and contents of knowledge must also exist, and are indicated in Fig. 4.1. Subdivisions of generalised knowledge can be identical in the theory and practice quadrants, for both the TSs and the DesPs.

A second basic model for design science, based on the concepts of systems engineering (*e.g.*, see Ref. [15]), is that of the transformation system; see Fig. 4.2. We have chosen to show a process as a rectangular symbol, and a real system as a symbol with rounded corners. All artificial processes can be *analysed* (abstracted) to conform to this scheme, and the corresponding elements and their properties can be found. They can be further analysed (decomposed) into constituent processes and operations. Those processes and operations in which a TS plays a major role are termed

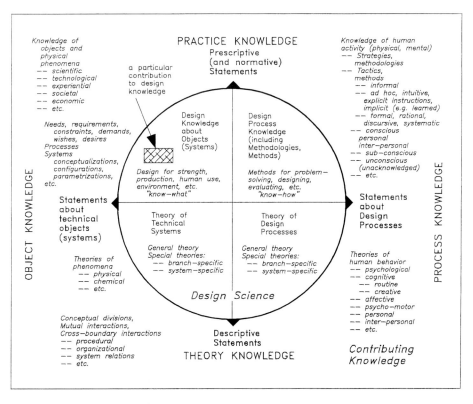

Fig. 4.1 Model (map) of design science – survey.

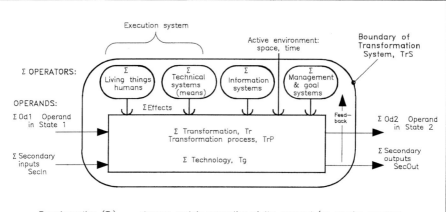

- Transformation (Tr) –– changes certain properties of the operand (as passive member of the process), by mutual interaction between object and means.

- Operand (Od) – WHAT is being transformed? –– object that is being changed in the transformation from an input state to an output state which should preferably be more desirable.

- State –– aggregate (vector) of values of properties (Pr) of a system at a certain time.

- Technology (Tg) – HOW is it being transformed? –– knowledge about the transformation, formulates what effects are needed.

- Effects (Ef) – WITH WHAT is it being transformed? –– means of transformation, actions exerted onto the operand, including the necessary energy, auxiliary materials, regulation and control.

- Secondary inputs (SecIn) –– (1) all necessary (desirable) further inputs to the process, and (2) all undesired inputs (disturbances, contaminants, products of the environment that enter as operands).

- Secondary outputs (SecOut) –– mostly undesirable opeand outputs of the process to the enviroment, their nature and composition depend on the chosen technology.

- Operators (Op) – WHO and WHAT delivers the necessary effects (as active member) to the operand:
    Op (Hu)   – living beings, particularly humans, but also animals, bacteria, etc.;
    Op (TS)   – technical (artificial) means, technical systems;
    Op (InfS) – information systems;
    Op (M&GS) – management and goal systems (directing, setting and achieving goals);
    Op (AEnv) – active environment.

- Active environment (AEnv) – WHERE is it being transformed? –– The part of the general environment that as operator influences the transformation (desired and undesirable effects).

- Space –– the main property of the environment (surroundings) of the transformation.

- Time – WHEN is it being transformed? –– time period during which the transformation occurs.

–––––––––– *** ––––––––––

- Types of effects acting on the operand, secondary inputs, secondary outputs, etc.:
    –– materials;   –– energy;   –– information (including signals).

- Types of operand, object being transformed:
    –– biological objects (humans, plants, animals, etc.);
    –– materials;   –– energy;   –– information.

- Structure of the transformation process:
    –– elements   = operations (O), or groups of operations;
    –– relationships = connections between the outputs of one operation (or group of operations) with the inputs of the immediately following operation (or group of operations);
    Operations can take place sequentially (in series) or simultaneously (in parallel).

Fig. 4.2 General model of the transformation process.

technical processes (TPs), a particular kind of transformation. Many cases show no direct involvement of humans, information, management, or the environment.

The model in Fig. 4.2 describes the complete transformation system and its elements, *i.e.*:

1. the operand (the object undergoing transformation within the TP) – the horizontal flow through the rectangle depicting the process in Fig. 4.2;
2. the technology and the structure of the process, those operations that are realised in the transformation process, and their connections;
3. the execution system, the operators "human" and "TSs (means)" that deliver the required physical effects (actions) to the TP – oval symbols in Fig. 4.2;
4. the other operators of the process, information, management and environment systems, often only acting indirectly;
5. the secondary inputs and outputs.

TSs deliver the *output effects* (necessary actions) that serve to perform the desired transformations of the operands within the TP. Machine systems, as special cases of TSs, use mainly mechanical modes of action to produce these work effects. Systems increasingly tend to become hybrids (*e.g.*, electro-hydraulic and computer-mechanical systems), particularly with respect to their propulsion and control organs.

Every TS exhibits several different sorts of *structure*, see Fig. 4.3 (first version published in Ref. [10]; compare "chromosome model"), which are carried by the TS whether they have been intentionally designed or not. These consist of elements, especially functions, organs or constructional parts, see Fig. 4.4 (first version published in Ref. [7], compare "domain theory"), and their relationships within the TS and across its boundary. The various types of effect delivered by a TS to transform the operand in the TP are generated by its structure(s).

Functions are the (verbal) descriptions of the actions that the TS is (or should be) capable of performing, analogous to the capability of a "human hand to grip a handle". The primary *working function* (capability for a TS-internal transformation effect) taking place *within* a TS is always accompanied by *assisting functions – auxiliary, propelling, regulating and controlling*, as well as *connecting and supporting* functions (also termed partial functions). These constitute classifying aspects for the function structure of TSs; see Fig. 4.4, level III.

Organs (sometimes called function carriers) of the TS, Fig. 4.4, level IV, act as the means to realise the functions. They consist of working and assisting organs – auxiliary, propelling, regulating and controlling, and connecting organs – that together form the complete organ structure of the TS. The connections between the TSs and the environment are the *receptors* and *effectors*, which are also respectively the start and end of the action chains formed by the organs. There is rarely a 1:1 relationship between functions and organs.

Every function and organ at any level of complexity may be realised by a *variety of constructional structures*; Fig. 4.4, level V. The possibilities of forming variations depend on the design characteristics – input, mode of action, and the design properties of the TS. There is rarely a 1:1 relationship between organs and constructional parts. Constructional structures exist at various levels of completeness, concreteness and detail: sketch layouts, dimensional layouts, detail parts drawings, *etc*.

Technical systems may be divided into four classes according to their *degree of complexity*, namely plant or equipment, machines, assemblies and parts. Each system of higher hierarchical order is composed of systems of lower order. The elements of

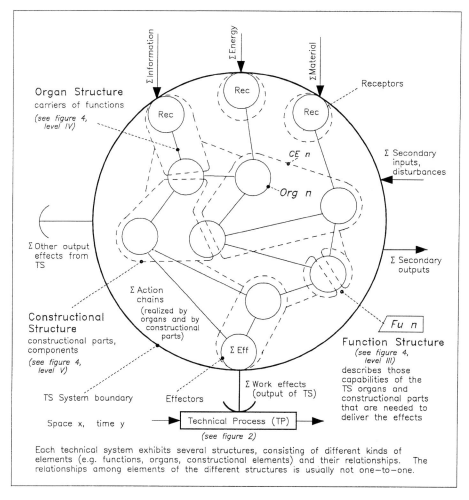

Fig. 4.3 Model of a TS – structures.

the constructional structure of TSs and their hierarchy are thereby determined. Examples are a car-manufacturing facility (industrial plant), a car (machine), a motor (engine) in that car (machine or assembly group, or module), a connecting-rod assembly in that motor (assembly group, or module), a threaded stud in the connecting-rod assembly (constructional part), *etc.*

Machine elements (MEs) are frequently occurring constructional elements of machine systems (TS with mainly mechanical modes of action). An ME can exist as a component of the first degree of complexity (*i.e.*, a detail part such as screw, key, split pin), or as a functionally composite group of parts (higher degree of complexity, *e.g.*, bearing, gear box, brake, coupling). In addition to the general class of MEs, there exist elements that are only found in some particular branches of engineering, such as a piston, valve, heat insulator, hook, resistor, capacitor, inductor, transformer, transistor, transducer, *etc.*

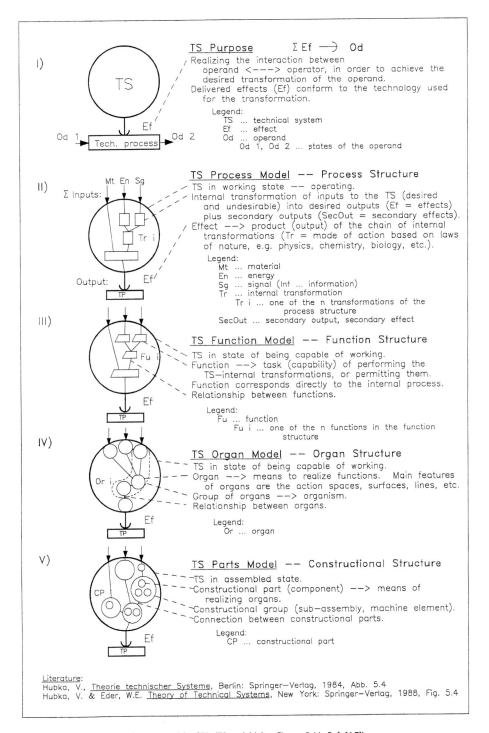

Fig. 4.4 Models of TSs (TS models) (see Figures 5.4 in Refs [6,7]).

The TS, as operator of its TP, has a set of properties that can be classified into a set of 12 property classes; see Fig. 4.5. Classes 1–11, the *external* properties that a customer can assess and observe: the working properties of the system. They summarise the customers' requirements in point form. Classes 1–7 cover the operational and life-cycle properties, classes 8 and 9 cover the immediate human interactions, and classes 10 and 11 cover the societal aspects. These properties exist whether they have been deliberately designed or not.

Class 12 describes the end result of designing: the *internal* properties. The elementary design properties are (conventionally) contained in the detail and assembly drawings, and any computer code and other instructions – these are the properties that are designed, the external properties result from the internal properties. The knowledge about the general design properties and design characteristics is: partly contained in the engineering sciences; partly in standards, guidelines, codes of practice, rules and regulations (*e.g.*, ISO 9000 and 14000, and the methods delivered by life-cycle engineering [1]); partly in prior experience of the designer and of the enterprise; and is augmented by creative thinking.

The phases of origination, operation (usage), maintenance, liquidation, *etc.* of TSs form the structure of the *life cycle* of the TS; see Fig. 4.6. Analysing the factors influencing the individual phases yields complete information about all situations in which the TS can exist.

Development of a TS during a period of time (innovation and further development process) leads either to small changes (variations, mutations) with basically unaltered work effects, or to larger changes when a new generation of products is created. The *state-of-the-art* in a branch of engineering is represented by those products whose properties have reached their highest levels of development at that time. Knowledge of this state is essential for engineering designers

4.3 Designing – general

Designing is a transformation process according to this model; see Fig. 4.7 (top). In this process, the future TS is generated. The DesP is an element in the process of generating TSs; Fig. 4.6. In the DesP, the *starting information* (in the form of requirements or descriptions of the assigned problem situation) is transformed into the *descriptions of the desired TS*; this is the main aim (goal) of designing. Description of the TS is performed by means of elementary design properties; see Fig. 4.5. Consequently, the search for suitable design properties is the main activity of design work.

As a TP, the DesP may be divided into a *finite number of partial processes* or operations. The majority of these belong to a few recurring classes of such operations, regardless of the type of DesP in question; see Fig. 4.7. *None* of the operational phases of the DesP is regarded as irrational. The DesP is composed of *rational* operations, although many of these operations are performed as subconscious actions. The DesP can be *neutral with respect to objects* (TSs, products) to be designed; a generalised model of the DesP exists that may be applied to designing *all kinds of TS* (and of MS). Level 3 shows the recurring *basic operations* (problem solving), which include evaluation as a major element.

The factors in the phases, stages and steps of engineering design that influence the optimal *quality* of a future product may easily be derived from the model of the DesP; Fig. 4.7 (top). This model shows the following components:

Theory of Technical Systems and Engineering Design Synthesis

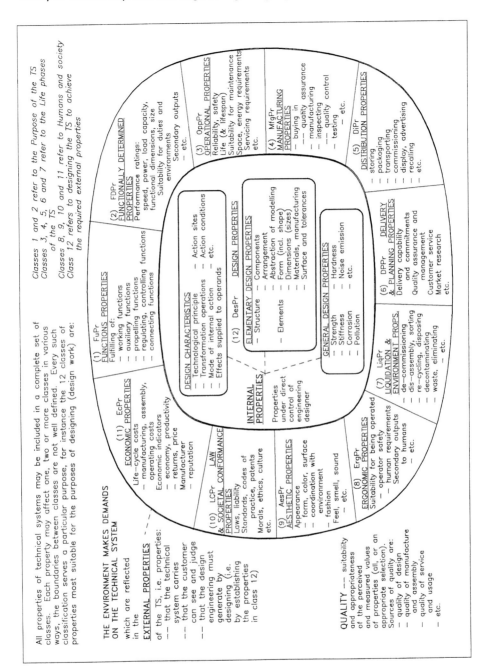

Fig. 4.5 Relationships among properties of TSs.

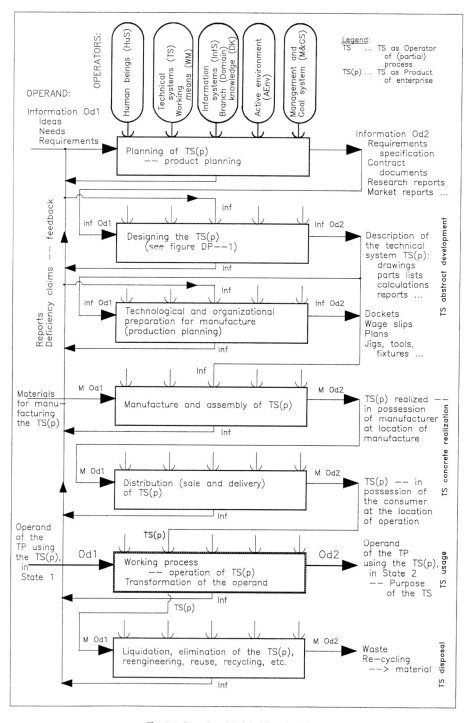

Fig. 4.6 General model of the life cycle of TSs.

Theory of Technical Systems and Engineering Design Synthesis

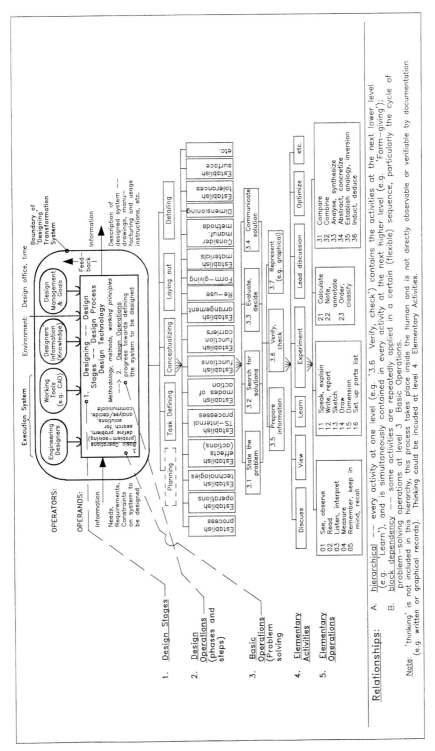

Fig. 4.7 Structure of possible activities in the design process.

(a) State 2 of the processed information: a full description of a TS that completely fulfils the given requirements, as the goal (output) of the DesP;
(b) State 1 of the information to be processed: the given requirements (customers' design specification) as the input to the DesP;
(c) a DesP, a transformation of information from State 1 (a) to State 2 (b).

This transformation is realised or influenced by six *operators* of the process:

(d) the engineering designers (usually as member of a design team), their professional profiles and personal characteristics, internalised knowledge, abilities, skills, attitudes, values, motivation, open-mindedness, adaptability, *etc.*;
(e) the working means available to the designer (tools, equipment, *etc.*, including computers and computer-aided design programs) and their usage;
(f) the existing technical knowledge (available information), particularly technical information, working methods, and techniques of representation, preferably externalised, collected and codified into a specialist information system;
(g) the guidance, leadership and control (management) of the DesP;
(h) the active environment (including time and space) in which the process takes place; and
(i) the DesP itself, the methods (procedures and techniques) employed by the designer, the *technology* of the process.

These operators decide whether the DesP can be completed successfully at all, what quality the result of the process will have (its output, the description of the designed TS), and what parameters of the DesP will be attained (*e.g.*, costs and duration of designing, committed costs of the proposed TS).

The most important factors are the designers (d), the available knowledge (f) and the design procedure (i). The TSs "working means" (e) that designers use (including computers) are useful, but cannot design anything by themselves. Therefore, a continuing need exists for descriptive/narrative/observational/protocol studies to clarify the human, psychological and sociological aspects of designing, including teamwork, differences between routine and non-routine (creative) work, critical situations, management, *etc.*

The strategy of methodical, systematic and planned procedure in design is generally defined in a *procedural model* that has been set up for an assumed state of the design situation consisting usually of generalised conditions. The concrete conditions that arise from a specific design problem and for a particular design engineer lead to mutations of the model, *i.e.*, *procedural plans, procedural manners*, personal working modes, *etc*. The plan for the DesP should be as highly *structured* as possible, the details reaching into "design operations" or "steps" where applicable. The complicated relationships between the desired (external) properties of the TS (the requirements) and the desired elementary design properties force design engineers to use a repeated *iteration* procedure based on the *progressive concretisation* of the TS being designed, *i.e.*, a movement from incomplete to complete information, and from approximate to definitive values.

Engineering design thus involves anticipating, *predicting*, and applying of existing (object) knowledge and the engineering sciences; see Fig. 4.1. Designing involves *synthesising*, going from abstract (principles, ideas, requirements) to concrete (description of a TS ready for manufacture); by bringing together selected parts into a whole, with equalisation (by compromise or avoidance, *e.g.*, using TRIZ, Invention

Machine or similar tools) of any conflicts or contradictions in the relationships among the individual parts (internal, external, cross-boundary and/or to the requirements of the problem), a new entity is created, for which the properties are optimal (at that time) for the envisioned task. It is important to note that the resulting properties are not merely the sum of the properties of the individual parts.

We will consider only such products where engineering design plays the dominant role. In products where industrial design (ergonomics, aesthetics, marketability, *etc*.) plays the dominant role, some of the following considerations will need other approaches.

The process of designing engineering products can be viewed in four main administrative phases: (A) *clarifying* the problem; (B) *conceptualising*, *i.e.*, generating several candidate solutions in principle, and selecting the most promising and optimal; (C) *embodying* in material forms, with proposed configuration and parameterisation; and (D) *detailing*, elaborating to the state in which the product can be realised, *i.e.*, implemented and/or manufactured. The interpretation of these phases is not uniform. The methods outlined in the following subsections are derived from design science [4–14], and include some indications for other methods where appropriate.

4.3.1 Starting designing – clarifying the problem – design specification

Many reports (*e.g.*, see Ref. [16]) and books have stated that engineering design should be started with a sufficiently clear definition of the required performance – a design specification. A design brief, as delivered by product planning or by a customer, is usually not sufficiently definite or complete for designing and designers.

At the start of a design problem, such a clear definition of performance (required properties of the system to be designed) is difficult to achieve, but an attempt should be made. Customers will frequently not be able to give such a definition of performance; they will request a piece of equipment according to their experience. Statements in a contract document are mostly not sufficient to define the design problem well enough for rational designing to begin, but, in general, give adequate definitions of the legal and financial commitments.

It is the responsibility of the engineering designers to formulate the design specification, in conformance with the customers' requirements, as agreed with the sales organisation, and anticipating, where possible any other concerns. This is a test of designers' capabilities for judgement. The responsibility to define the problem is currently rarely fulfilled. Any such design specification cannot be final; it must be continually reviewed and revised during designing in the light of progress. This demands periodic consultation between designers, sales and customers.

Theory: design science [4–14] delivers a model of the life cycle (Fig. 4.6) that delineates the main stages of how a system originates, is used, and ends its life. Each of these stages is a process (consisting of subprocesses and operations), with its operators, based on a general model of transformation processes; Fig. 4.2. For the working process, the process of *using* the TS, it is usually this operator – the TS(p) – that has to be designed. Sometimes, this working/usage process can also be changed (redesigned) to allow application of a different (*e.g.*, new) technology. The required (external) properties of the (future) TS are realised by establishing (defining) the elementary design properties, with the help of the other internal properties.

Method: an adequate first draft of a design specification can easily be drawn up by following the classes of external properties (Fig. 4.5), and by considering the life stages with their operators (Fig. 4.6), and, where possible, the secondary inputs and outputs of each process. This represents a useful scheme of classification for design specification statements.

This design specification should be sufficiently open to allow enough alternatives, but also sufficiently restricted to avoid finding too wide a range. It should be formulated as requirements and conditions of performance, preferably not as demands for particular hardware, firmware or software, *e.g.*, "snow should be removed from the roadway to leave a maximum depth of 30 mm". Consultations with the sales and customer representatives, life-cycle engineers, *etc.* should help to make this list of requirements reasonably complete and workable. *Priorities* must usually be stated and agreed with the customers, *i.e.*, whether a design specification statement must be fulfilled, or whether it is a stronger or weaker option. Quality function deployment (QFD) [17] can be useful as an entry into setting up a design specification.

4.3.2 Designing – design procedure – novel products

Experienced engineering designers working in their own field of products will normally use their own procedures (their tacit, internalised knowledge) to perform the design work. Considerations mainly stay in the constructional structure (components, details, subassemblies, assemblies). Many possibilities for optimisation, and even innovation, will be missed in this way.

Theory: design science [4–14] provides a full description of the abstract transformation process (Fig. 4.2), its technologies, and its operators. The transformation system also shows the logical relationships among the (primary and secondary) inputs and outputs of the process and its various operators. The effects produced by the operators (when they are in operation) cause the transformation to take place – a teleological sequence of causes and consequences known as *causality*. The (analytical) relationship of the existing external properties to the elementary design properties is also one of *causality*; external properties are caused by internal properties.

Conversely, if we can define what transformation we wish to anticipate and achieve, we can then deduce (analyse) the effects needed from the various operators, and find, select and combine (synthesise, establish) the available alternative means of achieving those effects – a many-faceted sequence of desired consequences and their possible causes known as *finality*. The relationship of the desired (and undesired) external properties of a future TS to the elementary design properties that are being designed is also one of finality. Appropriate internal properties must be generated (established) to fulfil the requirements for the future external properties: a synthesis relationship. In each step, a set of *goals* is defined, and suitable (alternative) *means* need to be found and established. This results in a sequence of *goals–means transitions* that eventually establish the elementary design properties; the goals can be operations in a TP, functions, organs or constructional elements – compare this with the "function–means tree".

The main operator of interest is the *TS*, TS(p), to be designed for its usage process, but it must also be suitable for all its other life-cycle processes (*e.g.*, sufficiently small environmental impacts), for societal, cultural and economic acceptance, *etc.* The TS is described in a set of theoretical models of different internal structures that may be recognised; see Figs. 4.3 and 4.4. Each structure consists of a different kind of element (*e.g.*, usage process operations, technologies, TS-internal functions, organs, and constructional parts), with the relationships among the elements. Structures can be "mapped" onto each other, but usually not in a 1:1 correspondence. These structures can also provide checklists for use during designing. Every designed (and realised) TS has all of these structures.

Method: as soon as a designer has understood the transformation process (as a black box, as a subject and its theory), he/she can apply the appropriate method. This method encourages consideration of all requirements, and of possibilities for generating alternative proposals (variations). Only on such a basis can an optimal solution be expected, though not guaranteed. The recommended methodology (and set of methods) for a novel system, iteratively and recursively applied to establish the future product, consists of the following.

(A) *Clarifying the task.*
- Establish a suitable design specification (see Section 4.3.1).

(B) *Conceptualising.*
- Determine the essential tasks of the transformation process – what the customer wishes to achieve by using the designed TS. *Model*: transformation process; Fig. 4.2. *Example*: "snow moved from road to side".

Several kinds of important subprocess (assisting processes) always appear as part of each transformation process: auxiliary processes; propelling (driving, energy delivering) processes; regulating and controlling processes; connecting and supporting processes; preparing, implementing/executing and finishing (checking, verifying, record-keeping) subprocesses. This represents a checklist to ensure that (eventually) all constituting partial processes (and their secondary inputs and outputs) have been taken into consideration.

- Choose a favourable technology for each of the partial processes, especially for those that are to be driven by a TS. *Example*: "scraping, pushing, throwing, melting".
- Establish all necessary output effects to be delivered by the operators (input effects to the operands needed for transforming them), especially by the TS(p), the product TS to be designed; Fig. 4.4, level I. *Example*: "accelerating snow at elevation angles 15–75°".
- Distribute these output effects optimally among the humans and the TS in the existing situation, *i.e.*, decide what will be done by humans, and what by the TS to be designed (which defines the TP).

This section of the method based on design science [4–14] is preferably used in the phase *clarification of the design task*, although (according to the theory) it should be part of the solution process (conceptualising) phases. It should be obvious that this procedure cannot be completed in such a linear fashion; iterative working is essential. Feedback from later stages to earlier ones will progressively drive the solution proposals towards an optimal state. A problem may need to be (recursively) broken down into simpler subproblems, and their proposed solutions reintegrated. If any opportunistic and intuitive step is taken outside this procedure, at least a check should be made to ensure that the results do not violate the procedural considerations and outcomes.

- Establish from this allocation of the TP what the operator "TS" (the one to be designed) must be capable of doing to transform its own inputs into the required output effects, *i.e.*, establish the essential (internal and transboundary) functions of the TS. Assess any effects from or on the operator "environment", as far as they can be anticipated at this stage. *Model*: function structure of TSs, and relevant explanations; Fig. 4.4, level III. *Example*: "lift snow, apply force at elevation angle, allow change of elevation angle".

Note that, for a novel system, the functions can be derived by synthesis from the TP. For a redesigned system, functions can be derived by analysis from the existing system structure. In both cases, the formulation of functions (in words) is done by the human designer, who can therefore change them to suit (*e.g.*, for an innovation), *e.g.*, to allow easier solving in organs. Attempting to combine functions in different arrangements, or decomposing them to simpler functions, may deliver impulses for different solution proposals. The function structure is preferably set up as a block diagram – a hierarchical tree, such as the "function–means tree", usually does not allow diagramming any relationships among tree branches.

- Establish with what means (in principles of operation, modes of action, and principles of construction – organs) each function can be solved and the requirements fulfilled. This may demand a reformulation of the functions to make them more easily solvable, or to distribute or combine functions. *Method*: morphological matrix, design catalogues. *Example*: "use centrifugal force on snow".
- Select and combine the most promising solution for each function; check that they are compatible (a) with each other, (b) with the design specification (or alter the design specification in agreement with the customer). Try different arrangements of these proposed solutions; choose the best combination. None of these can be finalised at this stage. *Method*: almost playful exploratory combination of organs in various forms and arrangements, with subsequent evaluation and decision to obtain a preferred solution-in-principle. *Model*: organ structure of TSs, and relevant explanations; Fig. 4.4, level IV. *Example*: "number, shape, dimensions, location (radius), angular velocity, *etc.* of blades to throw snow".

Note that this procedure will inevitably reveal the need for assisting functions (and organs, and constructional elements) – *evoked functions* (*etc.*) – to allow using different materials, permit assembly, transport, lubrication, maintenance, *etc.*

(C) *Embodying*.
- Place adequate material around the solution-in-principle, initially as a set of rough (probably sketched) proposals – a preliminary layout – select the classes of materials to be used, considering their energy consumption, toxicity, and by-products (secondary outputs), estimate life-cycle costs, *etc.* Progressively refine these proposals into a full definitive layout. Efforts will now concentrate more on manufacturability and assembly operations. *Method*: this is probably the earliest point in which conventional computer-aided (graphics, solid modelling or similar) design programs can be used. *Model*: constructional structure of TSs, and relevant explanations; Fig. 4.4, level V.

(D) *Detailing*.
- Prepare detail and assembly drawings (or their computer-resident equivalents), and all needed documentation, so that the future TS can be manu-

factured, assembled, adjusted, commissioned, transported, used, *etc.* to fulfil all required properties.

During all of these designing activities, the original design specification (Section 4.3.1) should be consulted, reviewed, and revised where necessary. At each of these stages, the possibility of formulating alternative proposals exists. Each proposal (for a particular structure) should be worked through to the same level of detail and completeness – just sufficient for the task of evaluating and selecting. All proposals can be evaluated, and the most promising taken forward into the next structure or stage. Initiatives for research to obtain future innovations should be noted. Combinatorial complexity (getting too many combinations for human capability of processing) can be controlled by this method of generating alternatives, evaluating, and selecting in stages as indicated. Good record keeping can allow engineering designers to retrench at any time, to follow a different proposal if sufficient difficulties arise. This documentation is also needed for design audits, life-cycle analyses, and possible litigation.

Experienced engineering designers (who do not know about these methods) will usually stay within the last two of these methodical rules, (C) and (D). Some of the methods mentioned in this methodology have been taken over from conventional procedures; some are also described in existing books (*e.g.*, see Refs. [18,19]). Iterative working and problem solving (with reflection [20]) is essential, but a recursive process (subdividing a problem, and recombining the solution proposals) is also necessary. Evaluation may be helped by a formal procedure, such as decision theory [21], but care must be taken that the non-measurable properties are not neglected in the decisions. Carrying the "customer's wishes" through the design specification and into the layouts and details may be helped by using QFD [18] as a form of design audit, especially before establishing the design specification, but also after establishing a layout, and after completing the details.

4.3.3 Designing – design procedure – redesigned products

Redesigning intends to achieve evolutions and innovations. If the aims and requirements can be recognised and achieved during designing, costs can be saved in developing and producing the redesigned system. It has been estimated that 95–99% of all design problems are concerned with redesigning. We claim that about 10–20% of all problems can benefit from a full procedure, as described in Sections 4.3.1 and 4.3.2, and probably 50% from adopting at least some of these recommendations. Yet, in many cases, in industrial (and other administrative) practice the designers (and their managers) never even formalise the requirements to be achieved by redesigning, *e.g.*, by setting up an agreed design specification.

Redesigning can rationally continue from the design specification by analysing (abstracting from) an existing product to determine the existing organs (and their structures), and/or the functions (and their structures) that these organs fulfil. This is an analytical reversal of the last three to five steps in the above full process (see Section 4.3.2), although with experience some of the analytical steps can be omitted. The higher the abstraction is taken, the more possibilities of variation are opened up, but the more effort must be expended. This is another classical trade-off situation between time and effort, and degree of innovation and optimisation.

Once the highest appropriate abstraction has been established, whether it is to the function structure, the organ structure, the subassemblies, or the details, and the results adapted to the new design specification, the systematic (methodical) procedure described can again be followed to complete the abstracted steps.

References

[1] Graedel TE. Streamlined life-cycle assessment. Upper Saddle River (NJ): Prentice-Hall, 1998.
[2] The Merriam-Webster dictionary. New York: Pocket Books, 1974.
[3] Klaus G. Kybernetik in philosophischer Sicht [Cybernetics in philosophical view]. 4th ed. Berlin: Dietz, 1965.
[4] Hubka V, Eder WE. Design science: introduction to the needs, scope and organization of engineering design knowledge. London: Springer, 1996.
[5] Hubka V, Eder WE. Einführung in die Konstruktionswissenschaft [Introduction to design science]. Berlin: Springer, 1992.
[6] Hubka V, Eder WE. Theory of technical systems. Berlin Heidelberg New York: Springer, 1988.
[7] Hubka V. Theorie technischer Systeme [Theory of technical systems]. Berlin: Springer, 1984 [second edition of: Hubka V. Theorie der Maschinensysteme [Theory of machine systems]. Berlin: Springer, 1974].
[8] Hubka V. Theorie der Konstruktionsprozesse [Theory of design processes]. Berlin: Springer, 1976.
[9] Hubka V, Eder WE. Engineering design. Zürich: Heurista, 1992 [2nd ed. of: Hubka V. Principles of engineering design. London: Butterworths, 1982].
[10] Hubka V. WDK 1 – Allgemeines Vorgehensmodell des Konstruierens [General procedural model of designing]. Zürich: Heurista, 1980.
[11] Hubka V, Andreasen MM, Eder WE. Practical studies in systematic design. London: Butterworths, 1988. [English edition of: WDK 4 – Fallbeispiele. Zürich, Heurista, 1980.
[12] Eder WE, editor. WDK 24 – EDC – Engineering Design and Creativity – Proceedings of Workshop EDC. Zürich: Heurista, 1996.
[13] Eder WE, Hubka V, Melezinek A, Hosnedl S. WDK 21 – ED – Engineering Design Education – Ausbildung der Konstrukteure – Reading Zürich: Heurista, 1992.
[14] Hubka V. Konstruktionsunterricht an Technischen Hochschulen [Design Instruction at Technical Universities]. Konstanz: Leuchtturm, 1978.
[15] Züst R, Matt D, Geisinger D, Winkler R. A change of paradigm for planning life-cycle optimized and innovative products. In: WDK 26: Proceedings of ICED 99 Munich. Zürich: Heurista, 1999;1081–4
[16] Feilden GBR. Engineering design, report of Royal Commission. London: HMSO, 1963.
[17] Bossert JL. Quality function deployment. New York: Marcel Dekker, 1990.
[18] Pahl G, Beitz W. Konstruktionslehre. 3rd ed. Berlin Heidelberg: Springer, 1993. Pahl G, Beitz W. In: Wallace KM, Blessing L, Bauert F, editors/translators. Engineering design. 2nd ed. London: Springer, 1995.
[19] Jones JCh. Design methods – seeds of human futures. 2nd ed. New York: Wiley, 1980.
[20] Birmingham R, Cleland G, Driver R, Maffin D. Understanding engineering design. London: Prentice-Hall, 1997.
[21] Schön DA. The reflective practitioner: how professionals think in action. New York: Basic Books, 1983.

A knowledge operation model of synthesis

5

Tetsuo Tomiyama, Masaharu Yoshioka and Akira Tsumaya

Abstract This chapter describes an attempt to formalise and model synthesis theoretically. It aims at establishing unified understanding of synthesis, beginning with an analysis-oriented thought process model and a synthesis-oriented thought-process model. These models are given logical interpretations to be performed on a multiple model-based reasoning framework. We then analyse design activities and show that design, including synthesis and analysis, is largely a knowledge-based activity. This results in a knowledge operation model that is decomposed into logical operations and modelling operations. The core of synthesis is considered to be abduction, and, within our framework, abduction is realised as model-based abduction. We outline an algorithm for model-based abduction. The knowledge operation model of synthesis was tested against an actual design case from which a reference model was built. The knowledge operation model was also implemented on the multiple model-based reasoning framework and the reference model was performed on it to perform the verification of the knowledge operation model.

5.1 Introduction

Engineering design consists of a variety of thought processes, but most of them can be categorised into either analysis or synthesis. Existing research attempts to model design processes have used such assumptions as "design as problem solving", "design as decision making", and "design by analysis", and have not explicitly addressed "design as synthesis". Compared with analysis, synthesis is less understood and codified as a model.

This chapter describes an attempt to formalise and model synthesis theoretically. By "theoretically" we mean an approach that tries to understand synthesis scientifically and to build its model based on this understanding. The model of synthesis should be used, as in other branches of science, *e.g.*, to understand the mechanism of design and to predict how design proceeds in a certain situation. If there is such a scientific model of synthesis, we can expect that designers could be guided with helpful suggestions even for a very difficult design task. We might also succeed in building an advanced intelligent design support system.

Driven by these motivations, the authors' group has conducted a series of research effort in design. During the 1970s, Yoshikawa proposed General Design Theory (GDT) [1]. GDT is a theory of design knowledge based on axiomatic set theory. The group also looked at experimental methods. For instance, a cognitive design process model was established through analysis of design protocols [2]. By logically formalising the cognitive design process model, a computable design process model was arrived at. Based on this, a design simulator was implemented [3].

This chapter integrates the results of these previous studies with a more recent project entitled "The Modeling of Synthesis" that primarily focuses on theoretical aspects. It aims at establishing a unified understanding of synthesis, beginning with our fundamental view on synthesis. In Section 5.2, we briefly review existing studies on design process modelling in the engineering design research community. Section 5.3 summarises our previous research on the synthesis-oriented thought process.

In Section 5.4, first we review synthesis as opposed to analysis. This leads to an *analysis-oriented thought-process model* and a *synthesis-oriented thought-process model*. We argue the logical reasoning aspect of these thought-process models. We then assume a *multiple model-based reasoning framework* that has features suitable for implementing an advanced design support system to verify our models of synthesis. This framework is based on the idea that synthesis is a knowledge-based activity. Therefore, we analyse design activities and show that design, including synthesis and analysis, can be interpreted as knowledge operations, which results in a *knowledge operation model*. A knowledge operation in this model is decomposed into *logical operations* and *modelling operations*. The logical operations are used for logical reasoning, and the modelling operations deal with design object models in the multiple model-based reasoning framework.

The logical operations inevitably include abduction. Since reasoning takes place within the multiple model-based reasoning framework that accommodates modelling operations to design object modellers, this abduction must be *model-based abduction*. We outline an algorithm for model-based abduction in Section 5.5. The knowledge operation model should be tested against actual design processes and verified through computer implementation of the multiple model-based reasoning framework. Section 5.6 deals with the test of the knowledge operation model against a reference model of a real design case and the verification through computer implementation, and Section 5.7 summarises the discussions.

5.2 Related work

The design process has mostly been studied in three ways: theoretically, methodologically, and empirically. The methodological, or "prescriptive", approach, which is typical among researchers in German-speaking countries, focuses on how design should be done (*e.g.*, see Refs [4,5]). It defines design and shows how to design as a procedure that begins with analysis of the required function followed by its decomposition into subfunctions to be embodied by mechanisms that exhibit physical phenomena to perform the required function.

In contrast, the empirical approach is bottom-up from experimental results, such as protocol analysis. There are a variety of approaches for analysing design protocol. Cross *et al.* [6] conducted the Delft protocol workshop. In that workshop, all of the participants used the same protocol data to easily compare different design process models (*e.g.*, see Refs [2,7]). Although these design process models explained phenomenologically how a design process proceeded, they failed to identify governing principles or driving forces of design.

The theoretical approach tries to set up a design process model in a top-down manner and aims at building "descriptive" design models. An example of the theoretical approach is GDT proposed by Yoshikawa's group at the University of Tokyo [1]. For modelling the design process as logical reasoning, many researchers (*e.g.*, see

Refs [3,8,9]) proposed to formalise design processes as abduction. Abduction is a logical reasoning mode first proposed by C.S. Peirce [10,11]. Abduction reasons out hypothetical explanations; *e.g.*, it reasons out fact p from rule $p \to q$ and premise q. In the design context, this abductive process is interpreted as reasoning out a "design solution" from "design knowledge" and "properties of the design solution". However, since abduction is not well formalised compared with deduction, these studies had to begin with formalising abduction itself and did not result in a clear understanding of synthesis.

Because all of these studies were more or less independently conducted, the engineering design research community, as a whole, has not yet arrived at a unified, scientific modelling of design processes. This paper is an attempt to arrive at such a theory of design processes with a particular focus on synthesis. Some efforts to overcome this problem include the following: Blessing [12] proposed a framework to compare design process models; Grabowski *et al.* [13] organised a workshop to discuss the possibility of establishing such a universal design theory by comparing different types of design theory and relating them with each other.

5.3 Design process modelling

A few research groups tried to understand scientifically, codify, and model design processes. Beginning with Yoshikawa's GDT, our group focused primarily on the theoretical and empirical approaches described in Section 5.2. This section briefly reviews our group's major achievement [14], which forms the foundations of the discussions in Section 5.4.

Takeda *et al.* [2,3] worked on modelling of design processes. The study began with design experiments in which design sessions were observed experimentally and protocol analysis was conducted. They proposed a cognitive design process model (Fig. 5.1) that viewed design as a repetition of unit design cycles. A unit design cycle has five steps, *viz.*, awareness of the problem, suggestion, development, evaluation, and decision. This cognitive design process model was logically formalised into a computable design process model (Fig. 5.2) in which the suggestion step was performed by abductive reasoning, and development and evaluation by deduction. Circumscription was used to revise knowledge when a contradiction happened. The model has two levels of reasoning: object-level and action-level (meta-level) to control the object-level reasoning.

A design simulator (Fig. 5.3) was implemented based on this computable model and it succeeded to play back design protocols obtained from design experiments. Although this simulator employed both deductive reasoning and non-deductive reasoning, such as abductive reasoning, circumscription, and meta-level reasoning, it could only "simulate" design processes rather than automatically design by itself. The result was far away from a model of synthesis.

5.4 A formal model of synthesis

In this section we describe our formal model of synthesis [15]. Though the focus is on synthesis, it does not necessarily only deal with synthesis. The theory, as a whole, covers the entire design activities, including both synthesis and analysis. Before we discuss synthesis and analysis, we here introduce the notions of fundamental

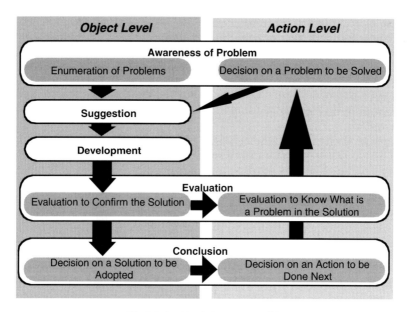

Fig. 5.1 Cognitive design-process model.

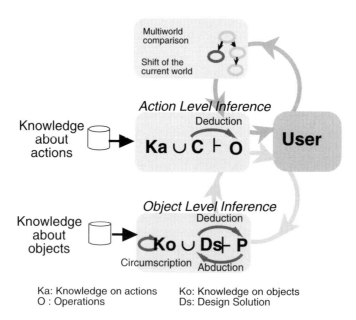

Fig. 5.2 Computable design-process model.

A Knowledge Operation Model of Synthesis

Fig. 5.3 The design simulator.

reasoning modes, such as deduction, abduction and induction, as mathematical preparation.

5.4.1 Mathematical preparation

Logic basically has two types of reasoning: deductive reasoning and reductive reasoning. Reductive reasoning includes induction and abduction. Before discussing abduction in design research, we briefly review these logical operations and their roles in the context of design.

Let us consider a logical scheme

$$A \vdash_\sigma Th, \tag{5.1}$$

which indicates Th (a set of theorems) is deducible from A (a set of axioms, *i.e.*, usually rules) using inference rule σ such as *modus ponens*. The symbol "$\vdash$" denotes "deducible". This is the simplest set-up of reasoning, in which deduction is a reasoning process to derive Th from A using σ and abduction is usually to derive A from given Th. The sets A and Th consist of logical formulae. We can introduce here semantic distinction of definitions Td that cannot be derived semantically from any other formulae. Axioms A can, therefore, be decomposed as follows:

$$A = K \oplus Fd,$$

$$Fd \in Td,$$

where K is knowledge and Fd is a set of "definition facts". Similarly, we can also introduce semantic distinction of observation To that are statements about observations in extralogic world. Therefore, we can consider observed facts Fo as part of Th, such that:

$$Th \supset Fo \in To.$$

Consequently, the reasoning scheme in Equation (5.1) is rewritten as:

$$K \cup Fd \vdash_\sigma Th, \tag{5.2}$$

which means Th (theorems) is deducible from knowledge K and facts Fd using inference rule σ. For example, K could be a set composed of rules in the form $P \rightarrow Q$. With facts in Fd such as P, we can derive Q using the inference rule.

This set-up is often assumed in artificial intelligence. Deduction is to obtain Th from K and Fd, whereas abduction could be performed in the following three reasoning modes [10,11]:

1. to obtain K and Fd from partially given Th;
2. to obtain K from Fd and partially given Th;
3. to obtain Fd from K and partially given Th.

Reasoning 1 is, in fact, the innoduction of Roozenburg and Eekels [16] and reasoning 2 is also called induction. These two are techniques often used in scientific discovery [17]. In this chapter, we argue reasoning 3 as abduction.

In order to clarify the relationship between these logical operations and reasoning in design processes, it is necessary to interpret these logical concepts with concepts related to design. One possibility is the following interpretations [3,9]:

- K – design knowledge;
- Fd – design solution;
- Th – property of design solution.

Under these interpretations, one may notice that derivation of design solutions can be conducted by abduction, derivation of the properties of design solutions by deduction, and derivation of design knowledge from a design solution and its property by induction. Note that deriving Th is intractable in standard logic. However, in many practical design applications, we do not need to derive all the deducible theorems, we usually only need to analyse a given set of goals. For example, in the case of Prolog, derivation of theorems boils down to confirmation only of goals of interest by refutation.

To understand the formalism above, let us consider a vehicle design case. Initially, design specifications will be described as "required properties" of design solutions in Th. Here, the initial specification of the design is "a vehicle that can transport more than four passengers at average cruising speed of $120 \, km \, h^{-1}$" (Fig. 5.4). First, the designer sets this specification as *Theorems*. To solve this problem, the designer uses knowledge about vehicles as *Axioms*. With these *Theorems* and *Axioms*, the designer

A Knowledge Operation Model of Synthesis

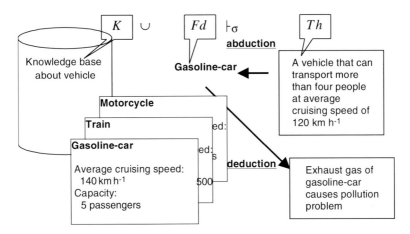

Fig. 5.4 Logical interpretation of design.

derives using abduction "gasoline-car" as a design solution candidate to be stored in *Facts*. Next, the designer derives properties of this candidate as *Theorems* by using deduction. In this case, "exhaust gas of gasoline-car" and "pollution problem" are derived. Modification of the specification, such as adding "avoid pollution problem" is extralogic.

Whereas the example above might be too simplified, one may notice that generating a design solution candidate is an abduction process, and abduction is the core of design; in particular, of synthesis. Although the importance of abduction is pointed out by many researchers, its formalisation and computing models are not clearly understood. This is the focus of Section 5.5.

5.4.2 Analysis versus synthesis

In classical science, it is widely accepted that a thought process is "rational" or "scientific" when the process takes the style shown in Fig. 5.5. It is a process to extract facts from observations and to try to give best explanations (theorems) of these facts from a set of axioms or hypotheses generated from the facts. According to Peirce, step (3) in Fig. 5.5 is an abductive process [10,11], whereas step (5) is often conducted deductively. In this sense, the thought process in Fig. 5.5 is largely analysis oriented.

Synthesis is considered an opposite thought process to analysis. In the context of design, analysis is a process to derive attributive descriptions about the design object's behaviour and function, whereas synthesis is to derive attributive descriptions about the design object from functional requirements to the design object. This justifies the claim that synthesis is the opposite of analysis. The problem here is that such a thought process as synthesis cannot be framed in the thought-process model described in Fig. 5.5. In addition, it is known to be extremely difficult to describe knowledge and its operations for synthesis explicitly in a deductive manner needed for step (3) in Fig. 5.5.

In contrast to the analysis-oriented thought-process model of Fig. 5.5, we here propose the synthesis-oriented thought-process model shown in Fig. 5.6. First, syn-

(1) Observation of phenomena
A phenomenon is observed as objectively as possible. Observations are recorded to form findings O.

(2) Extraction of facts
Observed facts Fo are extracted from observations O.

(3) Formation of hypotheses or selection of axioms
Observed facts Fo can be used to reason out hypotheses H. For instance, experiments can lead to formation of experimental equations. In obvious cases, a set of known axioms K is selected instead of H. The smaller number of axioms is preferred.

(4) Assuming definition facts
Initial definition facts Fd are assumed. Together with K (or H), this will be used to derive theorems Th. Usually, Fd contain such known facts as boundary conditions and initial conditions.

(5) Derivation of theorems from axioms
Theorems Th are derived from K (or H) and Fd deductively. This process may break down the original problem (i.e., derivation of theorems) into smaller subproblems (the "divide-and-conquer strategy").

(6) Verification of theorems against facts
The derived theorems Th are tested against the observed facts Fo to check the explicability of the theorems. This test checks if the derived theorems subsume and do not contradict the extracted facts (i.e., $Th \supseteq Fo$). If the test result was satisfactory, then the theorems are said to explain the extracted facts and the choice of K (or H) was appropriate. If $Th = Fo$, then K is complete. If $Th \supset Fo$, then $Th - Fo$ signifies unobserved facts or undiscovered facts in the future or past. If $Fo - Th \neq \emptyset$, then unexplained facts remain.

(7) Verification of theorems against other known axioms
The derived theorems Th are again tested against other known sets of axioms K'. This test verifies if the theorems are compatible with K' or at least if they do not violate K'. If the hypotheses obtained in step (3) pass tests (6) and (7), they become axioms.

Fig. 5.5 Classic analytical thought process.

thesis begins with describing requirements, rather than observation of phenomena to be explained. Then, axioms on which synthesis is based are selected. Synthesis involves multiple viewpoints as axioms, so the number of axioms is inevitably large. It is interesting to notice in this model that abduction plays a critical role in step (3) as well as deduction in step (4). In this sense, this model considers that synthesis is not just an opposition of analysis, but rather analysis and synthesis are complementary to each other. Just like analysis, synthesis also requires verification of solutions against axioms, which could be an analysis-oriented thought process. This model is similar to the design-process model for the design simulator, but this model captures the nature of synthesis: that synthesis inevitably involves multiple viewpoints as axioms.

5.4.3 Multiple model-based reasoning

Synthesis, in particular engineering design, is a model-based reasoning process. Figure 5.7 depicts a framework for model-based reasoning. First, the physical world is observed. By doing so, a designer creates a mental model through which models

A Knowledge Operation Model of Synthesis

---

(1) Describing requirements
Requirements for the synthesis R are described as theorems.

(2) Extraction of requirements of interest
From R, we only focus on interesting facts as Fo.

(3) Selection of axioms
Synthesis requires, by nature, various viewpoints to be considered. This means that the number or cardinality of K tends to be large.

(4) Derivation of solutions from requirements and axioms
Solutions Fd are derived as facts from K and Fo. The basic reasoning mode could be abduction, rather than deduction, unless there exists an algorithm to arrive at solutions. Just like analysis, the divide-and-conquer strategy might be used, but since the number (or cardinality) of K could be larger than analysis, trade-off and negotiation among different solutions are important.

(5) Derivation of theorems from axioms and facts
Theorems Th are derived from K and Fd deductively. This is very much the same as in the analysis oriented thought process. This process corresponds to the development subprocess in the cognitive design process model. Deduction and the divide-and-conquer strategy are central.

(6) Verification of theorems against requirements
The derived theorems Th are tested against the requirements of interest Fo to check if the derived Th subsume the initial requirements Fo; (i.e., $Th \supseteq Fo$). By doing so, we can check if the solutions Fd are satisfactory.

(7) Verification of theorems against other known axioms
The derived theorems are again tested against other known sets of axioms K'. This test verifies if Fd (and accordingly Fo) is compatible with not only K but also K' (for instance, not violating any constraints not considered at step (3)).

Fig. 5.6 Synthesis-oriented thought process.

in "media" are generated. These models include a drawing and a design object itself. Second, the designer abstracts these models and codifies them in a logical world. The designer first generates object-dependent models based on the set of axioms of his choice. These models serve as reference models for logical reasoning in the logical workspace. As Fig. 5.7 suggests, the designer's thought process controls logical reasoning in the workspace and the object-dependent models serve as "models" in the logic sense. In other words, the truth-value in the logical workspace is determined by the object-dependent models.

Whereas analysis is based on a single set of axioms, synthesis requires a number of sets of axioms. This requests that a model of synthesis must include features of multiple model-based reasoning (as opposed to single model-based reasoning). Each model is formulated based on a set of axioms (or models) of the identical design object. Since these models represent the same design object, they are dependent on each other and we need a mechanism to maintain consistency among them. This is the issue that the metamodel system exactly addresses [18].

To maintain consistency, we have to represent the relationship among models, and for this purpose we need a basic ontology. In this research, we define ontology as a basis for describing the recognition of the world for one purpose. This definition implies that all of the descriptions of the world are subjective in a certain sense and

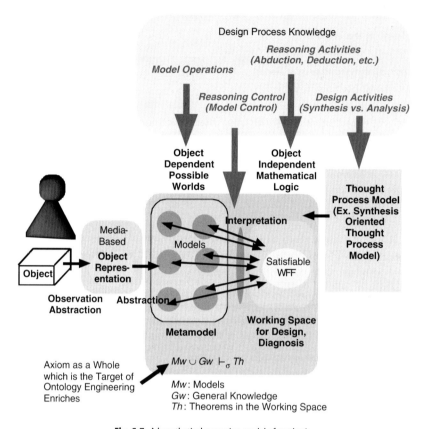

Fig. 5.7 A hypothetical reasoning model of synthesis.

there does not exist a neutral ontology. However, if the purpose is similar, we believe we can set up a more general and shareable ontology to guarantee interoperability among various kinds of models in the logical workspace.

To enable this multiple model-based reasoning, we have already proposed the pluggable metamodel mechanism that is a modelling framework for integrating multiple models [19]. In order to integrate multiple models, this mechanism uses physical concepts, including physical phenomena, attributes, mechanical components, *etc.* for general ontology. This general ontology is used for representing a design object as a network of relationships among concepts that appear in various models. Types of relationship include causal dependency among physical phenomena, arrangements of components, and attribute relationships.

5.4.4 Function modelling

As pointed out in Section 5.3, synthesis is subjective. This is most significant when dealing with the designer's intention. Function modelling is one approach to model the designer's intention. There are a number of reports on function modelling [20,21]. However, most of them, including our former research (function–

behavior–state (FBS) modelling), aim to model function from the viewpoint of design-object modelling and not so much of design-process modelling.

The FBS modelling represents a function as a subjective description of behaviours that are state transitions. Though the modelling scheme is effective as a post-mortem description of design results, it does not consider the characteristics of function design [22,23]. This is also true for German design methodology. Function analysis and functional decomposition always refer to entity descriptions. Though these methodologies advocate embodiment only after functional decomposition, observations of function design reveal that designers always need to have references to physical entities. Functional decomposition based purely on functional representation is almost impossible.

5.4.5 A reasoning framework of design

5.4.5.1 Knowledge operations in design

In this section, we discuss the knowledge operations in design based on observation of the designer's activities. Let us consider a day of a typical designer to write a technical proposal and specification. The following are detailed descriptions about each activity.

- Meetings and negotiation.
  The designer has discussions with other people to acquire new knowledge and information about the problem (**Knowledge/Information Acquisition**), or confirms information already obtained (**Confirmation of Information**), which sometimes results in revision of information, if confirmed information is different (**Information Revision**). From the viewpoint of knowledge, negotiation is conducted to resolve conflicts among participating sections' interest (**Conflict Resolution**).
- Coffee break.
  To use newly acquired knowledge or information effectively, it is necessary to relate new knowledge and information with that existing (**Knowledge Reorganisation**), which sometimes results in revision of knowledge if there are contradictions (**Knowledge Revision**). In addition, the designer reads magazines or chats with colleagues to obtain new knowledge and information (**Knowledge/Information Acquisition**).
- Information collection
  In this process, the designer intensively collects information related to the problem (**Knowledge/Information Acquisition**).
- Specification composition
  In this process, the designer articulates acquired information (**Knowledge/Information Reorganization**) and proposes candidate solutions for the problem by using different types of engineering tools (**Solution Synthesis**). After proposing the solutions, he also uses engineering tools for analysing them to check whether or not they solve the problem (**Object Analysis**).

In this example we identified seven different knowledge operations (Knowledge/Information Acquisition, Knowledge/Information Reorganisation, Information Confirmation, Knowledge/Information Revision, Conflict Resolution, Solution Synthesis, and Object Analysis), and these operations do not come in any particular order. This means that the designer's activities are not a straightforward process to solve a problem as the cognitive design-process model.

In addition, we found that the designer collects knowledge from various types of knowledge source. In particular, in detailed solution synthesis and object analysis processes, the designer collects and uses knowledge that is embedded in different engineering tools. These engineering tools act not only as working spaces, but also as knowledge sources. Therefore, we can formalise design processes as multiple model-based reasoning processes.

5.4.5.2 A hypothetical reasoning framework of design

To deal with the multiple viewpoints that is the nature of synthesis, we formalise knowledge operations and knowledge sources identified in the Section 5.4.5.1 by using the framework of the model-based reasoning system depicted in Fig. 5.7. In the knowledge/information acquisition process, a designer observes a physical world and collects knowledge and information about it. This process means the designer creates a mental model through which models in "media" are generated. These models in media include drawings and the design object itself. These models in media must be abstracted, captured, and codified as models on a computer. A different model can be generated for a different abstraction and codification system.

These codified models, as a whole, form a workspace for a designer by acquiring design information like design specification and his/her decision, and providing him/her with different perspectives about a design object. These codified models may correspond to object-dependent tools and systems, such as drawings, computer-aided design systems, computational tools, databases, and knowledge bases. Therefore, this framework, as a whole, needs to be a multiple model-based reasoning system. Each of these models describes the design object from a particular perspective and stores background knowledge about that domain.

The designers' actual thought process will take place in a logical working space that is object independent. Any reasoning in this object-independent logical working space must be controlled by the thought-process models. (There can be different thought-process models for analysis and synthesis, for instance.) The object-dependent level models serve as "models" for the object-independent model in the logic sense. In other words, the object-dependent models determine the truth-value in the logical working space of the object-independent model.

In the hypothetical reasoning framework of design (Fig. 5.7), object-dependent models deal with a variety of knowledge and information about design objects, and they are integrated with the multiple model-based reasoning system. In addition, object-independent models are controlled by thought-process models.

5.4.5.3 Modelling operations in the object-dependent models

The hypothetical reasoning framework in Fig. 5.7 has object-dependent models and an object-independent logical model. The object-dependent models are operated by such (modelling) operations as "Selection of a Modelling System", "Building a Model", "Modification of a Model", "Introducing a new modelling system", "Modification of the knowledge base of a modelling system", "Maintenance of models in different modelling systems", and "Reasoning about a model".

A Knowledge Operation Model of Synthesis

5.4.5.4 Logical reasoning operations in the object independent level workspace

We use the following ten reasoning operations of the synthesis- and analysis-oriented thought process (Figs. 5.5 and 5.6) as logic-level operations in an object-independent level:

- **Reasoning operations for the analysis-oriented thought process**
  - Observation of Phenomena
  - Extraction of Facts
  - Formation of Hypotheses or Selection of Axioms
  - Verification of Theorems against Requirement
- **Reasoning operations for the synthesis-oriented thought process**
  - Describing Requirements
  - Selection of Axioms
  - Derivation of Solution from Requirements and Axioms
  - Verification of Theorems against Facts
- **Reasoning operations for the analysis- and synthesis-oriented thought process**
  - Derivation of Theorems from the Axioms
  - Verification of Theorems against Other Known Axioms.

5.4.5.5 Formalising knowledge operations in design

We can now formalise knowledge operations in design by using the modelling operations, the reasoning operation, and model mappings between these two levels. Owing to space limitations, only the "Solution synthesis" operation is discussed. Other operations can be formalised similarly.

Solution synthesis is a knowledge operation in design to suggest a new candidate solution for a given problem. This process corresponds to the core part of the synthesis-oriented thought process (Fig. 5.6) and requires unit reasoning and modelling operations in the following order. First, a set of axioms for synthesis must be selected, which is formalised as **Selection of a Modelling System**. The description of a problem is formalised as **Building a Model**. The designer then proposes a new solution candidate by **Reasoning about a Model**. If the designer only uses the object-independent level, this process is formalised as **Describing Requirements, Selection of Axioms**, and **Derivation of Solution from Requirements and Axioms**. If there is a system that directly performs synthesis, such as the FBS modeller to support derivation of design solutions from functional requirements, **Derivation of Solution from Requirements and Axioms** can be performed by such a system. We have already started to formalise the concept of model-based abduction [24].

To select an appropriate modelling system, knowledge about modelling systems is necessary. After this, the designer has to map an object-independent model to a model in the selected modelling system. To do so, it is necessary to translate knowledge from the ontology for the object-independent model to the ontology for the selected modelling system.

5.5 The implementation strategy of the framework

In this section, we discuss how to implement the framework depicted in Fig. 5.7 with regard to our related research.

5.5.1 Multiple model-based reasoning system

Since each design-object model is formulated based on a set of axioms (or models) about an identical design object, and these models represent the same design object, they are dependent on each other and we need a mechanism to maintain consistency among them. As described in Section 5.4, we use the pluggable metamodel mechanism [19]. This mechanism needs to be equipped with an ontological system containing concepts that includeg physical phenomena, attributes, relationships, and mechanical components, *etc.*

5.5.2 Thought-process model

The framework depicted in Fig. 5.7 needs to have a thought-process model that controls knowledge operations. The synthesis- and analysis-oriented thought-process models (Figs. 5.6 and 5.5) can serve as the basis of a computational design-process model.

Since the purpose of this thought-process model is to control operations to design-object models, we also need to formalise knowledge for them. This knowledge was identified in Section 5.4.5. However, we need to associate design semantics with these knowledge operations, so that the process control does not lose the relevance to engineering design.

In addition, we have started to construct general ontology and knowledge for synthesis. We have formalised knowledge for building models [25].

5.5.3 Model-based abduction

5.5.3.1 Strategy for modelling of abduction operation

Though abduction is yet to be understood both logically and computationally, the following methods can be proposed to carry out abduction computationally [24].

- Use of analysis.
  - *Generate and test*: synthesis is a complementary process of analysis. Thus, abduction can be replaced by random generation of solution candidates and testing them with analytical methods. This method is simple but inefficient. However, if we do not have any knowledge available for abduction, this could be one possible method.
- Use of procedural knowledge.
  - *Catalogue retrieval*: when design knowledge that organises design solution candidates is available, the abduction process boils down to catalogue retrieval. This method is useful and efficient. However, obviously the availability of such organised knowledge is limited.
  - *Case-based reasoning*: an initial approximate design solution is looked up in the knowledge base. This case will then be modified to meet the requirements better. This is a useful idea if the knowledge base has sufficient cases, in such cases as routine design.
  - *Calculation model*: when a design problem can be formulated as inverse computation of analysis, we can reach a solution by mathematically solving formulae used for analysis. This method is useful and efficient. However, this method is also applicable only to limited design problems.

A Knowledge Operation Model of Synthesis

- Use of procedural knowledge and analysis.
  - *Optimisation technique (genetic algorithm, simulated annealing, etc.)*: most of these methods are similar to generate and test, but the procedure for generating new solution candidates is managed by the algorithms and an analysis method shall be given as an evaluation method. These methods are simple and not so inefficient as *generate and test*. However, codification of the design problem and evaluation criteria is not trivial for many design problems.

In this chapter, we take an approach that uses assistance of semantic information in the design context. By semantic information, we mean information about knowledge contents, *e.g.*, relationships between entities and their functions. This is confirmed by our finding from design experiments, that there is no knowledge directly coupling functional knowledge and attributive knowledge. Instead, the connection between functions and attributes is mediated by entities [2,26]. Because abduction is usually carried out to derive, for instance, an entity that performs the required function, it is natural to get assistance of such semantic information about the knowledge contents. In addition, as pointed by Bylander *et al.* [27], most of the abductive reasoning is non-polynomial hard. Introducing semantic information about knowledge adds another restriction to the abduction process and allows us to avoid the NP hardness problem.

Figure 5.8 shows the protocol data for designing a scale from the design experiment [2]. In this case, a standard scale and its components were utilised for mediating function (such as "measure weight") to attribute (such as "displacement").

From these observations, we propose to organise design knowledge based on entity concepts. In addition, abduction in design is a process to derive a design solution (entity) from the functional requirements and/or required properties. Therefore, to formalise abduction in design, we classify the design knowledge (*i.e.*, axioms) by the following two types, where e denotes an entity, f is a function, and p is a property (attribute is a kind of property):

- knowledge that describes an entity's functions: $e \to f$
- knowledge that describe an entity's property: $e \to p$.

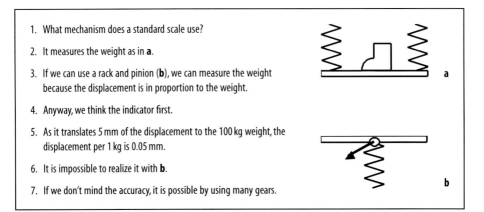

Fig. 5.8 Protocol data from the design experiment [2].

Both these types of knowledge can be used bidirectionally. Suppose there is an entity that fulfils the functional requirement f_i. This process should derive an entity e_i using the first type of knowledge using abduction. Then, the second knowledge is used to obtain property information p_j about the solution e_i.

5.5.3.2 Model-based abduction

Based on the discussion of semantic information in the design, we look carefully at the role of the entity in the design process. As Yoshikawa *et al.* [28] stated, designers use both existing (known) entities and virtual, non-existing, yet-to-be-created entities that are in their minds.

When we look at the protocol data (Fig. 5.8), we can find different roles of entity. In protocols 1 and 2, the designers extracted the basic configuration information from "standard scale". In protocol 3, the designers modified the configuration of the spring and the plate. In protocol 7, the designers point out the backlash of the gears and discussed the accuracy issue.

From the observation of these protocols, we find out that most of the entity concepts used in these protocols did not directly correspond to any particular existing entities. These entities are abstraction of existing entities obtained by selecting important ones. For example, let us compare Fig. 5.8a and b, which describes an existing scale. Both scales have a spring and plate; but Fig. 5.8a has two springs, whereas Fig. 5.8b has one spring. This example shows the designers used simplified virtual entities and they added more information by referring to the existing entities.

To represent these different roles of entity, we have previously proposed an *abstract entity concept* [29]. The abstract entity concept is similar to the class definition in object-oriented knowledge representation and has a finite number of attributes that characterise entities belonging to that class. In topological representation of design knowledge, such as GDT, an abstract entity concept corresponds to a representative element in the attribute space.

For example, an abstract entity concept "table" has two attributes, *viz.*, plate and legs. Suppose, with the reference to the particular table or more concrete abstract entity concept, the designer pays attention to the shape of the plate and the number of legs. He/she decides the value of each attribute by using abduction. Fig. 5.9 shows the evolution process of this abstract entity.

Based on this discussion above, we try to formalise abduction in design in a model-based reasoning environment that includes aspects necessary for design, such as attributes and function. From a comparison of Equation (5.2) and the formula in Fig. 5.7, axioms correspond to general background knowledge (Gw) and fact corresponds to model (Mw). Gw is a knowledge base that is related to a model-based rea-

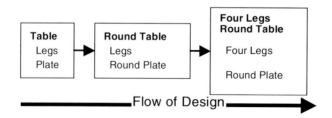

Fig. 5.9 Evolution of an abstract entity in design process.

soning system. Each knowledge base (Gw_i) forms a modeller from a specific aspect and it represents the relationship between entities and attributes and entities and functions. In addition, Gw_i should contain knowledge about the background theory of the modeller, how to use the modeller, and how to build a model for the modeller [28].

For example, a geometric modeller can handle the geometric information of entities stored in the modeller. Gw_{geo} should also contain knowledge about concepts used in algebraic geometry, such as point, edge, surface, cube, cylinder, *etc.*, as well as how to use this geometric modeller, and how to build a geometric model including input information. By using knowledge about the modeller, we can conduct relevance reasoning to select the proper modeller for solving the problem and guide how to use it from the design semantic point of view.

For handling functional concepts, a function modelling system, such as an FBS modeller, is used for managing the relationship between function and entity. The FBS modeller can represent the function of an entity and help the designer reason out an entity that is a solution candidate of the required function. In this framework, design requirements can be given in terms of function and attributes.

5.5.3.3 Algorithm of model-based abduction

Let us first outline the algorithm of model-based abduction. In this algorithm, we focus on how to deal with multiple axioms and concept categories, such as entity, attribute, and function, to obtain assistance from semantics information.

1. Derive the neighbourhood system of a solution candidate from one axiom set K_1.
   (a) Set requirements that can be treated by axioms K_1 as theory (Th_1) in Equation (5.2).
      (i) $K_1 \cup F_1 \vdash_\sigma Th_1$
      (ii) $\{e_1 \to p_1, e_1 \to p_2, \ldots, e_2 \to f_1, \ldots\} \cup F_1 \vdash_\sigma \{p_1, \ldots, f_1, \ldots\}$, where e_i is an entity concept, p_i is an attribute concept, and f_i is a function concept.
   (b) Derive design solution candidates, $F_1 = \{e_1, e_2, \ldots\}$, by abduction with the closed world assumption.
   (c) Analyse the neighbourhood system of F_1 in the attribute space and the function space by deduction using a modeller that corresponds to axioms K_1. This will enrich Th_i.
2. Apply previous procedures with another set of axioms and make the attribute information and function information richer.
3. Compute $F_n \cap F_{n+1} \cap \ldots$ for narrowing the solution space to reach a solution.

Now we can formalise an algorithm of model-based abduction as operations to the hypothetical-reasoning framework depicted in Fig. 5.7.

1. Set up the initial requirements in the working space (Th).
2. Select a knowledge base (or modeller, Gw) that can derive abstract entity concepts from Th (Relevance reasoning). If there is no more knowledge base to be used, then end.
3. Build a model (Mw) for the selected modeller.
   (a) Generate a solution candidate as an abstract entity concept. This can be done either by directly searching for a candidate using semantic assistance or by

indirectly identifying model generation operations that may arrive at an abstract entity concept. The former can be implemented with straightforward strategies, such as "generate and test", "catalogue search", "case-based reasoning", "calculation model", and "genetic algorithm". The latter can be implemented with the help from knowledge about model generation operations stored in Gw.
  (b) Build an aspect-specific model using knowledge stored in Gw. Transport information from Th and other modellers, if necessary. This corresponds to enriching information about the abstract entity concept and obtaining a specialised abstract entity concept.
4. Analyse the model (Mw) using knowledge stored in Gw. This is basically deduction, and the results (*i.e.*, the model's properties) are added to Th.
5. Confirm the validity of Th. This means comparison of the derived properties with the requirements.
  (c) If OK, the current Th is asserted. Go to step 2.
  (d) If No Good, do one of the following:
    (i) modify the specialised entity concepts resulting in a new model. Go to step 3b;
    (ii) modify the knowledge in Gw that might have caused the conflict. This operation is necessary to deal with the frame problem. Go to step 3a. Circumscription [30] could be useful for this purpose.

5.6 Verification of the model of synthesis

The model of synthesis described in Section 5.4 should eventually be verified by experimental data about design or by implementing a system that performs the algorithm depicted in Section 5.5. To do this, first a design case was selected and a reference model was built. This reference model was then used to verify the knowledge operation model of synthesis, and further used to run a design simulation on the implementation of the reasoning framework.

5.6.1 Selection of data for verification

Design cases for the verification of the model of synthesis should satisfy the following conditions:

- the design cases should not be toy problems nor unrealistically made-up cases; they must have "newness", which is the essence of synthesis;
- concrete data about the design process should be recorded;
- they should not be too huge and complicated; the data size must be reasonably small for further analysis.

Considering these, we paid attention to a machine design conducted at a university laboratory. The research was the development of a high-precision stereo lithography machine for micro-photoforming fabrication of micro-flexible mechanisms [31]. As the source of information, we used three bachelor theses in 1995, 1996, and 1997, a Ph.D. thesis in 1998, and summary reports written several times in the year. We also investigated weekly reports written by the students every week to analyse design

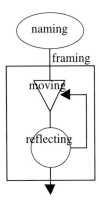

Fig. 5.10 Design activities of the frame cognition model.

activities. A typical weekly report consists of several lines about what a student did in a week.

From this design case, we have built a "reference model" of design that satisfies the three conditions above to verify the model of synthesis.

5.6.2 The reference model

To build a reference model, we employed the frame cognition model that identifies frame structure of design activities [32]. The frame cognition model classifies design activities (or design protocols) in to categories, *viz.*, "naming", "framing", "moving", and "reflecting". Figure 5.10 illustrates relationship among these four categories. The subject that relates to the present design conditions is named and articulated by activity "naming". The "framing" activity builds up a problem in a frame. The design proceeds and a solution is obtained by activity "moving". The solution is evaluated and refined by activity "reflecting". The "naming" and "framing" activities correspond roughly to the awareness of the problem of the cognitive design process model (Fig. 5.1). The "moving" activity corresponds to the suggestion and development stages, and "reflecting" corresponds to the evaluation and decision stages. Design protocols are analysed and categorised into these four activities.

By doing so, a design process is chronologically framed and a reference model is obtained. Figure 5.11 illustrates a part of the reference model displayed as a frame cognition model. Minor modifications were needed, however. First, the model was proposed from the analysis of design protocol data. Because the original protocol data came from a design experiment that dealt with conceptual design of a small and simple component [6], it basically does not consider the hierarchical nature of design very well. The definitions of the four design activities are not clear; in particular, "reflecting" was confusing. In addition, the model did not consider information flow at all. In particular, it totally lacks notions of how information acquisition was done, what kind of information was used, which knowledge source the designer had access to, *etc.* So, we had to improve the frame cognition model by taking the hierarchical nature of design and the acquisition process of knowledge and information during the design activities.

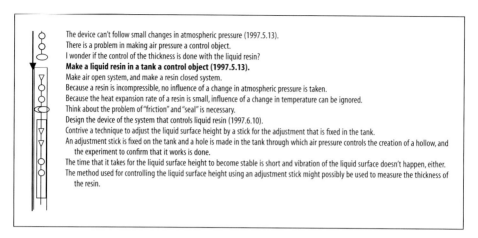

Fig. 5.11 Part of the reference model.

5.6.3 Verification of the knowledge operation model

This section discusses the verification of the knowledge operation model illustrated in Section 5.4.5 with the reference model. The knowledge-level operation model contains seven knowledge operations: "knowledge/information acquisition", "knowledge/information reorganisation", "information confirmation", "conflict resolution", "knowledge/information revision", "solution synthesis", and "object analysis". We check the correspondence of these seven knowledge operations with the design activities in the reference model. By doing so, the completeness of the knowledge operation model can be verified. We analysed a part of the reference model, which was research on the micro-photoforming fabrication method using a liquid hollow shaped by controlling liquid quantity. We confirmed that the seven knowledge-level operations cover the reference model, but randomly.

5.6.4 Vocabulary about design

We also conducted a comparative analysis between the knowledge-level operation model and the cognitive design-process model. However, this comparison resulted in the following problems.

1. Design activities are represented by short sentences that contain descriptions of concrete behaviour or thinking. It is difficult to relate design activities to one stage in the cognitive design-process model, because a design activity itself often contains various elements.
2. One design activity sometimes corresponds to more than one stage of the cognitive design-process model, because the granularities of design activities and stages of the design process-model are different.

We think that these problems happened, because we made direct correspondences between design activities of the reference model and stages of the cognitive design-process model. Rather, we need to introduce an intermediate abstraction level

A Knowledge Operation Model of Synthesis

Table 5.1 The results of the analysis using vocabulary.

Knowledge-level operation (no. of appearances)	Vocabulary (no. of appearances)
Knowledge/information acquisition (57)	investigation (26), knowledge acquisition (19), problem indication (16)
Knowledge/information reorganisation (33)	arrangement of the information (13), knowledge reorganisation (10), concretisation (7), drafting (3)
Information confirmation (10)	confirmation (10)
Conflict resolution (8)	conflict resolution (8)
Knowledge/information revision (9)	strengthening of the constraint (6), information revision (3)
Solution synthesis (44)	suggestion (24), selection (7), idea (8), improvement (3), association (1), decision (1)
Object analysis (58)	evaluation (27), experiment (9), trial manufacture (14), estimation (4), numerical analysis (3), derivation (1)

between design activities and the cognitive design-process model, or further the knowledge-level operation model. Therefore, we introduced standard terms (called "vocabulary") about design to represent design activities at both the more general and at the detailed levels.

For example, we can find the following design activities from the reference model:

- *"Because of the moving mechanism of the table at the top, the upper contact side of the tank deviates and a hollow warps caused by the movement of the table."*
- *"Should I fix a table (X-Y table) at the top of the tank?"*
- *"The X-Y table is changed to the bottom of the tank, and the top of the tank is unified with the tank."*

This series of design activities can be interpreted with the vocabulary as follows. The "estimation" of *"Because of the moving mechanism . . ."* and "knowledge acquisition" were performed owing to the contradiction in the experiment, and information is collected by the "investigation" to solve the problem. The "idea" of *"Should I fix . . ."* appeared in the process of the "arrangement of the information" of the collected information. According to the idea, "knowledge acquisition" and "knowledge reorganisation" are carried out, then "suggestion" of *"The X-Y table is changed . . ."* is made.

Likewise, each design activity can be explained by using terms from the vocabulary. The vocabulary included 23 standard terms that were obtained through brainstorming of the other team members of the project. All design activities in the reference model were interpreted, and, as a consequence, 102 design activities were converted into 223 that are correlated to only 23 kinds of standard term (see Table 5.1). These correspondences were one-to-one, thus indicating that the knowledge operation model of synthesis is a good interpretation of the reference model.

5.6.5 Verification through implementation of the reasoning framework

The model of synthesis described in Section 5.4 was then verified by implementing a system that performs the algorithm depicted in Section 5.5. The Knowledge Inten-

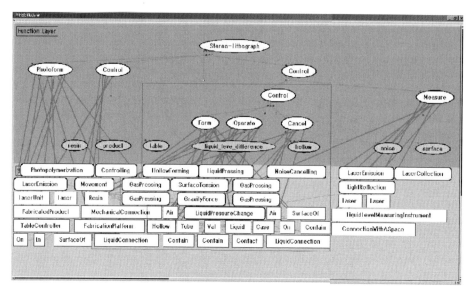

Fig. 5.12 An FBS model of a stereo-lithography machine.

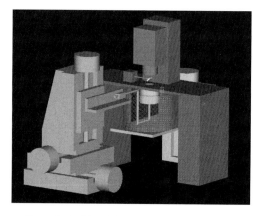

Fig. 5.13 A solid model of a stereo-lithography machine.

sive Engineering Framework (KIEF) [14,19,25] was used as a basis for this implementation. Figure 5.12 depicts an FBS model of the stereo-lithography machine design and Fig. 5.13 is its solid model representation created by the system. The system could follow the design of the reference model, automating many phases of this design.

5.7 Conclusions

This paper is a scientific attempt to build a model of synthesis and to formalise it. From our previous research results and observations of other researchers, we concluded that it is necessary to build such a model for facilitating unified understanding of synthesis, and it is possible to do so.

The first two models we discussed were the *analysis-oriented thought-process model* and the *synthesis-oriented thought-process model*. These models were given logical interpretations and the main findings were: (1) abduction plays a crucial role in both synthesis and in analysis; (2) synthesis and analysis are not opposite, but complementary to each other.

We then proposed the *multiple model-based reasoning framework* to be used as a platform for implementing an advanced design support system to verify our model of synthesis. We then identified knowledge operations that can describe designers' activities. These knowledge operations were decomposed into model operations, reasoning operations, and a model mapping operation between these two levels that can be performed in the framework. Among logical operations, abduction plays a crucial role for synthesis. We illustrated the algorithm for model-based abduction on the multiple model-based reasoning framework.

In Section 5.6, we tested the knowledge operation model against a reference model that was obtained from an actual design case. By introducing abstract standard terms that describe design activities, we could show good correspondences between the knowledge operations and the reference model. This justifies the completeness of the knowledge-level operation model. In addition, the knowledge operation model was implemented on the multiple model-based reasoning framework. The reference model was run on this model, indicating that the knowledge operation model could simulate the design process of the reference model.

Acknowledgements

This research was partially conducted within a project entitled "Modeling of Synthesis" conducted at the University of Tokyo, Osaka University, Nara Institute of Science and Technology, Tokyo Metropolitan University, and National Institute of Informatics, Japan. This project, JSPS-RFTF 96P00701, was financially supported by the "Research for the Future" Program of the Japan Society for the Promotion of Science.

References

[1] Yoshikawa H. General design theory and a CAD system. In: Sata T, Warman EA, editors. Man–machine communication in CAD/CAM. Amsterdam: North-Holland, 1981; 35–58.
[2] Takeda H, Hamada S, Tomiyama T, Yoshikawa H. A cognitive approach to the analysis of design processes. In: Rinderle JR, editor. Design theory and methodology – DTM'90, DE-vol. 27. New York (USA): ASME, 1990; 153–60.
[3] Takeda H, Veerkamp P, Tomiyama T, Yoshikawa H. Modeling design processes. AI Mag 1990;11(4):37–48.
[4] VDI-Gesellschaft. VDI-2222 Konstruktion und Entwicklung. Düsseldorf: VDI-Verlag, 1977.
[5] Hubka V. Theorie der Maschinensysteme. Berlin, Springer, 1973.
[6] Cross N, Christiaans H, Dorst K, editors. Analyzing design activity. Chichester (UK): Wiley, 1996.
[7] Ullman DG, Dietterich TG, Stauffer LA. A model of the mechanical design process based on empirical data: a summary. In: Gero JS, editor. Artificial intelligence in engineering design. Amsterdam: North-Holland, 1988; 193–215.
[8] March L. The logic of design. In: Cross N, editor. Developments in design methodology. Chichester (UK): Wiley, 1984; 265–76.

[9] Coyne RD, Rosenman MA, Radford AD, Gero JS. Innovation and creativity in knowledge-based CAD. In: Gero JS, editor. Expert systems in computer aided design. Amsterdam: North-Holland, 1987; 435-65.
[10] Hartshorne C, Weiss P, editors. The collected papers of Charles Sanders Peirce, vols I-VI. Cambridge (MA): Harvard University Press, 1931-1935.
[11] Burks A, editor. The collected papers of Charles Sanders Peirce, vols VII-VIII. Cambridge (MA): Harvard University Press, 1958.
[12] Blessing LTM. Comparison of design models proposed in prescriptive literature. In: Perrin J, Vinck D, editors. The role of design in the shaping of technology, COST A3/COST A4 International Research Workshop, Social Sciences Series, vol. 5. European Committee, 1996; 187-212.
[13] Grabowski H, Rude S, Grein G, editors. Universal design theory. Aachen (Germany): Shaker, 1998.
[14] Tomiyama T. From general design theory to knowledge-intensive engineering. Artif Intell Eng Des Anal Manuf 1994;8(4)319-33.
[15] Yoshioka M, Tomiyama T. Toward a reasoning framework of design as synthesis. In: Proceedings of the 1999 ASME International Design Engineering Technical Conferences and Computers and Information in Engineering Conference, DETC99/DTM-8743. New York (USA): ASME, 1999 [CD-ROM].
[16] Roozenburg NFM, Eekels J. Product design: fundamentals and methods. Chichester: Wiley, 1995.
[17] Valdes-Perez RE. Computer science research on scientific discovery. Knowledge Eng Rev 1996; 11(1):57-66.
[18] Tomiyama T, Kiriyama T, Takeda H, Xue D, Yoshikawa H. Metamodel: a key to intelligent CAD systems. Res Eng Des 1990;1(1):19-34.
[19] Yoshioka M, Tomiyama T. Pluggable metamodel mechanism: a framework of an integrated design object modelling environment. In: Bradshaw A, Counsell J, editors. Computer Aided Conceptual Design'97, Proceedings of the 1997 Lancaster International Workshop on Engineering Design CACD'97, Lancaster University, Lancaster, UK; 57-69.
[20] Chakrabarti A, Blessing L. Representing functionality in design [special issue] . Artif Intell Eng Des Anal Manuf 1996;10(4):251-3.
[21] Umeda Y, Tomiyama T. Functional reasoning in design. IEEE Expert Intell Syst Appl 1997;12(2):42-8.
[22] Umeda Y, Tomiyama T, Yoshikawa H. Function, behaviour, and structure. In: Gero JS, editor. Applications of artificial intelligence in engineering V, vol. 1: design. Southampton Boston/Berlin: Computational Mechanics Publications/Springer, 1990; 177-93.
[23] Umeda Y, Ishii M, Yoshioka M, Shimomura Y, Tomiyama T. Supporting conceptual design based on the function-behavior-state modeler. Artif Intell Eng Des Anal Manuf 1996;10(4):275-88.
[24] Yoshioka M, Tomiyama T. Model-based abduction for synthesis. In: Proceedings of the 2000 ASME International Design Engineering Technical Conferences and Computers and Information in Engineering Conference, September 10-13, 2000, Baltimore, MD, USA, DETC2000/DTM-14553. New York (USA): ASME, 2000 [CD-ROM, 9 pp.].
[25] Sekiya T, Tsumaya A, Tomiyama T. Classification of knowledge for generating engineering models – a case study of model generation in finite element analysis. In: Finger S, Tomiyama T, Mäntylä M, editors. Knowledge intensive computer aided systems. Dordrecht: Kluwer, 1999; 73-90.
[26] Tomiyama T. A design process model that unifies general design theory and empirical findings. In: Ward AC, editor. Proceedings of the 1995 Design Engineering Technical Conferences, DE-vol. 83. New York (USA): ASME, 1995; 329-40.
[27] Bylander T, Allemang D, Tanner MC, Josephson JR. The computational complexity of abduction. Artif Intell 1991;49:25-60.
[28] Yoshikawa H, Arai E, Goto T. Design theory by experiment – experiment method for general design theory. J Jpn Soc Precis Mach Eng 1981;47(7):46-51 [in Japanese].
[29] Yoshioka M. Theory of design knowledge operation. Ph.D. thesis, University of Tokyo, 1996 [in Japanese].
[30] McCarthy J. Circumscription – a form of non-monotonic reasoning. Artif Intell 1980;13:27-39.
[31] Xie T, Murakami T, Nakajima N. Micro photoforming fabrication using a liquid hollow shaped by pressure difference and surface tension. Int J Jpn Soc Precis Eng 1999;33(3):253-8.
[32] Valkenburg RC, Dorst K. The reflective practice of design teams. Des Stud 1998;19(3):249-71.

Part 2

Approaches

6. Two approaches to synthesis based on the Domain Theory
7. Using the concept of functions to help synthesise solutions
8. Design catalogues and their usage
9. TRIZ, the Altshullerian approach to solving innovation problems

Two approaches to synthesis based on the domain theory 6

Claus Thorp Hansen and Mogens Myrup Andreasen

Abstract The domain theory is described in this chapter. By a strict distinction between the structural characteristics and the behavioural properties of a mechanical artefact, each domain, *i.e.*, transformation-, organ-, and part-domain, becomes a productive view for design of mechanical artefacts. The functional reasoning within each domain and between the domains seems to be ruled by the function–means law (Hubka's law). On the basis of the domain theory and the function–means law we present two formal approaches to the synthesis of mechanical artefacts, namely a design-process-oriented approach and an artefact-oriented approach. The design-process-oriented synthesis approach can be seen as a basic design step for composite mechanical artefacts. The artefact-oriented approach has been utilised for the development of computer-based design support systems.

6.1 Introduction

In order to describe engineering design synthesis a theory should ideally encompass explanations with respect at least to three phenomena, namely the synthesis activities, the human aspect of synthesis, *i.e.*, creativity and reasoning from end to means, and the synthesised artefact, *i.e.*, the product being designed. Because of the different nature of these three phenomena, engineering design research has to rely on several sciences. Not even the core aspect of designing – synthesis – can be comprehensively explained by one theory. Therefore, current theories offer only partial descriptions and are only relevant and valid for certain aspects of engineering design.

The main sources for design theories are the human as problem solver, the artefact being designed, the design activities and the many factors influencing these activities, *e.g.*, knowledge, skills, instruments (like computer-aided design (CAD)), modelling, and process management. In this chapter we focus on engineering design synthesis from the artefact dimension.

Efforts, especially in German-language design research, have focused on the identification of a set of basic characteristics to be determined by the engineering designer through the design process in order to define the artefact. A comprehensive theory was formulated by Hubka and Eder [1] and has since then been recognised as a general theory of technical systems. Today, it is generally accepted in engineering design research that theories describing the design process have to rely upon theories related to the artefact being designed.

The domain theory was proposed by Andreasen in 1980 [2], and since then has been further developed. The core idea is to apply three views on the mechanical product that is to be designed, namely transformation, organ, and part views. These three views encompass substantial classes of structural definitions and behaviours of mechanical artefacts.

In this chapter we will present the domain theory and the function–means law. On this basis we will describe two approaches to engineering design synthesis related to two different design situations. If the artefact being designed can be based upon past designs, the engineering designer has more or less insight into the structure–behaviour relation. However, if the required functionality is new, the engineering designer can perform only a kind of trial-and-error approach, which consists of creating alternative structural proposals. For the design situation where the required functionality is new, we describe a design-process-oriented synthesis approach. The core of this approach is that the engineering designer synthesises an artefact by determining structural characteristics and making navigational manoeuvres within and between the three domains – transformation, organ, and part domains. For a design situation based on past designs we describe an artefact-oriented synthesis approach. The core of this approach is to analyse the past designs and organise the analysis results in accordance with the three domains. The artefact-oriented approach has been utilised by several research groups in their development of computer-based design support systems.

The structure of the chapter is as follows. In Section 6.2 we present the domain theory, and in Section 6.3 the function–means law. In Section 6.4 we describe the design object explained by the domain theory, and the two approaches to engineering design synthesis. In Section 6.5 we conclude on the nature and validity of the results described in the chapter.

6.2 The domain theory

The result of engineering design work is traditionally documented in the bill of materials and a set of drawings; *i.e.*, the production of the parts and the assembly process are specified. Thus, this type of documentation prescribes the production of the product, and, based on this documentation, it is difficult to reason about the product's functionality and properties with respect to other product life phases: distribution, sales, use, *etc*. Therefore, we expect that the introduction of other types of view on the artefact that support the engineering designer in reasoning about functionality and behaviour will improve design work. In this section we present the domain theory [2], which offers three synthesis-oriented views on the mechanical artefact being designed.

6.2.1 Systems theory

Contributions to a general artefact theory can be found in cybernetics, systems theory, and systems engineering [3–5]. A core element in these theories is the distinction between structural *characteristics* (German = *Merkmale*) and behavioural *properties* (German = *Eigenschaften*). A system's structure is designed by determination of its characteristics, whereas the behaviour of a system is how it reacts to input and how human beings perceive its properties.

In engineering design research, several behavioural views upon mechanical artefacts are utilised: man–machine interaction, transformation, views showing functionality or specific views on strength, dynamics, reliability and other properties. The systems theory approach may be seen as a modelling view: the system designer's view determines what he/she sees as elements and relations, and the system model predicts the behaviour in accordance with the view.

In the design methodology theories (explicitly by Hubka [6], but also partly articulated by Pahl and Beitz [7] and others), four different structures or structural system views have been proposed for designing mechanical artefacts:

- a transformation system;
- a system of functions, often termed a function structure;
- a system of organs or function carriers;
- a system of machine parts.

According to systems theory, a system's function is a class of behaviour, *i.e.*, function is a property not a characteristic of the system. Thus, it adds to blurredness in the design methodology theories to see functions as structural elements. The explanation behind this unfortunate approach may be that several engineering design researchers use function expressions for labelling organs or subsystems, leading to a misinterpretation that functions are having both a structural and a behavioural meaning. However, strictly speaking, a function structure of a mechanical artefact does not exist, but one may label the organs being synthesised by their functional expression and show the structure of these organs.

Now the reader might well ask why one does not accept function as a distinct description of artefact, owing to it not being a characteristic, yet still accept transformation as a characteristic even though it too may be seen as a property of the man–machine system. Our answer is that in the transformation domain, the main characteristic is technology. Thus, we could decide to name this domain the technology domain, but we prefer the name transformation domain owing to tradition in German literature.

6.2.2 The domain theory

According to the domain theory, the mechanical artefact to be designed can be seen in three domains (see Fig. 6.1):

- a transformation domain, where focus is on the purpose-oriented transformation of operands like material, energy, and data, which occur when the operator and mechanical or mechatronic product cooperate;
- an organ domain, where focus is on the mechanical product's active elements, *i.e.*, organs that create effects, and their mode of action;
- a part domain, where focus is on the allocation or distribution of the organs into machine parts, which can be produced and assembled so that every machine part solves its tasks and contributes to the totality.

6.2.2.1 The transformation domain

The purpose of any product is to support a transformation or process; see Fig. 6.1. In interplay between the mechanical product and the human operator the effects delivered are those that are necessary for the transformation of an operand.

The operand may be material, energy, data, or biological objects. Normally the end result of the transformation, *i.e.*, the operand in its end state, fulfils the purpose

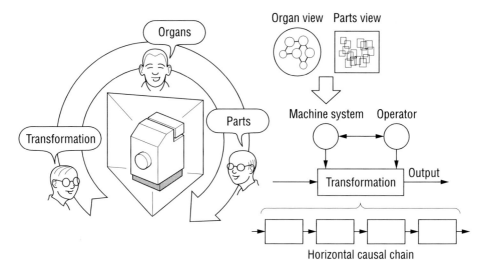

Fig. 6.1 Three views on the mechanical artefact, and three patterns of explanation [8].

or covers the human's need. Each individual step in the transformation is termed an operation, and an operation alters one or more characteristics and properties of the operand. In order to obtain the required end state of the operand more operations are normally needed. Hubka and Eder [1] call this the *horizontal causality chain*.

One may see the structure of operations and for each operation the characteristics of the man–machine system's interplay with the operand as the structural parameters in the transformation domain. The overall structural characteristic in this domain is the *technology*. The transformation output, *i.e.*, the operand in its end state, may be seen as the *functional* aspect in the transformation domain. Other behavioural aspects are the transformation's qualities, termed "universal virtues" by Olesen [9] and divided into the following pragmatic classes: cost, quality, time, efficiency, flexibility, risk, and environmental effects.

6.2.2.2 The organ domain

The active elements, which create the required effects in the mechanical product, are the organs. The mode of action of an organ is based upon a physical effect. Some organs are in direct physical contact with the operand, whereas others deliver effects in a chain structure, which Hubka and Eder [1] call the *vertical chain of causality*.

An organ of the mechanical type consists of wirk elements. We use the German word for effect, namely *Wirkung*, for the elements creating effects (from the theory proposed by Ersoy [10]). For a substantial class of mechanical artefacts we focus upon three types of wirk element, *viz.*, wirk surface, wirk volume, and wirk field. Firstly, an interaction between two surfaces may create an effect, *e.g.*, the interaction between the teeth of two gear wheels constitutes a gear as an organ. Secondly, an organ may be carried by a wirk volume, *e.g.*, a shaft can be seen as an organ, which transmits rotation and torque. Thirdly, a volume bounded by material surfaces can be a wirk field, *e.g.*, the cavity of a linear hydraulic motor, *i.e.*, the cylinder, constitutes an organ, which in interaction with a piston creates a linear movement.

In the organ domain, structure is defined as a structure of organs, and for each organ the structure of wirk elements. Behaviour in the organ domain may be divided into functions and properties. The function of an organ is its ability to create an effect. Organs carry many types of property, which may be classified after Pahl and Beitz [7] or Hubka and Eder [1], in the classes: functional, operational, ergonomic, aesthetic, distribution, delivering, planning, law conformance, manufacturing, design, economic, and liquidation properties.

6.2.2.3 The part domain

In a mechanical product the wirk elements are distributed or allocated among the parts in such a way that the parts can be produced and assembled respecting the requirements given by the wirk elements. One organ will normally need several wirk elements distributed on more parts, whereas a machine part often contributes to the realisation of more than one organ.

Although it is easy to identify the structural characteristics of the part domain, the behaviour causes trouble. Structure in the part domain consists of the parts and their assembly relations, and for each part the characteristics are form, material, dimension, surface quality, and tolerance. But what is the behaviour of parts?

Mortensen [11] defines *task* as behaviour in the part domain. His line of argumentation is as follows. The results of an organ-oriented design activity may ideally be seen as the determination of the organs' wirk elements. These wirk elements and their characteristics may be seen as the *tasks* of the machine parts; for instance, if a wirk element is being specified by the engineering designer as a wirk surface, which is plane, smooth and with certain hardness, this specification means that one of the machine parts shall contain the surface; see Fig. 6.2. Thus, a system model in the part domain is a structure of parts with assembly relations, and the behaviour is fulfilment of tasks. This behaviour is also based on physical effects, but normally they are more basic (material properties, geometry, positions) than in the organ domain.

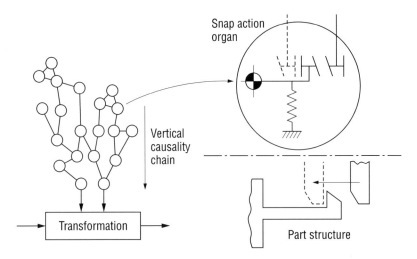

Fig. 6.2 The vertical causality chain of organs. One organ, a snap-fit joint, and its related part structure are shown [8].

Jensen [12] claims that there is no behaviour in the part domain. His line of argumentation is as follows. The structural characteristics in the part domain reflect the production of the mechanical product. However, during production a part is an object being shaped through a sequence of operations forming a piece of material into a machine part, *i.e.*, the part is an operand being transformed. A part being an operand does not actively deliver any effect and does not satisfy a purpose. Therefore, there is no behaviour in the part domain. Thus, as a consequence of this argumentation, the part domain is not a system view of the mechanical artefact.

In our opinion, both Mortensen [11] and Jensen [12] have a relevant and valid line of argumentation. Mortensen has three system views on the mechanical artefact being designed, but has problems in defining a part's task. Jensen skips behaviour in the part domain at the cost that this domain is not a system view. Although these two lines of argumentation result in a contradiction, this contradiction does not invalidate the domain theory. In the authors' opinion it is a matter of taste and purpose as to which definition to choose. We leave further considerations regarding this question open for the reader.

6.2.2.4 Visualising the domain theory

The domain theory can be represented in a graphical model as shown in Fig. 6.3. Each domain can be drawn as a two-dimensional plane with two axes: abstract/concrete and undetailed/detailed. Along the axis undetailed/detailed the number of identified characteristics vary, and along the axis abstract/concrete the number of characteristics, which have been given a value, vary. Design models of a mechanical artefact can be positioned in one of the three planes, depending on the domain to which they belong, and on the actual level of abstraction and degree of detail [2,13]. As an example, three sketches of a hinge are shown in Fig. 6.3. These three design models belong to the organ domain. The first model is the most abstract and undetailed representation of the hinge. The second model is on the same level of abstraction, but with more detail, *i.e.*, more characteristics have been identified. The third model is the most detailed and concrete, *i.e.*, most characteristics have been identified and given a value.

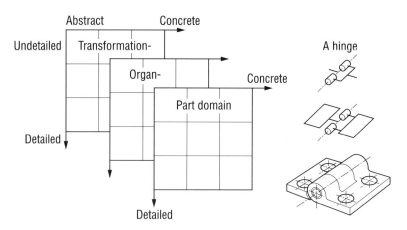

Fig. 6.3 A graphical representation of the domain theory. Three organ models of a hinge [2].

6.3 The function–means law

It is well known that goals and means can be arranged in a hierarchical structure, where what is means seen from the top down becomes goals when seen from the bottom up. In the domain theory we need another type of hierarchy, namely a hierarchy where the goals, *i.e.*, required behaviour, are linked to the means, *i.e.*, the structural characteristics of the mechanical artefact, realising the goals.

The domain theory, presented in Section 6.2.2, introduces three structural-behavioural system views upon the mechanical artefact being designed. It was pointed out that behaviour is different in the three domains, namely purpose, effect, and task, and similarly for the structural elements, namely transformation, organ, and part. Now we shall have a closer look at the function–means law.

The function–means law links the structural and behavioural aspects of a mechanical artefact. It was published in 1967 by Hubka [14], but did not get an explicit treatment until 1980 [2] when it was termed Hubka's law. Today we recognise the law as a fundamental explanatory pattern of mechanical artefacts with many applications.

The function–means law may be expressed as follows [2]: in the hierarchy of effects (the functions), which contribute to realisation of the mechanical artefact's overall purpose function, there exist causal relations, determined by the organs (the means), that realise the effects.

The law says that a means synthesised by the engineering designer for solving a required function seldom is sufficient in itself, but calls for additional functionalities (like energy, control, support, and auxiliary functions) to be realised by additional means.

6.3.1 The function–means tree (F/M-tree)

The function–means law may be modelled as a tree structure: the F/M-tree. The principle in the set-up of the F/M-tree is a hierarchical arrangement of function levels and means levels, connected with lines that correspond to the causal relations between the functions and means. As an example, upper levels of a F/M-tree for a tumble dryer are shown in Fig. 6.4. Please note that we use here the so-called effect-type of function [13], and that the means are of type organ. We return to the F/M-tree in the following.

6.4 Engineering design synthesis

Neither the domain theory nor the function–means law explains or shows an approach to engineering design synthesis. They merely mirror some fundamental traits of the nature of mechanical artefacts. In this section we will take a closer look at the synthesis activity itself and relate it to these traits.

Engineering design synthesis, meaning to find the structural characteristics of a design, which is believed to posses the required properties, is a complex and difficult activity. If the artefact can be based upon past designs, the engineering designer has a more or less complete insight into the structure–behaviour relation. However, if the required functionality is new, the engineering designer can perform only a kind of trial-and-error approach. In this approach it is necessary to create structural proposals.

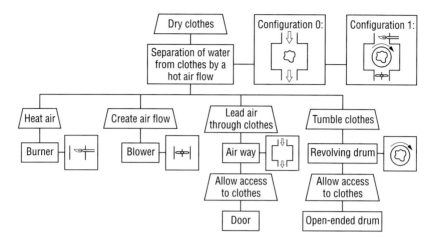

Fig. 6.4 The upper levels of a F/M-tree for a tumble dryer [15].

Literature proposes many types of synthesis method. Some methods are based upon principles or more embodied solutions with insight into the structure–behaviour relation, whereas others are blind or based on exploratively proposing design variants. Examples of methods are the use of design catalogues [16], morphology [17], variation methods [10,18,19], reasoning about physical principles [18] or utilising basic problem-solving principles [20].

This section contains three subsections. In Section 6.4.1 we focus on the design object and its synthesis as it can be explained based on the fundamental traits of the nature of mechanical artefacts expressed by the domain theory and the function–means law. Then, we focus on the two different design situations outlined above and describe approaches to engineering design synthesis. In Section 6.4.2 we focus on the design situation where the required functionality is new, and we describe a design-process-oriented synthesis approach. The core of this approach is that the engineering designer synthesises an artefact by determining structural characteristics. In Section 6.4.3 we focus on the design situation based on past designs and we describe an artefact-oriented synthesis approach. The core of this approach is to analyse the past designs and organise the analysis results in accordance with the three domains. The artefact-oriented approach has been utilised by several research groups in their development of computer-based design support systems.

6.4.1 The design object and its synthesis

In Section 6.2 we saw that a mechanical artefact is dominated by two patterns: the horizontal and the vertical causality chains. The relation between the man–machine system and the transformation is shown in Fig. 6.1; the man–machine system delivers the necessary effects for the transformation.

Focusing on the mechanical artefact we see the organ domain and the part domain. The organ structure creates effects based upon the wirk elements. These wirk elements are allocated or distributed among the machine parts. The organ view and the part view are *complementary*:

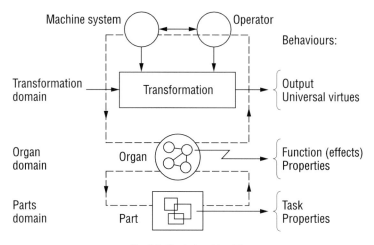

Fig. 6.5 The design object [8].

- the organ view explains the creation of effects, coupling together wirk elements, based on physical effects;
- the part view explains each wirk element's realisation by machine parts, leading to a part structure, which respects the organ structure.

By applying the domain theory's definitions the design object can be illustrated as shown in Fig. 6.5, and we have obtained the following:

- in each domain the engineering designer is looking on behavioural aspects and their structural realisation;
- there exists *design causality* between the domains, *i.e.*, decisions made by the engineering designer in one area result in conditions and constraints for other areas.

The design activity may progress in all three domains simultaneously, respecting the causal relations. Thus, the design object is threefold: *man–machine system* characterised by technology, and the mechanical artefact seen in the *organ domain* and in the *part domain*.

Figure 6.6 shows three principal synthesis steps according to the domain theory. There are two types of step within each domain, namely detailing and concretisation, and a synthesis step jumping from one domain to another. A step of detailing means defining more characteristics, whereas concretisation means giving values to characteristics, *e.g.*, length is 15 mm. As an example, Fig. 6.3 shows a stepwise determination of an organ, a hinge, from an abstract and undetailed description into a description, which may be transformed from organ domain into a part structure definition. This stepwise determination consists of two synthesis steps: a step of detailing and a step of simultaneous detailing and concretisation.

6.4.2 Process-oriented synthesis

The F/M-tree has been known and utilised in our research and education since early 1970s [2,14], but has been developed in several directions since then. In this section

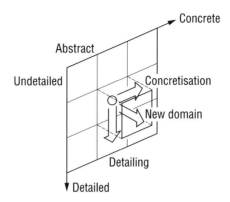

Fig. 6.6 Synthesis within a domain and from one domain to another [21].

we will describe an approach to engineering design synthesis of composite mechanical artefacts based on the work of Svendsen and Hansen [22–24].

According to Hansen [24] the decomposition of an artefact into manageable chunks or modules is a very common activity in design work for at least two reasons. Firstly, chunks are more easily designed than the composite artefact, because of their lower complexity. Secondly, the decomposition makes it possible for teams of designers to work in parallel on different chunks, so reducing the product development time. Although decomposition is a very frequently performed activity, a study of literature shows that there does not exist a theory that explains decomposition and the laws on which it could be based [24]. According to literature, decomposition of a composite artefact is carried out in a pragmatic and part-oriented manner.

Svendsen and Hansen propose a basic design step based on the function–means law. When the overall purpose function of the composite mechanical artefact has been identified, the basic design step consists of the following substeps:

1. search alternative means to realise the function (synthesis step: detailing)
2. assign values to the alternative means (synthesis step: concretisation)
3. select "the best" means between the alternatives (decision-making step)
4. fit the selected means into the totality of the composite artefact (composition step)
5. identify subordinate functions to support the selected means (decomposition step).

The engineering designer's point of departure when carrying out a basic design step is a required function. In the first substep the engineering designer searches for alternative means to realise the function, *i.e.*, a step of detailing. For these means the engineering designer gives values to their characteristics, *i.e.*, a step of concretisation. The engineering designer selects "the best" means between the alternatives. The selected means is fitted into the currently designed totality of the composite artefact, *i.e.*, composition. In the last substep the engineering designer identifies subordinate functions, which have to be realised in order to ensure that the artefact realises the required overall purpose function.

From the description of the basic design step we observe that:

- to carry out one design step is to proceed down one level in a F/M-tree, see Fig. 6.7;
- the basic design step is to be carried out recursively, starting from a required overall purpose function, via organ structures and organs, and will, in principle, end up with wirk elements at the bottom level of the F/M-tree.

We observe that our basic design step has features similar to the Roozenburg and Eekels model of the basic design cycle [25] and to Cross's model of the engineering product-design process [26,27].

- Our basic design step is cyclic, like the basic design cycle. The "stopping criterion" for the basic design step is the arrival at a lowest function–means level, where the means are of type wirk elements and where the required overall purpose function is being realised by the composite artefact. This criterion corresponds to the criterion of the basic design cycle, being the arrival at an acceptable design.
- In Cross's model a symmetrical relationship is assumed between problem and solution, and between subproblems and subsolutions. For the basic design step this feature corresponds to the function–means relation at each level within the F/M-tree.
- In Cross's model there is a hierarchical relationship between problem and subproblems, and between solution and subsolutions. For our basic design step this feature corresponds to the decomposition step, *i.e.*, to identify subordinate functions, and the composition step, *i.e.*, to fit a selected means into the totality of the composite artefact. We see this to mirror the recursive nature of designing.

Thus, based on the domain theory and the function–means law, we have described a basic design step for designing a composite mechanical artefact. However, a proper graphical model of the basic design step has not yet been developed.

In order to apply the F/M-tree to control the progression of the design process, Svendsen and Hansen [23] have proposed a number of enhancements of the F/M-tree, namely to include configurations, to include alternative boxes, and to extend function expressions.

Svendsen and Hansen propose that the entities of a means, *i.e.*, organs and/or wirk elements, and their relations be described in a configuration. The configuration of a means can be expressed in a graphical model; see Fig. 6.7. As the design process proceeds, more function–means levels are set up, and (as a consequence of the domain theory) the means are gradually determined, *i.e.*, their characteristics identified and assigned values. Thus, for any means, there exists a sequence of configurations mirroring this gradual determination. The configurations will change from being a sketch that shows the concept of the means, via configurations with information content and expression similar to a layout drawing, to the final configuration being a detailed drawing of wirk elements; see Fig. 6.7. Tjalve [19] has described a number of visual variation methods to create alternative organ structure solutions.

To ensure that a F/M-tree in which alternative means are included does not become extremely large, Svendsen and Hansen propose listing the alternatives below one another in the means box, and the chosen solution for the means rewritten in a field at the bottom of the box; see Fig. 6.7.

In order to clarify the formulation of a required function, Svendsen and Hansen extend function expressions from verb–noun pairs to include a description of the

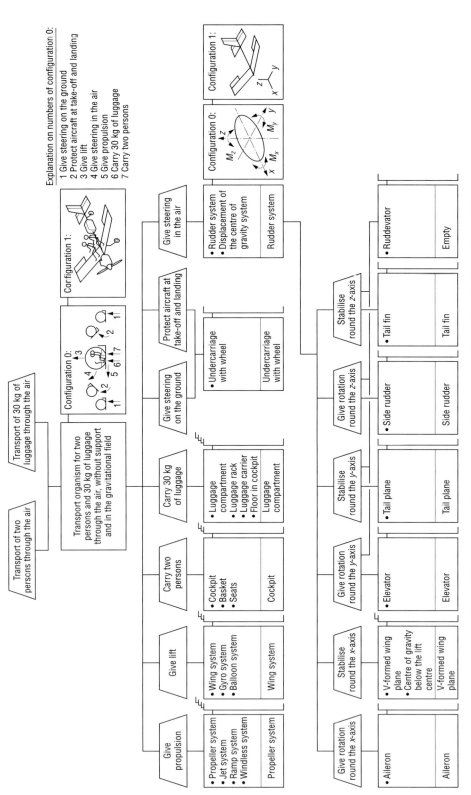

Fig. 6.7 F/M-tree for an ultralight airplane, with configurations shown (adapted from Svendsen and Hansen [23]).

object or the environment. For example, for the airplane shown in Fig. 6.7, the functions "give steering on the ground" and "give steering in the air" require different effects, and are realised by different means.

6.4.2.1 Problem analysis

The first activity in the basic design cycle of Roozenburg and Eekels is to analyse the problem [25]. What does the analysis activity of our basic design step looks like? The F/M-tree is often used for identifying the functions and means of the mechanical artefact in its normal, primary use situation, as it is the tradition in German language literature [6,7]. However, required functions may also stem from necessary operations in other product life phases, like installation, maintenance, or disposal [28]. Therefore, the engineering designer has to add branches to upper levels of the F/M-tree, and link these required functions to the means for the basic functionality.

Andreasen *et al.* [28] show how a mechanical artefact may be seen as consisting of a set of superimposed organ structures, where each organ structure corresponds to a specific property of the artefact. These superimposed organ structures are allocated or distributed among the machine parts in an often-opportunistic manner, where machine parts synthesised for other means are utilised. In the F/M-tree this may lead to parallel branches, when the means, when transformed into machine parts, are integrated. When describing an existing artefact by help of the F/M-tree, neglecting this fact may lead to misinterpretations of the design.

6.4.3 Artefact-oriented synthesis approach

As mentioned above, we distinguish two types of contribution to a synthesis theory in this chapter, namely a design-process-oriented approach and an artefact-oriented synthesis approach. The artefact-oriented approach encompasses many types of computer support possibilities, and several developments have been created based upon the domain theory and the function–means law, as we shall see in the following.

It is a well-known fact that a very high percentage of product development is based upon past designs. The amount of take-over may be up to 80–90%, and the company still sees the design as "new". Utilising patterns from past designs is a very powerful way of designing: it leads to a high rationalisation potential, and it can also create many promising solutions to the synthesis problem.

6.4.3.1 Utilising the F/M-tree

Past design may be crystallised in design catalogues [16,18], and dynamic design support systems based upon knowledge about design principles may be created. One such is Schemebuilder [29], which is actually utilising the function–means tree for capturing and linking decisions made about organ solutions.

The design group at Chalmers University of Technology has worked intensively with enhancements of our approaches, for finding a methodics for early design [30], and for handling design specifications [31] aiming at a more comprehensive support of designing.

The design group at Linköping University is, in a similar way, working on an approach for conceptual design of complex products, investigating the combination of F/M-tree, interaction analysis, and specifications [32], and investigating the design of part structure [33].

6.4.3.2 Developing a product model

In our research work on a Designer's Workbench [11,12,22], a digital and semi-formal product model showing all domains and mirroring both the F/M-tree and the structural relations was created. We call this basic modelling concept the chromosome model [34]. We focus here on the structural aspects in each domain, but allow a non-formal specification of the elements in the form of drawings, photographs, sketches, text or CAD-defined designs.

In this form, the chromosome model is currently used for a knowledge-based system for aluminium design (ALULIB) [11], for a support system for environmental product-life-oriented designing, for cost calculation (HKB-System) [35], and for prototypes of support systems.

A more formal modelling of the design need a "spelling" of the product in all three domains, for instance a spelling of the organ structure and each organ constituents; see Fig. 6.8. Jensen [12] and Mortensen [11] have made contributions in this area,

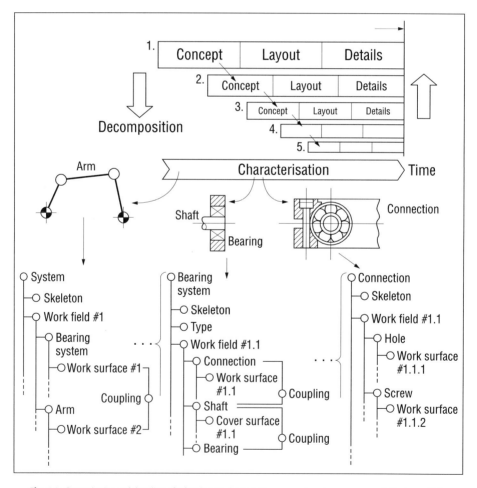

Fig. 6.8 Concretisation and detailing of a four bar mechanism. The progressing chromosome model is shown [11].

and Mortensen has shown that no general design language exists in the organ domain. So Ersoy's [10] theory about design characteristics is valid, but limited to a certain type, or gender, of organ.

Thus far, we have succeeded in creating a type of product model that shows all three domains and their relations, and which explains the relations between the elements and the constitual characteristics of the elements in the three domains. The product model has been enhanced into a so-called product family master plan [11], which is able to serve for modularised product families and for configuration systems. This model covers the product life phases, the function and organ aspects, and the modular product structure; in this way it serves as a powerful structure for knowledge related to the designing. This product model concept has been tested in several industrial applications.

6.5 Implications and conclusion

Both the domain theory and the function–means law belong to the attempts to contribute to a general theory of technical systems. They are phenomena explanations based upon certain concepts, *i.e.*, they are theory elements.

The uniqueness of the domain theory is its ability to explain the link between a product's purpose and the reasons for its details. The domain theory links the engineering designer's considerations about the man/machine system delivering effects for the purposeful transformation, via considerations about organs creating effects, to considerations about the parts being produced and assembled.

The F/M-tree, which can be seen as a graphical representation of the function–means law, may, on the one hand, be applied to explain designs and, on the other hand, be applied as an element of a design synthesis method. The F/M-tree gives both support for the engineering designer's creative thinking and support for information technology modelling and computer manipulation of product models. We believe that the multi-aspect modelling of artefacts for developing computer-based product models is of immense importance.

Our research attempts described in this chapter may be seen as a proposal for supporting the design process based upon an engineering designer's understanding of the mechanical artefact.

References

[1] Hubka V, Eder WE. Theory of technical systems. New York: Springer, 1984.
[2] Andreasen MM. Machine design methods based on a systematic approach – contribution to a design theory. Dissertation, Department of Machine Design, Lund Institute of Technology, Sweden, 1980 [in Danish].
[3] Klir J, Valach M. Cybernetic modelling. London: Iliffe Books, 1967.
[4] Chestnut H. Systems engineering methods. New York: Wiley, 1967.
[5] Hall AD. A methodology for systems engineering. Princeton: Van Nostrand, 1962.
[6] Hubka V. Theorie der Konstruktionsprozesse. Berlin: Springer, 1976.
[7] Pahl G, Beitz W. Engineering design – a systematic approach. Berlin: The Design Council/Springer, 1988.
[8] Andreasen MM. Conceptual design capture. In: Proceedings of the Engineering Design Conference, Brunel, 1998.
[9] Olesen J. Concurrent development in manufacturing – based on dispositional mechanisms. Dissertation, Institute for Engineering Design, Technical University of Denmark, 1992.

[10] Ersoy M. Wirkfläche und Wirkraum, Ausgangselemente zum ermitteln der Gestalt beim Rechnergestützten Konstruieren. Dissertation, Braunschweig, 1975.
[11] Mortensen NH. Design modelling in a designer's workbench – contribution to a design language. Dissertation, Department of Control and Engineering Design, Technical University of Denmark, 1999.
[12] Jensen TAa. Functional modelling in a design support system – contribution to a designer's workbench. Dissertation, Department of Control and Engineering Design, Technical University of Denmark, 1999.
[13] Buur J. Theoretical approach to mechatronics design. Dissertation, Institute of Engineering Design, Technical University of Denmark, 1990.
[14] Hubka V. Der grundlegende Algorithmus für die Lösung von Konstruktionsaufgaben. In: XII Int. Wiss. Kolloquium der TH Ilmenau "Konstruktion", 1967.
[15] Hansen CT. Towards a tool for computer supported structuring of products. In: Riitahuhta A, editor. Proceedings of the 11th International Conference on Engineering Design in Tampere, Tampere University of Technology, vol. 2, 1997; 71–6.
[16] Roth K. Konstruiren mit Konstruktionskatalogen. Berlin Heidelberg: Springer, 1982.
[17] Zwicky F. Entdecken, Erfinden, Forschen im morphologischen Weltbild, München, 1966.
[18] Koller R. Konstruktionsmethode für den Maschinen-, Geräte- und Apparatebau, Springer-Verlag, Berlin/Heidelberg, 1976.
[19] Tjavle E. A short course in industrial design. Newnes–Butterworth: 1979.
[20] Altschuller GS. Creativity as an exact science. Glasgow: Gordon and Breach, 1988.
[21] Andreasen MM. The theory of domains. Report from a Workshop on Understanding Function and Function-to-Form Evolution, CUED/C-EDC/TR 12, 1992; 21–47.
[22] Svendsen K-H. Discrete optimization of composite machine systems – a contribution to a designer's workbench. Dissertation, Institute of Engineering Design, Technical University of Denmark, 1994.
[23] Svendsen K-H, Hansen CT. Decompositon of mechanical systems and breakdown of specifications. In: Proceeding of the International Conference on Engineering Design (ICED'93), The Hague. Zürich: Heurista, 1993.
[24] Hansen CT. An approach to simultaneous synthesis and optimization of composite mechanical systems. J Eng Des 1995;6(3):249–66.
[25] Roozenburg NFM, Eekels J. Product design: fundamentals and methods. Chichester: Wiley, 1995.
[26] Cross N, Roozenburg NFM. Modelling the design process in engineering and in architecture, J Eng Des 1992;3(4):13.
[27] Cross N. Engineering design methods, strategies for product design. 3rd ed. Chichester: Wiley, 2000.
[28] Andreasen MM, Hansen CT, Mortensen NH. The structuring of products and product programmes. In: Tichem M, editor. Proceedings of the 2nd WDK Workshop on Product Structuring, Delft University of Technology, 1996; 15–43.
[29] Bracewell RH, Bradley DA, Chaplin RV, Langdon PM, Sharpe JEE. Schemebuilder, a design aid for the conceptual stages of product design. In: Proceeding of the International Conference on Engineering Design (ICED'93), The Hague, vol. 3. Zürich: Heurista: 1993; 1311–8.
[30] Schachinger P. Methods for early design phases – a product modelling perspective. Thesis, Chalmers University of Technology, 1999.
[31] Malmqvist J, Schachinger P. Towards implementation of the chromosome model – focusing the design specification. In: Riitahuhta A, editor. Proceedings of the 11th International Conference on Engineering Design in Tampere, vol. 3, 1997; 203–12.
[32] Liedholm U. On conceptual design of complex products. Thesis, Linköping University, 1999.
[33] Ringstad P. Early component design – theory and a procedure. Thesis, Linköping University, 1996.
[34] Ferreirinha P, Grothe-Møller T, Hansen CT. TEKLA: a language for developing knowledge based design systems. In: Proceedings of the 1990 International Conference on Engineering Design, vol. 2. Edition Heurista: 1990; 1058–65.
[35] Ferreirinha P, Della Rossa A. Knowledge based integration of process- FMEA and cost analysis. In: Proceeding of the 12th International Conference on Engineering Design, Technische Universität München, vol. 3. 1999; 1559–64.

Using the concept of functions to help synthesise solutions

7

Gerhard Pahl and Ken Wallace

Abstract Using the concept of functions to help synthesise solutions provides many advantages over more intuitive approaches. By adopting the approach described, a broader solution field is generated, and designers are less likely to be constrained by existing solutions. The approach starts by defining the problem in a solution-neutral way, and this helps identify the overall function for the proposed product. From this starting point, a function structure is developed that consists of linked subfunctions representing the flows and conversions of energy, material and signals. Guidelines for handling the concept in practice are provided, along with examples. Using function structures assists the management of the design process, as subtasks are easily identified and assigned to individual team members. Planning time scales and milestones is also facilitated.

7.1 Introduction

New solutions are frequently found intuitively and spontaneously. The starting situation is usually an unsatisfactory existing solution or a competing product. By analysing existing solutions, their deficiencies usually become apparent and additional requirements that new solutions must fulfil are identified. Then, on many occasions, by carefully analysing existing solutions, new solution ideas are stimulated, and through appropriately modifying form designs and layout designs, or both, an improved product results. There is nothing to be said against this frequently practised approach if the new solution satisfies the requirements and is found quickly.

If, however, one is searching for a new solution in order to develop a successful long-term product for the market, it is recommended that the search for solutions is conducted at a generally applicable, that is more abstract, level. This is so that a broader and more favourable solution field is considered. This type of approach increases the chances of identifying favourable solution principles because it encourages the designer's mind to roam freely and not be fixed on well-known principles or designs. Such an abstract and unbounded reflection can be achieved by using the *concept of functions* to help synthesise solutions.

A *function* defines intended purpose in a solution-neutral way. Thus the aim of the design, or the goal to be achieved, is clearly stated. Such a formulation leaves the solution and design characteristics completely open, and thus avoids any early fixation of thoughts. Starting with the appropriate functions makes the current problems and intentions easier to recognise, and opens the way for an unbiased and broader solution search. In addition, the consideration of functions provides a better basis for identifying faults and disturbing influences. Using functions, therefore, provides an important and many-sided aid to the synthesis of new and effective solutions.

A simple example demonstrates the advantages of using the concept of functions. In the drive train between a piston engine and the equipment it is driving, torsional vibrations arise. From similar situations, the use of a rotationally elastic coupling is a solution known to reduce the vibrations. The instruction to the designer might, therefore, have been simply: "use a torsional coupling". With hindsight, this approach is unlikely to provide an optimal solution, as it is too strongly focused on components. A much better approach is to focus on the crux of the task and, using the functional approach, to describe the purpose of the task in a solution-neutral manner, namely: "reduce torsional vibrations". From this perspective, the following solution possibilities emerge:

- alter the shaft stiffness;
- change the mass distribution in the drive chain;
- build in two balancing elements; and
- incorporate a rotationally elastic balancing coupling, with or without damping.

So, at least five different solutions are possible. Which one is selected will depend on the spatial and operational constraints, as well as the project budget and time scale. By using a solution-neutral formulation, the designer is encouraged to think more deeply about the fundamental principles and to consider more solution possibilities. In this way, individual creativity is enhanced and, at the same time, any fixation on previous solutions and designs is reduced.

In order to use the concept of functions effectively, a methodical approach and certain guidelines are helpful. The approach and guidelines set out below are based on those in Chapter 2 of the book *Engineering Design* by Pahl and Beitz [1].

7.2 Functional interrelationship

For the purpose of describing and solving design problems, it is useful to use the term *function* for the general input–output relationship of a definite and limited system, whose purpose is to perform a task. The type of system under consideration can be different. It can either be the whole of a technical system or only part of it, *i.e.*, a sub-system. The choice of the system boundary depends on the overall complexity and the scope of the influencing factors.

For static processes, it is sufficient to determine the inputs and outputs. For dynamic processes, which change with time, the task must be defined further by a description of the initial and final magnitudes. The function thus becomes an abstract formulation of the task, independent of any particular solution.

Functions are usually defined by statements consisting of a verb and a noun, for example "increase pressure", "transfer torque" and "reduce speed". They are derived from conversions of energy, material and signals (information). In most mechanical engineering applications, a combination of all three types of conversion is usually involved, though one will often dominate.

If the overall task for the selected system has been adequately defined, *i.e.*, if the inputs and outputs of all the quantities involved and their actual or required properties are known, then it is possible to specify the *overall function*. The overall function is the relationship between the inputs and outputs of the system, *e.g.*, plant,

Using the Concept of Functions to Help Synthesise Solutions

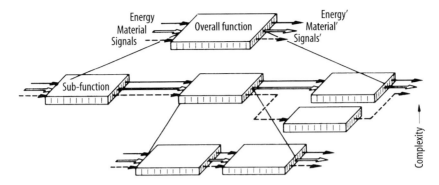

Fig. 7.1 Breaking down an overall function into subfunctions.

machine or assembly, and specifies the overall purpose of the system. This purpose is determined by the overall requirements.

An overall function can often be divided directly into subfunctions corresponding to subtasks; see Fig. 7.1. The relationships between subfunctions and the overall function, and between the subfunctions themselves, is often governed by certain constraints, inasmuch as some subfunctions have to be satisfied before others. On the other hand, it is usually possible to link subfunctions in various ways, and in this way create solution variants. In all such cases, the links must be compatible. The meaningful and compatible combination of subfunctions into an overall function produces a so-called *function structure*. This function structure can be systematically varied to satisfy the overall function, and is the starting point for producing several new solutions.

It is useful to distinguish between main and auxiliary functions. *Main functions* are those subfunctions that serve the overall function directly. They determine, in a fundamental way, the overall solution, and it is not possible to dispense with them. *Auxiliary functions* are those subfunctions that contribute to the overall function indirectly. They have a supportive or complementary role and are often determined by the specific nature of the particular solution being considered. The determination of auxiliary functions, therefore, changes with the development of a particular solution.

As an example, consider the packing of carpet tiles stamped out of a length of carpet. The overall function is "pack carpet tiles". Regarding the subfunctions, the first task is to introduce a method of control so that the perfect tiles can be selected, counted and packed in specified lots. The main flow here is that of material shown in the form of a block diagram in Fig. 7.2, which, in this case, is the only possible sequence. All the subfunctions shown are main functions.

On closer examination we discover that this chain of subfunctions requires the introduction of auxiliary functions, see Fig. 7.3, because:

- the stamping-out process creates off cuts that have to be removed;
- rejects must be removed separately and reprocessed; and
- packing material must be brought in.

It is not always possible to make a clear distinction between main and auxiliary functions, especially if either the boundary of the system or the scope of the chosen

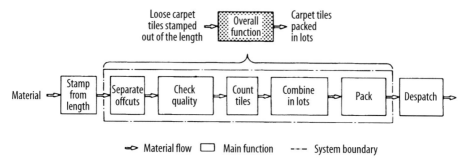

Fig. 7.2 Function structure for the packing of carpet tiles.

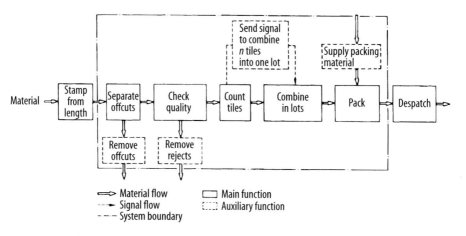

Fig. 7.3 Function structure for the packing of carpet tiles with auxiliary functions added.

solution changes. However, the terms are useful, both to help structure the solution and to organise the designer's work. A flexible approach should be adopted.

Next, we identify all the required subfunctions, *i.e.*, the subtasks and subproblems. From this analysis, we obtain our first impression of the structure of the solution. The design task has been simplified, as it is now possible to focus on finding subsolutions for each of the subfunctions (subproblems) in turn.

Task-specific functions are those derived directly from the current design task. They are formulated using task-specific words. In the carpet tile example above, all the functions are task-specific functions. This formulation provides a strong engagement with the current problem. Since these functions are very directly expressed, they stimulate and guide, in a problem-orientated way, the creative search for solutions. For this reason, we recommend that one should always use task-specific functions during the initial search for solutions.

Various authors [2–4] of design methods have put forward the concept of *generally valid functions*. In theory, it is possible to define functions so that the lowest level of the function structure consists exclusively of generally valid functions that cannot be subdivided further. This is the highest possible level of abstraction. Because of

Using the Concept of Functions to Help Synthesise Solutions

Table 7.1 Generally valid functions.

Characteristic input (I)/output (O)	Generally valid functions	Symbols	Explanations
Type	Change		Type and outward form of I and O differ
			I < 0
Magnitude	Vary		I > 0
			Number of I > 0
Number	Connect		Number of I < 0
			Place of I ≠ 0
Place	Channel		Place of I = 0
Time	Store		Time of I ≠ 0

their simplicity of application, Pahl and Beitz recommend the proposal of Krummhauer [5]. His proposal for generally valid functions is shown in Table 7.1. These functions apply equally to the flows and conversions of energy, material and signals. Generally valid functions are an excellent aid for organising systematic overviews of design solutions, such as those found in design catalogues and classifications schemes, since they provide a generally applicable and ordered basis for such classifications.

In cases where no specific functions are easily recognisable at the outset, generally valid functions can often be used. They provide a stepping stone to the accurate formulation of appropriate task-specific functions. Figure 7.4 shows the function structure for the overall function "pack carpet tiles" shown in Fig. 7.2 represented using the generally valid functions defined in Table 7.1. In the sequence shown in Fig. 7.4, the main flow is that of material, with secondary flows of energy and signals. The first generally valid function in the sequence, placed outside the system boundary, is "connect energy and material" (stamp from length). The sequence of steps

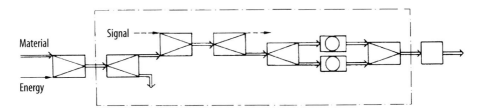

Fig. 7.4 Function structure for the packing of carpet tiles using generally valid functions.

inside the boundary, matching the five functions inside the boundary in Fig. 7.2, is as follows: (1) "separate material" (separate off cuts); (2) "connect material and signal" (check quality); (3) "separate material and signal" (count tiles); (4) "separate material" (combine in lots); and (5) "store material" and "connect material" (pack). Finally, the last generally valid function, placed outside the boundary, is "channel material" (dispatch). It can be seen that a disadvantage of generally valid functions lies in their very abstract nature – it is difficult to link them to concrete embodiments. In practice, using them to help in the search for solutions requires some interpretation using task-specific characteristics.

Logical relationships, referred to as *logical functions*, need to be established between the inputs and outputs of particular subfunctions. These relationships are based on binary logic using the statements: true/untrue, yes/no, in/out, fulfilled/not fulfilled. These are represented by AND-functions, OR-functions and NOT-functions, along with their combinations NOR-functions (OR with NOT), NAND-functions (AND with NOT), and storage functions. Such logical functions are often necessary in function structures, and in such cases must be included. Some function structures, such as those for control systems, can be built entirely out of logical functions. Such function structures already represent *principle solutions* that can be realised through the appropriate selection of standard control elements and their connections. For examples of the use of logical functions, see Chapter 2 of *Engineering Design* [1].

With the help of task-specific functions in particular, it is possible to proceed directly from the requirements list to an appropriate formulation of the overall task and the required subtasks in terms of subfunctions. Designers will be able to obtain a good overview of where new solutions should be sought and where existing solutions can be applied. This formulation also provides an excellent planning tool, making it apparent just how much work there is still to do. The derivation of the functions (subtasks and subgoals) provides the opportunity to plan for a goal-orientated creative search for solutions and achieve results in a shorter time. The handling of functions, in practice, requires practice and experience. Section 7.3, therefore, concentrates on the practical application of the concept of functions.

7.3 Handling the concept of functions in practice

After the crux of the overall task has been identified, it is possible to define the overall function using the basic flows of energy, material and signals. This expresses the relationship between inputs and outputs independently of any solution. At this stage, it is important to focus on the main flow, which could be that of energy, material or signals. This main flow must be specified as precisely as possible. Depending on the complexity of the problem, the overall function will, in turn, be more or less complex. A complex overall function must then be broken down into subfunctions of lower complexity. The object of breaking down complex functions is:

- to determine simpler subfunctions that will facilitate the subsequent search for solutions; and
- the combination of the subfunctions into a simple and unambiguous function structure.

The function structure is clearly represented by a block diagram. Function structures should be kept as simple as possible in order to encourage simple and economical solutions.

If no clear relationship between the subfunctions can be identified, it can be helpful for the search for a first solution principle to start with the mere enumeration of important subfunctions, without considering their logical or physical relationships.

The next step in the search is to develop preliminary function structures and identify the relationships that exist within them. By doing this the first definitive function structure will emerge, which will not necessarily be complete. Working systematically, the gaps in the structure and superfluous functions can be identified until, finally, the function structure is completed, or simplified, as appropriate.

Once all the necessary subfunctions to fulfil the task have been identified, the next step is to concentrate on the search for solutions for the main function that defines or characterises the overall solution. Starting from this main function, and its appropriate solution principles, one then proceeds to identify, test and fix the other important main functions identified in the overall relationship.

As an example of the systematic development and variation of a function structure following the guidelines given above, the function structure used during the design of a fuel gauge to measure the contents of a motor vehicle's fuel tank is shown in Fig. 7.5. The development starts with the overall function "measure and indicate quantities of liquid". The main flow is a flow of signals. The main function, which will characterise the overall solution, is "receive signal", followed by the functions "channel signal" and "indicate signal".

The development of the function structure in Fig. 7.5 then proceeds as follows. Having received a signal, this may need to be changed before channelling it to the indicator. The requirements list requires that measurements of the quantity of fuel are made in tanks of different sizes and different shapes. The first of these two requirements requires the introduction of an auxiliary function "adjust signal", and the second the introduction of "correct signal". As an external source of energy may be required, a further auxiliary function "supply external energy" is then introduced. Depending on the scope of the solution required, different system boundaries are chosen. First, a system boundary is shown if the required solution only needs to provide an output signal to be channelled to an existing indicator. The final function structure in Fig. 7.5 shows the system boundary if the solution needs to include the means of indicating the measurements. The search for solutions should start with the key task-specific function "receive signal", since the solution selected for this subfunction not only influences which other subfunctions are included in the final function structure but also the particular solutions selected for these other subfunctions.

The required auxiliary functions frequently emerge from the selected solution principle, so it is not normally possible to identify all of them at the beginning of the development of a solution. For main functions and auxiliary functions that are similar, it is, as a rule, possible to adopt similar solutions.

Especially in adaptive design, existing solutions and existing function structures can provide a stimulus for finding the required subfunctions. It is certainly not wrong to use the function structures of existing solutions in order to stimulate ideas. It is essential, however, that studying previous solutions is not allowed to cause a fixation of the thought processes. It is essential always to check any ideas found in this way for their legitimacy on the basis of the particular task in hand and its requirements.

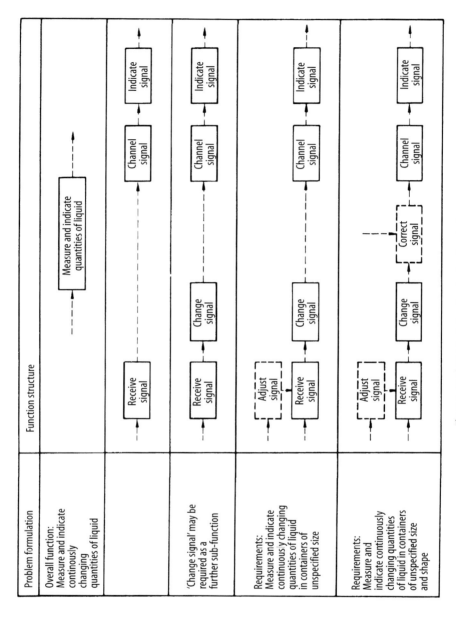

Fig. 7.5 Development of a function structure for a fuel gauge.

Using the Concept of Functions to Help Synthesise Solutions

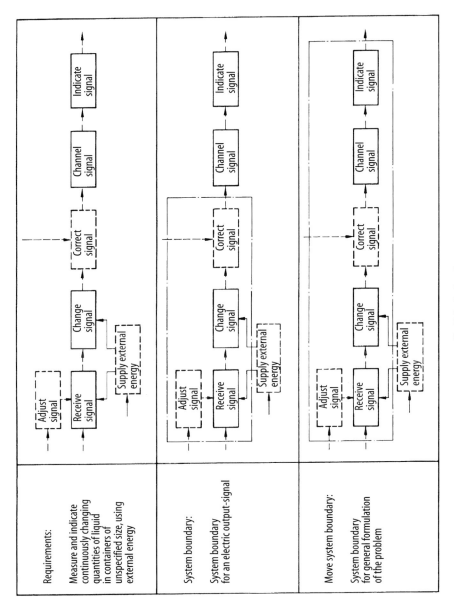

Fig. 7.5 *Continued*

An accurate function structure is the basis for planning design work, since it is possible to identify all the necessary subtasks required in the search for a solution and assign these tasks to individual design-team members. With the help of the function structure, it is easier to plan the time scales and milestones for the overall design and development of the product.

In addition, function structures are an excellent basis for generating the structures of modular systems. They can also be used, along with the selected subsolutions, for developing production modules and assembly procedures.

Finally, functions are the starting point for fault finding using a fault-tree analysis. In a fault tree, the fulfilment of individual functions is systematically negated. By doing this, possible failures of the overall solution can be identified, along with the possible sources of these failures [1].

7.4 Inappropriate use of the concept of functions

In the literature, and in practice, one frequently comes across the terms "fault function" and "disturbing function". According to design theory, a function is defined as the *desired* purpose. A fault or a disturbance is never desired. It follows, therefore, that such ideas should not be linked to the concept of functions. It is better to replace "fault function" with "fault suppression" and "disturbing function" with "disturbing influence" or "disturbing effect".

In addition, one often comes across the notion that auxiliary functions are unimportant functions. In technical systems there can be no important and unimportant functions. All functions are important if they are required for the product to function. Unnecessary or superfluous functions should be eliminated. It is clear, however, that in order to reduce the work required in the search for solutions, designers initially focus on some functions in preference to others. However, for the product to function satisfactorily, solutions must eventually be found for all necessary functions, both main and auxiliary.

7.5 Summary of the approach and its advantages

A function defines intended purpose. When solving design problems, where the aim is to create a technical system to perform a specified task, functions are represented by verb–noun pairs, with their inputs and outputs clearly identified. All technical systems can be modelled as a block diagram showing functions linked by the flows of energy, material and signals.

The approach for applying the concept of functions can be summarised as follows:

- prepare a solution-neutral statement that identifies the crux of the task at the appropriate level of abstraction;
- from that statement, identify the *overall function*, the system boundary and the overall inputs and outputs;
- break down the overall function into subfunctions with their individual inputs and outputs;
- using physical, logical and embodiment relationships, develop a function structure, *i.e.*, a block diagram linking the subfunctions (main and auxiliary);

- systematically vary the function structure to provide a range of possible solution variants, aiming for the clearest and simplest function structure;
- search for solutions for each of the subfunctions in the definitive function structure, using known solutions where appropriate;
- combine these subsolutions into a compatible overall solution;
- analyse the solution for possible faults and disturbing influences.

The approach appears logical and simple when described, but, in practice, it can be very difficult to apply. A flexible and iterative approach, along with trial and error, should be adopted, and several attempts may be required before a suitable function structure begins to emerge. If getting started is proving difficult, simply listing functions, without any attempt to link them, can be a useful approach. Beginning from both extremes can also be helpful. One can begin at a very concrete level by analysing the functions of an existing product; or at a very abstract level by using only generally valid functions that have no concrete embodiments.

The advantages of systematically applying the concept of functions are:

- more successful in synthesising novel solutions than more intuitive approaches;
- a broader solution field is considered;
- any tendency to focus on known solutions is avoided;
- the problem solving is broken down into more manageable chunks;
- function structures provide a foundation for creating modular systems;
- planning design work and assigning subtasks to individual designers is facilitated;
- possible faults and disturbing influences are easier to identify; and
- planning production modules and assembly procedures is simplified.

Evidence from applying the approach in practice shows that more innovative solutions frequently emerge, and the logic underpinning these solutions is more clearly understood. However, applying the approach can sometimes be difficult, and a flexible approach should be adopted. It is worth persisting.

References

[1] Pahl G, Beitz W. Engineering design – a systematic approach [Wallace KM, Blessing LTM, Bauert F, Transl. Wallace KM, editor]. 2nd ed. London Berlin: Springer, 1996.
[2] Koller R. Konstruktionslehre für den Maschinenbau. Grundlagen für Neu- und Weiterentwicklung technischer Produkte mit Beispielen. 3rd ed. Berlin, Springer, 1994.
[3] Rodenacker WG. Methodisches Konstruieren. Konstruktionsbücher, vol. 27. 4th ed. Berlin: Springer, 1991.
[4] Roth K. Konstruieren mit Konstruktionskatalogen, vol. 1, Konstruktionslehre. 2nd ed. Berlin: Springer, 1994.
[5] Krummhauer P. Rechnerunterstützung für die Konzeptphase der Konstruktion. Dissertation, TU Berlin, D 83, 1974.

Design catalogues and their usage

Karlheinz Roth

Abstract Design catalogues not only allow optimum design solutions or appropriate elements for their realisation to be found very quickly, but also provide optimum variation operations. This chapter describes the creation of design catalogues, their form and how to work with them.

8.1 Purpose of design catalogues

Design catalogues are a very effective aid for methodical designing. They are not only collections of technical objects, normally elementary, or basic technical function solutions, but also provide groupings of variation operations used to change technical conditions. In accordance with Ref. [1], design catalogues represent "... knowledge storage present outside human memory, normally in tabular form, that is created using methodical criteria, largely *complete* and *systematically* structured within a certain framework. They permit precise *access* to their contents and consist of *classifying criteria*, a *main part*, *selection characteristics*, and possibly also an *appendix* ...".

Because design catalogues contain the required elements, frequently used functions [2–5] and even variation operations, they simplify the design process. Indeed, because the design catalogues are largely complete, sometimes they make conscious that there are more effective variants to reach the goal other than those considered.

8.2 Types and structure of design catalogues

The previous discussion indicates that there must be three types of design catalogues [1]:

- object catalogues (see Fig. 8.1);
- solution catalogues (see Fig. 8.3);
- operation catalogues (see Fig. 8.5).

8.2.1 Object catalogues

These are design catalogues that contain task-independent details of physical, geometric, and technological nature that are important for the design process [1].

Classifying criteria		Main part				Selection characteristics		Appendix
Number of teeth		Side view	Transverse profile	Pinion z_1	Wheel z_2	Transmission ratio	Efficiency	Reference
	No.	1	2	3	4	5	6	7
1	1	1.1 $\alpha_n = 20°, \beta = 30°$	1.2 $1.967 \cdot m_n$	1.3 $h_{aP1} = 0.383 m_n$ $(0.281 m_n)$ $h_{fP1} = 1.1 m_n$ $x_1 = +1.013$ $(+1.031)$	1.4 $h_{aP2} = 1.1 m_n$ $h_{fP2} \geq 0.38 m_n$ $x_1 = -0.6$	1.5 about until 1:48	1.6 about 88%	1.7
2	2	2.1 $\alpha_n = 20°, \beta = 25°$	2.2 $5.046 \cdot m_n$	2.3 $h_{aP1} = 0.481 m_n$ $(0.431 m_n)$ $h_{fP1} = 1.1 m_n$ $x_1 = +0.947$ $(+0.961)$	2.4 $h_{aP2} = 1.1 m_n$ $h_{fP2} \geq 0.48 m_n$ $x_2 = -0.6$	2.5 about until 1:20	2.6 about 90%	
3	3	3.1 $\alpha_n = 20°, \beta = 20°$	3.2 $6.092 \cdot m_n$	3.3 $h_{aP1} = 0.554 m_n$ $h_{fP1} = 1.1 m_n$ $x_1 = +0.892$	3.4 $h_{aP2} = 1.1 m_n$ $h_{fP2} \geq 0.55 m_n$ $x_2 = -0.6$	3.5 about until 1:16	3.6 92%	See Ref. [6]
4	4	4.1 $\alpha_n = 20°, \beta = 20°$	4.2 $7.216 \cdot m_n$	4.3 $h_{aP1} = 0.659 m_n$ $h_{fP1} = 1.1 m_n$ $x_1 = +0.822$	4.4 $h_{aP2} = 1.1 m_n$ $h_{fP2} \geq 0.66 m_n$ $x_2 = -0.6$	4.5 about until 1:12	4.6 94%	
5	5	5.1 $\alpha_n = 20°, \beta = 20°$	5.2 $8.320 \cdot m_n$	5.3 $h_{aP1} = 0.752 m_n$ $h_{fP1} = 1.1 m_n$ $x_1 = +0.753$	5.4 $h_{aP2} = 1.1 m_n$ $h_{fP2} \geq 0.75 m_n$ $x_2 = -0.6$	5.5 about until 1:10	5.6 96%	

Fig. 8.1 Object catalogue: "evoloid gearwheels" for pinion gear teeth counts from $z_1 = 1$ to 5 for producing a large transmission ratio in a single-stage gear unit [6], where α_n is the profile angle, β is the angle of inclination, h_{aP} is the reference profile addendum, h_{fP} is the reference profile dedendum, x is the addendum modification, m_n normal module with $m = 25.4/\text{pitch}$.

Figure 8.1 shows a typical object catalogue with *classifying criteria, main part, selection characteristics*, and *appendix*. It contains all evoloid gearwheels [6] (involute gearwheels for very small numbers of pinion teeth) from $z = 1$ to 5 to reduce the speed ratio. Every user can immediately check in the classifying criteria that no gearwheel is missing or superfluous. Because the classifying criteria are prepared using the so-called ordinal scale [7,8] (which permits a sequence of unique integers), the structure is unique and does not exhibit any overlapping.

Evoloid pinions and the usual spur gears can be used, for example, to increase greatly the transformation ratios per gearwheel pairing. This allows the number of gear stages to be reduced and/or the module to be increased and thus also to reduce the total size of a gearbox.

Figure 8.2 shows an application. Using a hand drill as an example, Fig. 8.2a illustrates working with this catalogue. The usual **two**-stage gear unit of a conventional machine has been replaced by a **single**-stage gear unit whose pinion has $z = 3$ teeth,

Design Catalogues and Their Usage

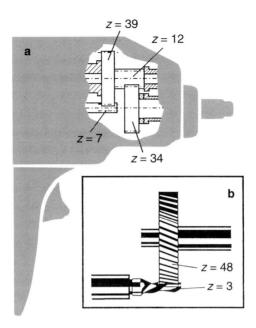

Fig. 8.2 Use of the design catalogue from Fig. 8.1. A three-toothed pinion from this catalogue produces the stage reduction and size reduction of the gearbox housing of a hand drill. (a) Gearbox with two gearwheel stages and the usual involute gearwheels with $i = (34/7)(34/12) = 15.8$; (b) gearbox with evoloid gear teeth has **one** gearbox stage with a transmission ratio $i = 48/3 = 16$.

Fig. 8.2b and as in Fig. 8.1, the design catalogue, rather than $z = 7$ teeth. Module m has also been increased at the same time [9].

8.2.2 Solution catalogues

These are design catalogues that contain an almost complete range of solutions for the task [1]. The solutions can apply to various design phases [7], and thus to function combinations, effect assignments, design structures, part combinations, and manufacturing possibilities.

Figure 8.3 shows a typical solution catalogue, namely *mechanical power multiplication*, without external energy [1]. This is a basic design catalogue, because almost every mechanical design that produces an effect needs power. The available power is normally too small and must be increased.

As in the object catalogue, Fig. 8.1, the four columns *classifying criteria*, *main part*, *selection characteristics*, and *appendix* are easy to recognise.

The classifying criteria have primary importance for producing a design catalogue. To avoid the need to make subsequent changes, it must be largely complete. However, this assumes a complete study and understanding of the content. Thus, for example, no mechanical power may be forgotten in the solution catalogue, Fig. 8.3. Therefore, it is shown here in its complete length. The almost completeness mentioned above is described in detail below.

Classifying criteria			Main part			Selection characteristics		Appendix
System	Effect	Force direction	No.	Principle sketch	Example	Force multiplication	Condition of force multiplication	Reference
1	2	3		4	5	6	7	8
1.1	1.2	1.3 intersecting	1	1.4 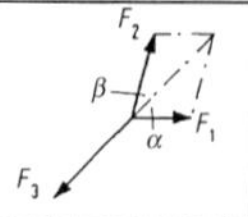	1.5 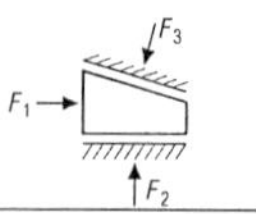	1.6 $\kappa = \dfrac{F_3}{F_1} \approx 5\ldots10$	1.7 $\sin\alpha > \sin\beta$	1.8
	Force splitting	2.3 parallel	2	2.4 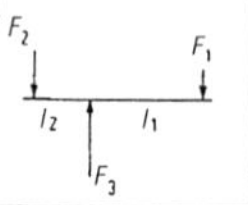	2.5 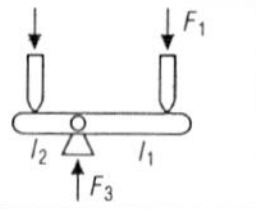	2.6 $\kappa_{max} = 50\,000$	2.7 $l_1 > l_2$	
3.2 Energy conducting systems	3.2	3.3 intersecting	3	3.4 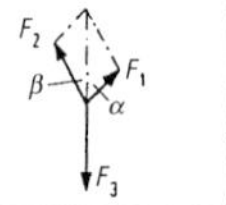	3.5 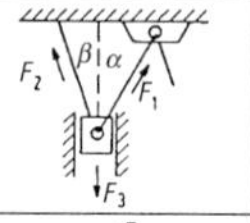	3.6 $\kappa \approx 2$	3.7 $2\cos\alpha > 1$	
	Force addition	4.3 parallel	4	4.4 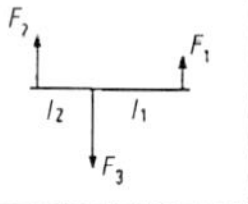	4.5 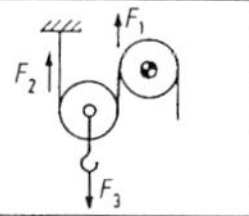	4.6 $\kappa_{max} = 800$	4.7 $F_2 > 0$	
	5.2 Pressure extension	5.3 at will	5	5.4 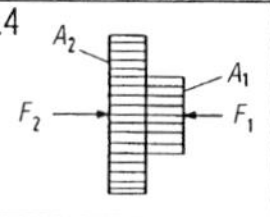	5.5 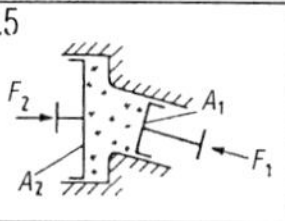	5.6 $\kappa_{usual} \approx 2\ldots10$	5.7 $A_2 > A_1$	See Ref. [1]
6.1 Energy connecting systems	6.2 Friction	6.3 right-angled	6	6.4 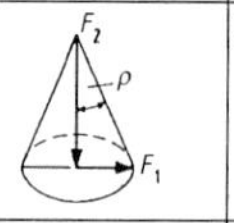	6.5 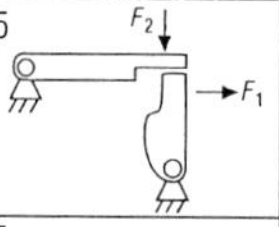	6.6 $\kappa = 5\ldots10$	6.7 $\mu < 1$	
	7.2 Different spring stiffness	7.3 intersecting	7	7.4 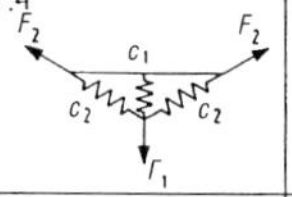	7.5 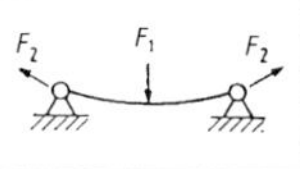	7.6 $\kappa \approx 5$	7.7 $c_1 \ll c_2$	
	8.2 Pulse transmission	8.3 in line	8	8.4 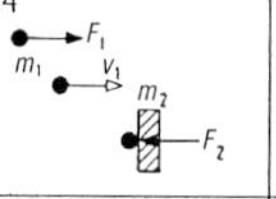	8.5 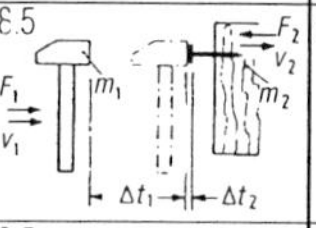	8.6 $\kappa \approx 10\ldots100$	8.7 $\Delta t_2 < \Delta t_1$	
Energy storing systes	9.2 Recoil	9.3 at will	9	9.4 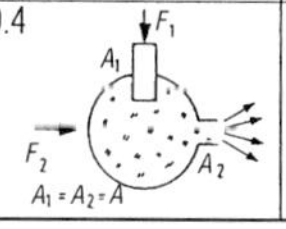	9.5 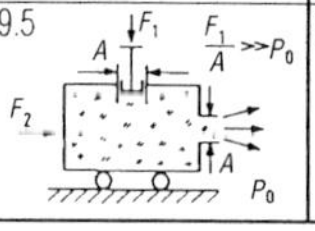	9.6 $\kappa < 2$	9.7 $\dfrac{F_1}{A} \ll p_0$	

Fig. 8.3 Solution catalogue for the function *"replicate mechanical force"* without the addition of any supplementary energy. Representation as block diagram and as form. The design catalogue contains *classifying criteria*, a *main part*, *selection characteristics*, and an *appendix*.

1. "System" column. As a result of the classification in system theory [7,10], the energy can be *stored*, *transferred* (i.e., forwarded with or without conversion), and *combined*. These energy states have been used for the "general functions" in design theory [7]. Because the power change is considered only as result of transferring, saving or combining, the assignment has been made based on the normal scale [8]. One action here excludes any of the others.

2. "Effect" column. The structure in column 2 is similar. Only *one* effect acts for the power multiplication. Because all the effects are known, you can immediately determine whether any effect is missing and, if so, which. Rows 7–9 also contain the "volumetric spring stiffness", "pulse transmission", and "recoil action" effects. In contrast, the design catalogue does not contain possible magnetic and electrical power multiplications. They are excluded, because only "*mechanical power multiplications*" are considered here.
3. "Force direction" column. Geometric, qualitative considerations are also desirable, easy to handle and test. It is immediately possible to determine and test whether the forces intersect, run parallel, are right-angled or in all directions.

The structuring considerations discussed in the example are based not only on the "yes" or "no" decision, but also on a larger repertoire. However, the principle solutions require that the choice of *one* property excludes the other. This means non-unique considerations, such as "large–small", "hot–cold", "expensive–cheap", *etc.*, are not suitable for the classifying criteria.

This clean structure, which guarantees the near completeness, systematics, and freedom from redundancies, produces the contents automatically, and one receives, for example, all mechanical multiplications of a force by wedge, lever, rope and pulley, differential pulley block, pressure multiplier, friction multiplier, spring multiplier, and also by hammer and recoil. The main part contains a typical representative of each of these. The selection characteristics of the solution catalogue, Fig. 8.3, allow the selection of a force multiplier from one of the properties mentioned there. The optionally added appendix contains here the documentation of the contents. Whereas the selection characteristics and the appendix can be changed or added subsequently, the classifying criteria are fixed; in the main part, only the examples can be changed, and formula and sketches added.

The example in Fig. 8.4 shows the use of a commonly used jointed connection [11] in which a small force must produce a large force in order to remove the cork from a wine bottle. This represents a corkscrew [1].

Row 1 of Fig. 8.4 lists again the capabilities for principles of the force multiplication from Fig. 8.3; row 2 shows their use to produce corkscrews.

There is a form of corkscrew for almost all of the effects from Fig. 8.3, where, however, those in boxes 2.1 and 2.8 are new. The form in box 2.1 uses the knee lever. The form in box 2.8, in which the bottle is held with applied spiral and the weight forcibly hit against the upper shoulder, operates with the impulse principle. The cork is literally "hammered out".

Row 3 shows variants that result from the use of rotary wedges (threads), levers and change of friction. The principles can also be used for large forces; in particular, when two force multipliers are used successively [1] (bolt cutter, car jack, block and tackle, rope and pulley, door handle, pneumatic drill, *etc.*).

8.2.3 Operation catalogues

These are design catalogues that contain not only process steps, processes or rules, but also their application conditions and usage criteria [1]. In contrast to the other catalogue types, they consist only of classifying criteria and a main part.

Figure 8.5 shows such an operation catalogue (two-dimensional). It contains the measures (operations) that can change the structure, namely through change of the

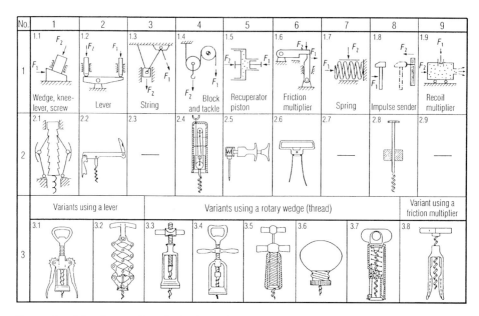

Fig. 8.4 Use of the solution catalogue for the design of corkscrews, in which a small manual force can produce a large cork removal force. Row 1 again shows the general force multiplication solutions, realised with the various effects from the design catalogue, Fig. 8.3. Row 2 shows the resulting corkscrews. Boxes 2.1 and 2.8 show the still unknown solutions with knee lever and weight force. Row 3 shows various solutions that use levers, rotary wedges (threads), and a change of friction [1].

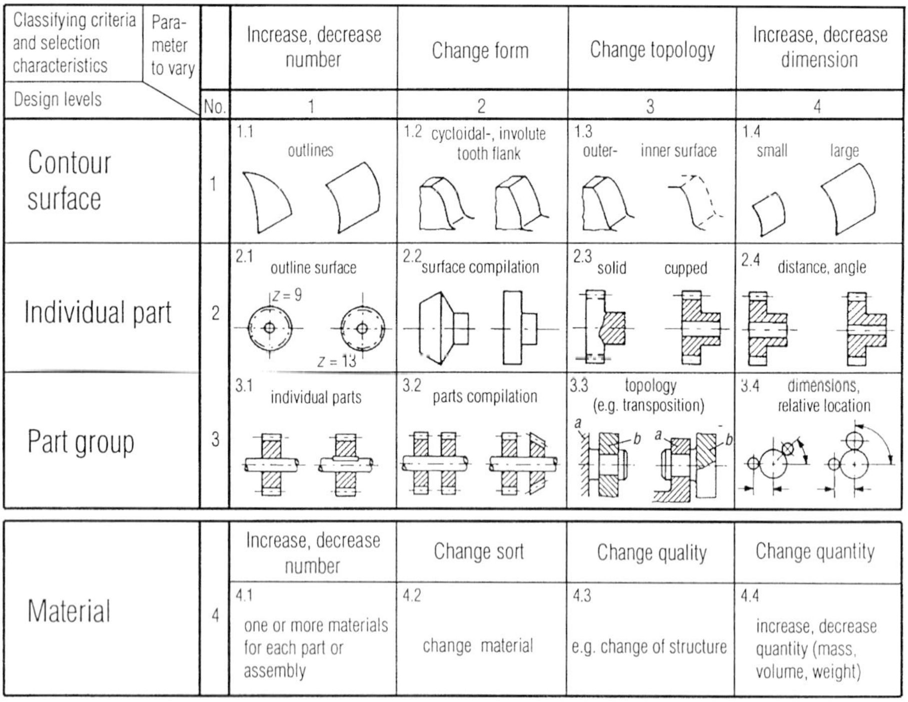

Fig. 8.5 Operation catalogue for the possible form variation operations. It shows four fundamental possibilities at a level of *contour surface*, *individual parts*, *part groups*, and *materials*.

number, form, topology, and dimensions. These changes can be made to the contour surface, individual part, parts unit or material.

It is clear from the figure which alterations can be worked out on surfaces and component parts, *e.g.*, box 2.1: changing the number of teeth.

8.3 Requirements placed on design catalogues

As the purpose of working with design catalogues is to obtain an overview of the processed subarea that is as complete as possible, a design catalogue must guarantee almost full **completeness**. In addition, no redundancies should occur between the individual forms. The contents should also be **free of redundancies, expandable** and provide good **access capabilities**. The expansion is always possible for the *selection characteristics* and the *appendix*. However, it is available for the *contents* only when the "white boxes" (boxes without contents), which occur when required, can be filled later. This was originally the case for the periodic table of elements used in chemistry. The periodic table for the chemical elements is a prototype for all design catalogues. The "white boxes" that could not be filled there led to the postulation and subsequent discovery of new elements. The design catalogues could still have unknown solutions that belong in the "white boxes".

8.4 Desirable forms of design catalogue

Design catalogues provided in table form on paper have two major forms: **one-dimensional** catalogues, Fig. 8.6 (and also Figs. 8.1 and 8.3), and **two-dimensional** catalogues, Figs. 8.5 and 8.7.

For *one-dimensional* design catalogues, Fig. 8.6, each new category is set in a separate row. Figure 8.1 shows an example in which every involute gearwheel with a different number of teeth is set in a separate row.

Classifying criteria				Main part			Selection characteristics			Appendix
I	II	III		Equ	Sk	Ex	F	v	Pr	Doc
1	2	3	No.	4	5	6	7	8	9	10
A	a	α	1							
		β	2							
	b	α	3							
		β	4							
	c	—	5							
B	a	—	6							
	b	—	7							
C	a	—	8							
	b	—	9							

Fig. 8.6 Structure of a one-dimensional design catalogue with *classifying criteria, main part, selection characteristics,* and *appendix.*

Classifying criteria and selection characteristics		a	b	c	d	e
	No.	1	2	3	4	5
A	1	A.a	A.b	A.c	A.d	A.e
B	2	B.a				
C	3	C.a				
D	4	D.a	design catalogue content (main part)			
E	5	E.a				
F	6	F.a				

Fig. 8.7 Structure of a two-dimensional design catalogue with *classifying criteria* and *selection characteristics*.

Two-dimensional design catalogues, Fig. 8.7, provide two structures in the heading column and in the heading row. The contents lie at the intersection of the structures, as in Fig. 8.5. The classifying criteria must be used as selection characteristics at the same time.

It is also possible to have *three-dimensional* catalogues when the contents are contained in two-dimensional catalogues and a third structure is present that spreads over several pages.

The systematic representation is ideally suited for storing the design catalogue contents in a computer in order to be able to fetch them for use at any time. There have also been isolated attempts to produce complete catalogue hierarchies in which, for example, a row of the overview catalogue can be extended to provide a complete detailed catalogue [12].

8.5 Use of design catalogues

It is favourable to use existing design catalogues [1].

1. Object catalogues: consists of systematically structured variants of objects.
2. Solution catalogues: selection of a functional solution. Fitting this solution with the demands of the special case afterwards.
3. Operation catalogues: these describe the different methods to vary the objects.

References

[1] Roth K. Konstruieren mit Konstruktionskatalogen, vol. II, Konstruktionskataloge. 3rd ed. Berlin Heidelberg: Springer, 2001.
[2] Roth K. Franke H-J, Simonek R. Aufbau und Verwendung von Katalogen für das methodische Konstruieren. Konstruktion 1972;74(11):449–58.
[3] Roth K. Aufbau und Handhabung von Konstruktionskatalogen. VDI-Berichte Nr. 219, 1974; 89–99.

[4] Roth K. Design models and design catalogues. In: ICED-Conference, Boston, 17–20 August, publication series WDK 13. Zürich: Heurista, 1987; 60–7.
[5] VDI 2222-2 Konstruktionsmethodik, Erstellung und Anwendung von Konstruktionskatalogen [Design engineering methodics. Setting up and use of design catalogues]. Berlin: Beuth Verlag, 1982.
[6] Roth K. Zahnradtechnik: Evolventen-Sonderverzahnungen. Berlin Heidelberg: Springer, 1998.
[7] Roth K. Konstruieren mit Konstruktionskatalogen, vol. I, Konstruktionslehre. 3rd ed. Berlin Heidelberg: Springer, 2000.
[8] Orth B. Einführung in die Theorie des Messens. Stuttgart Berlin Cologne Mainz: Verlag W. Kohlmann, 1974.
[9] Roth K. Zahnradtechnik: Stirnrad-Evolventenverzahnungen. Berlin Heidelberg: Springer, 2001.
[10] MacFarlane AGJ. Engineering systems analysis. London: George G. Harrap, 1964.
[11] Roth K. Konstruieren mit Konstruktionskatalogen, vol. III, Verbindungen und Verschlüsse. 2nd ed. Berlin Heidelberg: Springer, 1996.
[12] Diekhöner G. Erstellen und Anwenden von Konstruktionskatalogen im Rahmen des methodischen Konstruierens. Dissertation, TU Braunschweig, 1981. Fortschrittberichte der VDI-Zeitschriften, Reihe 1 No. 75. Düsseldorf: VDI-Verlag, 1981.

TRIZ, the Altshullerian approach to solving innovation problems

Denis Cavallucci

Abstract In the current socio-economic context, companies are required to pursue an on-going reduction in terms of the time-to-market process while increasing their innovation capacity. Yet this objective is hindered because the relevance of the solutions depends on human creative, intrinsic and random skills, which means that any technological and economic forecasts related to the new product (or technological system) are necessarily uncertain. The basic notion behind systemics suggests, therefore, that a certain distance should be taken from these situations and that they should be observed according to a systemic time/level reference, which should enable the company to forecast the development of technical systems. There are an impressive number of studies that present analyses of developing the technique. Most of these studies are backed by the formalisation of laws. The emergence of a theory on innovation proposed by Altshuller also offers an advanced study of these laws. This chapter aims to give a brief presentation of TRIZ (from the Russian acronym for theory of inventive problem solving) from a methodological and structural angle, and to put forward how it might be integrated into current Western approaches.

9.1 The genesis of a theory

9.1.1 Introduction

Traditional methods of design all involve a phase at some stage in the process when the idea is triggered off. However, they all delegate this ability to come up with ideas to the intrinsic creative skills of mankind. The problems are reformulated and analysed from various angles; the conditions that are most conducive to generating ideas are set up for people, but the result – that flash of genius – always and ever stems from a person. The account of TRIZ presented here [1–3] does not call into question your intrinsic creativity, but lifts your abilities to have ideas to a higher plane by guiding you in the right direction. Figure 9.1 gives a graphical representation of the TRIZ model, where, after a phase of reformulation (evolving from a problem to a model of the problem), the approach offers you generic solutions linked to the context of your problem (evolving from model of problem to model of solution) so that your creativity can then take up the work and interpret these beginnings of ideas by transposing them to your specific problem (evolving from a model of solution to a solution). It should be noted that the role of TRIZ begins when reformulation starts with one or several models of the solution. Then the creative skills of the designer are dedicated to interpret these models with their industrial realities in order to build a concrete given solution.

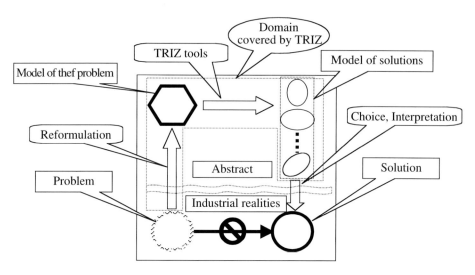

Fig. 9.1 Solving a problem using TRIZ.

Traditional approaches to problem solving are often based on methods such as trial-and-error. The heuristic or haphazard advance of the problem towards a solution may take a long time, depending on how complex the problem is. TRIZ is a theory that helps conquer the blank-page syndrome by guiding the designer in a narrow prospective direction. Then, backed by a systematic process, it addresses the problem by bringing in tools at each stage of the solution-finding process. These tools guide you in generic directions that other people have followed in similar problem configurations.

Two advantages in this approach very rapidly become apparent.

- The analyses do not take into account the industry of origin. Because of this the process becomes transdisciplinary, thereby multiplying the opportunities of finding solutions to problems tenfold by searching for them in other domains. For example, a solution from the chemical industry to solve a problem of mechanics: the chemical effect on the cavity surface on plastic injection tools to create a marbled surface finish for moulded parts.
- The analyses highlight simple fundamental principles and chemical, physical and geometric effects that can be found in specialised literature. For example, Archimedes' principle, the Coana effect, Seebeck effect, *etc*. An application of the Seebeck effect led to heat produced by a domestic oven being used to generate electricity in a house [4].

9.1.2 Altshuller: evaluation of a life dedicated to others

When we speak of a life dedicated to progress in design, we can hardly refrain from quoting its author and the work he conducted at key moments of his life. Nevertheless, only very few publication are dedicated to describing Altshuller's life through the key phases of the genesis of TRIZ [5,6]. Here, we will simply honour the memory of the ideological inspiration of a man of genius who moved away from solitary, clandestine research work towards a theory that, today, is universally acknowledged and forms the basis for the industrial and research activities in hundreds of establishments throughout the world.

9.1.3 Opening to the West gives TRIZ the opportunity to develop

Owing to its informal nature, the work conducted by Altshuller and his team is often termed by his former colleagues as being "behind the scenes". At that time, the Soviet system did not want to acknowledge the fundamental value of his work. Some people even say that it was by moving to the West that TRIZ really managed to develop to its full potential. This is probably true, and the emergence of associated computer tools [7,8], which sometimes benefit from tremendous back up in terms of communications, has no doubt contributed in part (or in any case speeded up the process).

9.2 An approach to classifying Altshuller's work

9.2.1 Introduction

The picture that TRIZ sends us back is actually linked with patent analysis. Nevertheless, a deeper analysis of the history of TRIZ and Altshuller's life obviously shows that patents only represent a reduced view of his work. The following representation (given in Figure 9.2) gives us a quick and synthesised overview of the various fields forecast by Altshuller and of a close to accurate classification of his referenced and known results.

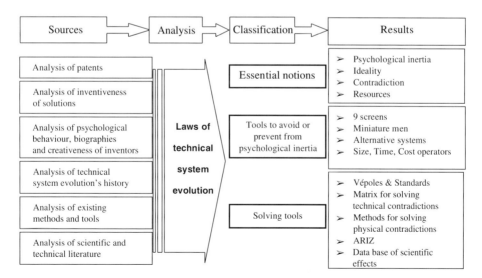

Fig. 9.2 Proposed classification of Altshuller's work.

9.2.2 Basic notions

When putting TRIZ into practice, Altshuller developed certain notions that he claimed were basic to any problem concerned with idea seeking. We must, therefore, be familiar with these notions and follow certain rules in order either to avoid the pitfalls of complex problems, or else to go straight to the essentials by ridding ourselves of stigma inherent in idea seeking, such as the "blank-page" syndrome or the satisfaction that can be experienced with compromise solutions. The essential notions highlighted by Altshuller are not merely quoted, but analysed in depth in TRIZ. You are invited to consider them as "hot lines" for reflection and keep them in mind throughout the life of a project, so as to go faster, further and with greater relevance towards the solution.

- The notion of contradiction: here, we touch on the key point in TRIZ. Indeed, most new actions are based on the principle that an in-depth analysis of the problem must enable us to identify its inherent contradictions. To be solved, all problems must be reformulated so that the contradiction clearly stands out. According to Altshuller, there are three types of contradiction:
- *organisational* (the initial vision of the problem) – this is vague and does not help us find the right direction towards a solution;
- *technical* (the first level of detail) – this sets two parameters linked to the same object (or system) in correlative opposition;
- *physical* (the ultimate level) – this offers a key parameter to the problem, which must subsist in two diametrically opposed states (values).
- The notion of psychological inertia: this is the main obstacle to creativity – in TRIZ, tools have been formalised to overcome this hurdle.
- The notion of the final ideal result: all situations and all systems can be defined in terms of the ideal. By following the recommendations, which help to define the ideal, another step has been made towards formalising the direction towards the solution. According to Altshuller, following the notion of ideality means minimising a system's cost functions and harmful functions while optimising the useful functions of that same system.

9.2.3 Spotting Altshuller's original idea: the laws (or regularities) of developing technical systems

Initially, Altshuller defined a law as being a logical trend in development [3,9]. As with all laws, it is destined to be followed or breached. The norm induces us to follow the law, whereas the alternative of breaching it may turn out to be dangerous. Analysis and formalisation of these laws form the basis of TRIZ, which stipulates that the emergence of a new product is no more than the fruit of the accumulation of all mankind's knowledge. Knowing this development logic, therefore, implicitly helps to bring the designer closer to the right solution, thereby broadening his aptitude to catch a glimpse of it. Altshuller classified the development laws for technical systems under three headings: static, cinematic and dynamic.

9.2.3.1 The "static" laws

The static laws give a motionless vision of the system at a given instant t. Their purpose is to check the structural and functional wholeness of the system.

Law 1: wholeness of parts. For a system to ensure its main function, it must have four fundamental parts ideally fulfilling their role in the functioning of the system.

These four main parts are:

- the driving force (engine), whose function is to generate the energy required to ensure the main function;
- the element of transmission (transmission), which will channel this energy towards the working element;
- the working element (work), which, within the limits of the system under study, will ensure the physical contact between our system and the physical element it acts upon;
- the control element (control), whose main function is to react to the variations in the functioning of the system by adapting automatically to its form, structure and informational output.

The corollaries to this law are as follows:

- each element must participate fully in the good working order of the system;
- at least one of the parts must be controllable to adapt to the variations of the control element.

A diagrammatic representation of law 1 is given in Figure 9.3. Its purpose is to specify the interrelationship between the parts of the system, and it also serves to limit the physical outline of the system under study.

Law 2: conductible energy flow. One condition that is essential to the functioning of a system is the free and efficient circulation of energy through its four main parts. Furthermore, all technical systems act like a power converter. Consequently, the energy must be transferred faithfully without any loss of driving force via the transmission to the working and control parts. The transmission of energy from one part to another may be material (camshaft, gearwheel, lever, fluid, gas, *etc.*), a field (magnetic, electric, thermal) or a combination of the two.

The evaluation of energy losses in the ratio given by input energy (developed by the driving force) over output energy (delivered by the working element) is an important indicator of whether law 2 is being breached.

Law 3: coordination of the rhythm of the different parts. An essential condition for the optimised functioning of a system consists in establishing coordination in the rhythm (frequency, vibrations, periodicity, resonance) of all the parts. Any discrepancy between the functioning rhythm of one part and that of another inevitably generates a loss of efficiency that is harmful to the overall performance of the system. Thus, it is important to set up a form of harmony between the parts (or their com-

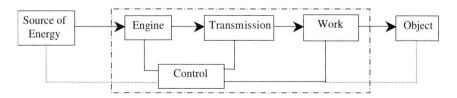

Fig. 9.3 Diagrammatic representation of law 1.

ponents) with the aim of achieving more efficient functioning. Conversely, the latent coordination of the system parts can sometimes generate non-optimised functioning. Under these conditions, the objective will be to develop towards the discordance of the functioning rhythm of the system parts.

9.2.3.2 The "cinematic" laws

Under the cinematic laws, the system is observed in a broader time–space system. This means the analysis no longer bears on the current observation of the system, but is extended towards its past (several stages previously) in such a way as to spot any discrepancies with the cinematic laws.

Law 4: increase in ideality. The development of any technical system strives to attain the highest level of perfection (ideal). By definition, an ideal technical system is a system where the weight, volume and surface area are reduced to the minimum (or even disappear altogether) without changing the working capacity. In other words, an ideal technical system is a virtual system that retains and fulfils its own specific functions. In practice, systems develop towards the ideal when their functional performance improves, while their costs diminish.

Law 5: the unequal development of the parts. The parts of a system develop at unequal rates. The more unequal the development of the parts, the more complex the system becomes. The result is that physical and technical contradictions arise, which, in turn, will generate problems for the future development of the system. The system can then only develop logically if these contradictions are solved.

Law 6: transition towards a supersystem. A system that has exhausted all its development potential can be merged into a supersystem as one of its component parts. Any further development of this subsystem will, therefore, involve the development of the supersystem. It should be noted that, within this law, it is possible to extend the notion of supersystem to the adjacent system, which, if it can take on the functions of the system under study, eliminates the need for its very existence. The analogy with the notion of function integration is clearly formalised here and also contributes to the development of our system towards ideality.

9.2.3.3 The "dynamic" laws

This family of laws is different from the previous two owing to the fact that a given system must opt between developing towards law 7 or law 8. The laws are, in fact, a means of projecting our system into the future, there being, logically speaking, only one of the two directions open to it. The choice governing whether the system will follow either law 7 or law 8 is not, however, a complex one, since the two directions are totally opposed. The designer can, therefore, easily observe which direction is probable by adopting the logic behind the law.

Law 7: transition from the macro-level to the micro-level. The development of the system's working components first transits via the macro-level and then develops to the micro-level. The notions of macro-level and micro-level are directly linked to the structural level observed (solid, granulated, powder, liquid, fields). This law reflects the development trends of technical systems towards the miniaturisation of system components, such as microelectronics, micro-instruments and mechatronics.

However, there are four types of transition open to us:

- from macro-level to macro-level;
- from macro-level to micro-level;

- from micro-level to micro-level;
- from micro-level to macro-level.

The macro-level to micro-level trend is one of the most fundamental trends in the development of modern technical systems, whereas the transition from micro-level to macro-level is not recommended, since, in most cases, it is tantamount to a step backwards. The other two cases are not particularly innovative.

Law 8: increase in dynamics and controllability. The development of "monobloc"-type technical systems always strives towards the highest level of dynamics and controllability. According to this law, rigid systems move towards segmentation while underlining a more efficient controllability via a transition in terms of control fields going from mechanical fields to electrical fields and then magnetic fields and, finally, electromagnetic fields.

This development only occurs if increased dynamics do not generate a loss of controllability; consequently, the compatibility of system components must also be increased.

9.2.3.4 Summary of the laws

To sum up, all systems inevitably transit by this development logic. When a problem arises in the development of a system, this is no doubt due to the fact that one or more laws have been breached, that a blockage in one of the directions for development is an underlying cause thereof and hinders the transition to the next generation. The suggested attitude in the face of these laws is to observe and spot the positioning of the system under study in relation to the three static laws and then the three cinematic laws, so as to identify one of the two dynamic laws as the logical development for our system through time. Then, logic requires the system to develop towards a level of ideal perfection (more effective respect of the laws). The development of the system should, therefore, be directed towards a more effective respect of whichever laws have been noted as being deficient.

9.2.4 Case study: improving the performance of an intake manifold [10]

9.2.4.1 Description of the problem

The intake manifold (Figure 9.4) is an important element in the gas inlet zone of the cylinders. Its shape and the lack of space available under the bonnet mean that, today, it is an element that has undergone little optimisation in terms of its structure and shape. With the aim of improving its performance in use, the intuitive method developed in the previous analysis was applied at MGI Coutier (French automotive supplier) to generate innovative concepts that may improve its performance.

9.2.4.2 Positioning the manifold in relation to the laws of evolution

In TRIZ, Altshuller formalised eight laws of a technical system's evolution. This fundamental aspect of a system's study has been formalised in our case in order to locate the intake manifold in these laws. This study has been performed by grading from 0 to 4 each law in order to predict the logical evolution of the system.

Law 1: taking the parts as a whole (Grade: 1). The main function of the manifold is to feed the cylinders with a mixture:

Fig. 9.4 The subject of our project: the intake manifold.

- engine element – vibrations;
- work element – the volume of air;
- transmission element – the inside walls;
- control element – none.

Comment: law 1 stipulates that the absence of one of the four components of the system leads to overall malfunction. In the case in point, a control element would appear essential for the manifold to function smoothly.

Law 2: free circulation of energy through the parts (grade: 2). In terms of energy, the manifold causes some loss as the energy moves around its major parts. Consequently, this law is not optimised.

Law 3: operational rhythm of the parts (grade: 1). Some parts of the system have not reached their operational peak. The coordination of the rhythm is therefore imperfect.

Law 4: law of increasing the ideal function (grade: 3). The total elimination of the manifold is a case used in Formula 1 engines. In our case, the analysis will examine each component of the manifold. Has the optimum level been reached in terms of the definition of the elements making up the manifold (shape, material, *etc.*)? The answer is yes, insofar as the measurement takes into account technological feasibility and cost factors.

Law 5: unequal development of the parts (grade: 3). All the parts of the manifold have been developed. None has been developed to a lesser degree *than* any other. The grade is therefore good. It should, however, be noted that the absence of one of the 4 parts of the system gives us a biased view of the situation with relation to law 5.

Law 6: transition to the supersystem (grade: 2). Can we get rid of the manifold and transfer its functions to the supersystem or adjacent systems? The answer to this question is basically no, since the current level of system maturity is relatively optimised. A manifold integrated into the cylinder block or the extension to the air filter should, nonetheless, be borne in mind.

Dynamic choice regarding law 7 or law 8. On a micro-level, the transition law would appear utopian, since a manifold made up of a liquid substance is inappropriate. We will therefore focus our attention on law 8.

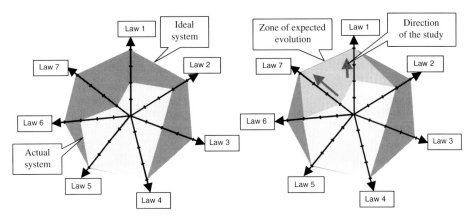

Fig. 9.5 Representation of the manifold in relation to the laws of development.

Law 8: increasing the dynamics and controllability (grade: 0). We can clearly see that this law is not respected at all, since the manifold is a rigid, fixed element and, therefore, devoid of any dynamic properties or possibility of control.

9.2.4.3 Interpreting the positioning

Interpreting how the manifold is positioned in relation to the laws of development (Figure 9.5) leads us to conclude this study as follows:

- firstly, it is necessary to examine the need to have a control element in the system (law 1);
- secondly, it is necessary to provide the manifold with dynamic properties (law 8);
- thirdly, the manifold fulfils useful functions on adjacent systems (or supersystems) with a symmetrical structure and operation; the manifold must, therefore, be symmetrical itself to give optimum response (law 3).

9.2.4.4 Findings of the study

On the basis of the work involved in generating the concepts by the study group and the team at the computational centre, who digitally validated the solution, the innovative nature of the solution, together with the need to protect itself from competition, led MGI to initiate a patent application. The concept given in Figure 9.6 will be the basis for drawing up the patent. It shows that a dynamic element (a piston moved by a spring) is moving directly in relation with the speed of the engine's pistons. These various speeds create a variable airflow in the plenum chamber; this actuates the piston and then modifies the overall length of airflow fluctuation.

To summarise this upstream study on concept research, the following points have been deciding factors in reformulating the problem.

- Highlighting the utility of applying TRIZ when no obvious contradiction arises.
- The absolute need to project the system in the radar diagram of the law of evolution. This enabled us, in particular, to demonstrate:

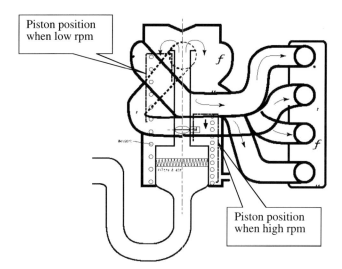

Fig. 9.6 Drawing of the final solution for patent.

- the limits of the system's operation due to its non-symmetry;
- in which direction concept research should progress – the dynamics of the manifold.
- Finally, the principles of separating physical contradictions constitute the most appropriate tools in our situation.

To conclude, the necessary complementarity between the specialist in the technical field under study and the TRIZ specialist who plays a "facilitating" role at key moments in the project has once more been highlighted in this case.

9.2.5 Tools for breaking down the blockages of psychological inertia

The aim of the tools for breaking down the blockages of psychological inertia is to force you to view your problem from a different angle while removing this hindrance to creativity. By applying these tools at the right moment in the project, you are distancing yourself from your problem, so as to see it from a different angle in order to unblock the situation or better perceive its contradictions.

- The nine-screen (or multi-screen) approach: the method traditionally used to observe systems is a single screen. Using the nine-screen tool, Altshuller proposes that the field of vision should be broadened by observing the system under study through several screens laid out over a time scale (past, present, future) and at systemic levels (subsystem, system and supersystem).
- The miniature people method: this tool allows a derivative of Gordon's synectics [11] to be used by specifying the role of the substances present at the micro-level. Miniature people, whose characteristics are clearly defined, enable the engineer to observe a problematic situation whilst maintaining a certain distance.
- Dimension, time and cost (DTC) operators: applying this tool leads to a distortion of the problem. This distortion works by giving infinite variations of the

system's dimension, time and cost. The problem appears under a different angle and the contradictions are clearer, if not patently obvious.

9.2.6 Problem-solving tools

Problem-solving tools are the most important fruits of Altshuller's work. They should be applied once the contradiction has been clearly identified. Which tool proves to be most suitable will depend on the type of contradiction.

- The 11 methods for solving a physical contradiction: the basic laws of dialectics visualise a physical contradiction as a wall that is impossible to surmount. How can a single parameter meet two opposing characteristics? Using his analysis of patents as a basis, Altshuller formalised 11 methods that can solve a physical contradiction.
- The matrix for solving technical contradictions: in his findings, Altshuller stresses that the fundamental principles used by inventors from all industries do not exceed 40 all in all. These "problem-solving" methods were then divided into subgroups to obtain a finer degree of definition. His work then consisted in drawing up configurations of standard problems and expressing them in the shape of conflicts between parameters. The number of main families of conflicting parameters when seeking a contradiction stands at 39. From this point, it is possible to represent these conflicting situations in the form of a matrix by indicating for each one of them which principle(s) was/were used by others in similar configurations.
- The "Vepoles" (from Russian – a contraction of the words substances (*vechestvo*) and field (*pole*)): one of the key phases of problem solving is modelling. With the aim of using a synthesis of fundamental effects, a form of modelling of "substances–field" type has been formalised in TRIZ. This means of representation, with its graphic rules and symbolic form, makes it a powerful modelling tool, and can guide the engineer towards standard problem-solving methods.
- The standard methods: probably the most powerful tool developed by Altshuller. He summarised the standard methods used by inventors in their work and then built up a definition dialectic for each one. Today there are 76 of them, classified in five different classes. The purpose of each standard is, after the "Vepole" modelling stage, to set your problem in a standard solution context in order to follow the TRIZ model for problem solving as presented in Figure 9.1.

9.2.7 ARIZ, the algorithm for applying TRIZ

Some years after TRIZ appeared, the need for a framework approach arose, destined to structure the development of the application of developed concepts and tools. Altshuller and his team members designed ARIZ (from the Russian acronym for algorithm of inventive problem solving) to meet this need [12]. ARIZ claims to be a series of steps with clear objectives:

- to open the project up to a broad potential of ideas;
- to serve as a "frame" for the TRIZ project;
- to yield optimum results;
- to enable the user (phase 1) to develop from an organisational contradiction to a clear and exploitable physical contradiction using the TRIZ tools in the

following phase (phase 2) and then in the final phase (phase 3) to check and apply the solutions according to the schedule put forward.

The history of ARIZ demonstrates its sound structural progression (from 1956 to 1985), with iterations going from versions 56 [13], 59 and 61, which were no more than a programmed series of steps, towards versions 64, 65 and 68, where the problem-solving tools appear for the first time in the process. The term algorithm then takes on its full meaning in versions 71, 75 and 77 [1,14], where the structured analysis (with its new rules and recommendations for use) is added to the quality of the tool. Finally, versions 82 and 85-B and 85-C [15] (for the final version to come out of the work of Altshuller and his team) reflect the achievement of years of development and progress in the versions of ARIZ, which are now used as a basis for teaching the TRIZ approach in a number of institutions.

9.3 TRIZ's contribution to integration in the design process

In this section, three theories for integrating TRIZ into other design methods will be compared [16]. This comparison will be led (in Figure 9.7) from the intuitive design model's vision, which gives an abstraction of a designer's main activities centred around four essential verbs (Collect, Create, Construct, Produce). Our aim is not to favour one or other of these approaches, but to demonstrate that each one has its advantages and drawbacks that prevent our deciding on any one of them.

9.3.1 Using TRIZ in an approach consisting of applying a series of tools

Several articles (especially in the USA) have been published on this subject [17,18]. Integration in a series of tools (Figure 9.7a) offers two advantages: benefiting from

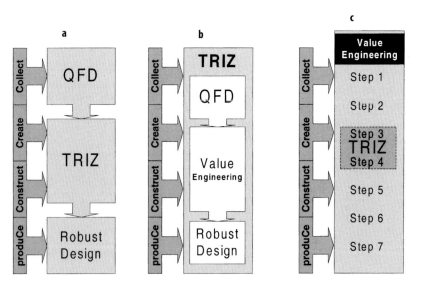

Fig. 9.7 a TRIZ in an approach where a series of tools is applied. **b** TRIZ as a meta-method. **c** TRIZ as a component of an existing method.

the strong points of complementary tools, and providing a broader frame for the project by extending its limits beyond the design habits inherent in a single method. This logical form of integration does, however, present some drawbacks:

- the need to design interfaces between methods;
- the increased complexity of the learning process;
- a longer time span for the project.

9.3.2 TRIZ as a "meta-method"

Integration could also be taken from the angle of considering TRIZ as the basic method for the project, and the strong points of other methods could be integrated during the project (Figure 9.7b). This type of integration would involve all the limitations linked to TRIZ [19] and also all the drawbacks quoted in Section 9.3.2.

The only advantage in this type of integration would be that the project would conform to a certain logic and a formalism specific to a method, thereby getting rid of the vagueness generated by moving from one method to the next.

9.3.3 TRIZ as a component part of an existing method

Finally, in the strategies of integration, another approach would lead us to integrate the strong points of TRIZ (creation phase) into a project conducted using another method (Figure 9.7c). Let us take value engineering (VE) as an example. Here, TRIZ could be used to strengthen phases 3 and 4 of VE (creation phase), where some of its tools could be applied and thereby make the entire project more relevant.

The advantages of this integration are twofold.

- VE is a widely used method in industry and teaching, and it has the benefit of being highly structured. This means it can provide the project with a functional orientation, which is sometimes lacking with TRIZ.
- The amount of training required as a result of this integration could be low, insofar as only some TRIZ tools have been applied.

The only drawbacks to this type of integration are:

- integration does not reach the maximum potential inherent in combining methods;
- all designers who have not previously used a specific method (excepting a personal method) must make substantial efforts in terms of training.

9.3.4 The intuitive design model approach to methodological integration

The logical conclusion of this section is that none of the three approaches gives absolute satisfaction in terms of relevance and simplicity of integration. To fulfil these two criteria, we must, from the beginning of the integration strategy, take into account the fact that it is important to reach the maximum level of relevance in the use of the methods, while basing this integration on the existing know-how of the company's designers [10,20,21].

This conclusion opens a gateway to reflections on the definition of a generic integration strategy specific to a given company.

The four phases, Collect, Create, Construct and produCe (namely the four Cs), shown in Figure 9.7a–c clearly pinpoint the essential notions for all the methods in our studies. Since they are highly generic, they provide a structural approach to the project and in no way consist of a rigid, fixed structure. This abstraction phase will enable us to put forward hypotheses as to the use of one or more "strong points" associated with a method and to link them up to one or more strong points in another.

We should remember that no method takes the methodological history of the company into account. Its structure is fixed in relation to this, and any company that operates one method or another (or sometimes merely rules) is, therefore, obliged either to train themselves regarding the new method they wish to adopt, or to adapt it to what already exists in their company by acting intuitively in terms of its integration.

Backed by this statement of fact, our analysis is therefore based on a prior survey of what exists on a methodological level [22]. This survey is then used in a logical fashion to build up the integration strategy best suited to the company. Our conclusion is that, ideally, the designer should be able to increase the relevance of the project with a minimum number of changes to his design habits. In order to do this, we offer to draw up a survey of the strong points of each method with the aim of detecting any deficiencies in terms of relevance in the project and to bridge these deficiencies, not by integrating a method wholesale, but by integrating its strong points alone.

Once the state of the art of the strong points in design methods has been formalised, and since the methodological and technical knowledge of the company is a known factor, it becomes possible to deduce which intuitive design method may be most appropriate for a given company.

Here is a summary of the key points in our approach:

1. avoiding straying too far from design habits;
2. maximum use of latent knowledge;
3. minimising efforts in skill-building;
4. fostering the acquisition of the methods' strong points;
5. taking stock of the methodologies without preconceived ideas;
6. drawing up a development plan for the design activity based on an accurate data summary.

9.4 Potential development of the theory in research

On the basis of an in-depth analysis of current activities linked to TRIZ around the world, there are three major areas for future contributions.

9.4.1 Contributions to integrating TRIZ in one or more existing methods

As was pointed out in Section 9.1, the strong point of TRIZ is its relevance in the creative phase, whereas current Western methods are rarely interested in this aspect of design. Therefore, obvious complementarities can be envisaged with most of these methods [23,24]. To quote a few: VE, quality function deployment (QFD), design for manufacture (DFM), axiomatic design, the Pahl and Beitz approach, concurrent engineering, and robust design. For each of these methods, one or more interfaces would need to be built up in order to formalise the integration. Thus, we can easily conceive that TRIZ can significantly improve the efficiency of VE in the creative phase by making solution seeking systematic and efficient. Likewise, if we have to accept the best compromise in QFD, TRIZ can refuse this compromise effectively by proposing to solve the contradiction at the root thereof.

9.4.2 Contributions to the development of TRIZ itself

In practice, TRIZ gives us the feeling that the way in which it was drawn up was in correlation with the industrial, scientific and political context of the ex-USSR. The arrival of TRIZ in the West in this form obviously gives rise to a quantity of incoherencies, which mean that, as it stands, the theory cannot be used to its full potential. Therefore, readapting TRIZ to Western reality is necessary, and this can be done, especially from the angle of modelling substances–field (Vepoles) where the graphic formalism needs to be reviewed. It should be noted that concept evaluation is also lacking in the theory, and that formalising a matrix of relevance would seem to be indispensable. This matrix can, moreover, significantly increase its relevance by integrating cognitive and economic data.

Finally, at the current time, the image of TRIZ leaves us with the impression that the tool is in its finished form as far as the content is concerned. Its operationalisation and the computer versions of some of its facets help to convey this image. In fact, discussions with representatives of the former Soviet bloc who contributed to its creation have demonstrated that this is far from true. For example, work on the foundations and developments in terms of principles and standards could be linked with research in the field of capitalising on knowledge.

9.4.3 Contributions to other fields of activity

Finally, let us note the developments in some research laboratories of the former Soviet bloc that have resulted in the use and interpretation of Altshuller's precepts in disciplines linked with industrial fields (marketing, management, publicity, politics, education, *etc.*). The main purpose of this work is to develop research in specific areas by using TRIZ as a theoretical basis for calling into question the certitudes in these fields. It should be noted that, in most cases, the notion of contradiction is at the root of the questioning.

9.5 Orchestrating the work in Altshuller's wake

The death of the genitor of TRIZ on 24 September 1998 poses a significant problem here: whereas in his lifetime no-one could have contested his vision of how the theory should develop, the current state of affairs with its multitude of TRIZ derivatives offers both positive and negative aspects. Indeed, the wealth of the myriad developments of TRIZ in various countries can but make the debate more exciting, but it will also make it vaguer – a point that is already a negative aspect of TRIZ. In such a context, the novice can but experience difficulties in discerning what is really hidden behind this enigmatic acronym. Perhaps the answer can be found in the recent setting-up of the world TRIZ (MATRIZ) association in Petrosavodsk (Russia), the mythical home of TRIZ. This brings together a large number of international associations with the aim of sharing the experiences of those involved in TRIZ worldwide, while benefiting from the visions of those who worked closest to Altshuller, in terms of validating the theory's development.

Three "developmentalist" visions linked to TRIZ should be noted.

- The SIT vision (capturing the key points): developed in Israel in cooperation with the Ford group. Its main purpose is to preserve the basic elements of TRIZ while trimming away what could seem to be complex in terms of acquiring the theory. The aim is to enable a minimum amount of knowledge to be communicated to the largest number of people so that they may benefit from a significant part of TRIZ's potential.
- The TechOptimizer vision by Invention Machine [8] (computerise the basics, benefit from the software potential of data bases). First developed in the former Soviet Union by a specialist in artificial intelligence and an ex-pupil of Altshuller's distance-learning course, this approach to TRIZ aims to set down the key points in the hidden mass of TRIZ (some of its problem-solving tools) and to offer a computerised version, while inviting the designer to use a conventional functional analysis approach as the main structure of the project (so that the user can keep his reference points). Then (always on the basis of ideas developed by Altshuller), once the company has developed a database oriented to "functions to be carried out" for the major scientific effects, the software invites the designer to benefit from this structured, progressive synthesis of basic scientific know-how in his search for solutions.
- The Ideation vision (the so-called "classical" TRIZ era is a thing of the past, make way for I-TRIZ). A former TRIZ school (Kichiniev), which emigrated virtually entirely to the United States, proposes a vision that combines the user-friendliness of software with methodological force, wherein we bear in mind the initial precepts of TRIZ, but we also attempt to structure the process of contradiction emergence more systematically. The idea of simplifying TRIZ is also relevant here in that the user perceives very few of the mechanisms that give rise to proposals stemming from what they have called "classical" TRIZ tools [25].

9.6 Conclusions

9.6.1 Industrial integration strategies

The difficulty of integrating a new approach into a company's competence is a well-known fact. Starting from the principle that an initial contact has been established and that there is a real interest in acquiring this competence, other important obstacles have to be overcome.

Firstly, the psychological difficulty regarding the people involved. People fear the unknown; it puts the designer in a situation of doubt, which sometimes makes him reluctant to admit the interest of building up competence. For example, the sheer volume of content in TRIZ or the complex terms in axiomatic design can be off-putting.

Secondly, the difficulty linked with the time available for this type of investment must be controlled. In all companies (especially small ones) the time/man ratio is tight and there is little room for a methodological building up of competence.

Thirdly, the difficulty linked with the different visions of the theory is also well and truly apparent. Points of view abound in all companies. The technical manager will not have the same interest in the new theory as the designer or even the specialised workshop technician.

Finally, the difficulty of adapting the approach to the size of the company is also a crucial point. In our experience of theory integration into companies, we have encountered the obvious difficulty of company size. This means a large industrial group will encounter problems in internal communication, in decision-making for the department that pilots this type of competence, and several cases will need to be conducted in-house to convince the general management. On the other hand, a medium-sized company will run into the hurdle of the financial aspect of mobilising the rare polyvalent executives it has, as well as the heavy financial investment needed at the outset to build up competence and dedicate a certain time to training. Small or medium-sized companies can find financial resources in state or regional assistance programmes to meet this requirement, but they do not possess the technological and scientific resources to bring in the know-how they do not have in terms of applying solutions.

9.6.2 Creativity and innovation: the missing (or forgotten) link in the design process

Let us not forget that creativity is a thought process that generates ideas. Innovation results from the application of these ideas. However, only a small number of ideas actually end up as an innovation (it is estimated that, on average, you need 60 ideas to achieve an innovation). Indeed, the usefulness of an idea does not need to be demonstrated, whereas an innovation must yield concrete results. The ideas generated in the creativity phase must, therefore, be submitted to evaluation criteria, the ultimate criterion, of course, being consumer satisfaction. In the evaluation of the systematisation of the design process, our encounter with industrial designers led us to state that only the phases involving a creative act remained little, or even not at all, controlled. In our research, a questionnaire targeting designers and other people involved in innovation in the company was aimed at understanding their difficulties as linked to the design process. To this effect, the following question was put to nearly 60 people involved in design in around 40 companies: How do you evaluate your aptitude to overcome the various obstacles in product development?

By the term "aptitude", we mean:

- the time forecast to overcome this obstacle;
- the relevance of overcoming this obstacle (certitude of attaining the best possible outcome).

The synthesis of answers to this two-sided question was eloquent indeed. Only the downstream phases of design seemed to give satisfaction. The upstream phases remained dependent on the company's creative potential, the chance factor or the perspicacity of someone involved in the design process to put the new concept on the market.

9.6.3 An asset for product design

The fundamental sciences are all too often ignored in design. This is probably due to the fact that courses in engineering schools (especially in Western countries) rarely touch on physics and chemistry. The result is that engineers in companies turn away from these resources and fall into design routines centred on the know-how of their own discipline. Reducing mechanical systems to simple principles of physics may, however, contribute significantly to the development of the design process. We conclude that work that aims to reduce the statement of the problem to a set of functions linked with physical effects represents a fundamental asset for the design process. This asset comes under a logical idea of providing the designers with the state-of-the-art know-how on a given subject so as to propose all possible leads to finding a solution.

Investment in the creative phases also seems to give a significant competitive advantage. Indeed, industrial balance is a fact of life, and it is often through the creativity of its designers that the design performance of a project team is acknowledged. Efforts made in this direction by a number of companies attest to the difficulty that lies in controlling these competencies. To us, TRIZ seems to go towards improving this factor.

A final statement must be made on the fact that the context of TRIZ is part and parcel of the culture of the former USSR. This statement is backed up by the fact that 95% of the bibliographical references (scientific articles, scientific reviews, works, *etc.*) are in Cyrillic, which does not make it any easier to understand the basics of the theory. Likewise, virtually most of the researchers working on the subject of TRIZ are still in Russia and can only communicate in Russian. The few works of reference that we were able to use at the beginning of our research were written in English (mostly from the USA, Sweden and Israel). But, very quickly, we realised that very little of what TRIZ is really about appeared in these works. We therefore found it only logical to translate and summarise some of the reference works from Russian so as to grasp a theoretical state-of-the-art that goes beyond a simple transcription of what had already been interpreted in the published information on the subject. Nevertheless, in view of the quantity of literature still available today in Russian, extensive work in summarising, translation and operationalisation remains to be done.

References

[1] Altshuller GS. Creativity as an exact science. New York (USA): Gordon & Breach, 1988 [ISBN 0677212305].
[2] Altshuller GS. To find an idea: introduction into the theory of inventive problem solving. Novosibirsk: Nauka, 1986 [in Russian].
[3] Altshuller GS, Zlotin BL, Sussman AV, Filatov VI. Search for new ideas: from insight to technology (theory and practice of inventive problem solving). Kishinev: Karte Moldaveniaske, 1989 [in Russian].
[4] Killanders A. Generating electricity for families in northern Sweden. Report from the Department of Manufacturing Systems Royal Institute of Technology, 1996.
[5] Lerner L. Altshuller, an outstanding destiny. Ogoneck, 1991 [in Russian].
[6] Khomenko N, Cavallucci D. Genrich Altshuller, un inventeur pas comme les autres. Info'TRIZ: Bulletin d'information de TRIZ-France 2000;(0):2.
[7] Souchkov V, Alberts L, Mars N. Innovative engineering design based on sharable physical knowledge. In: Gero JS, Sudweeks F, editors. Artificial Intelligence in Design'96. Kluwer: 1996;723–42.
[8] Tsourikov V. Inventive machine: second generation. AI Soc 1993;7(1).
[9] Salamatov YuP. A system of laws of engineering evolution. In: Selutsky AB, editor. Chance for adventure. Petrozavodsk (Russia): Kareliya, 1991.
[10] Cavallucci D, Lutz P. Intuitive design method, a new approach to enhance design process. In: 1st International Conference on Axiomatic Design (ICAD 2000), June 21–23, MIT, Cambridge, USA, 2000.
[11] Gordon W. Synectics: the development of creative capacity. New York: Harper, 1961.
[12] Altshuller GS. The innovation algorithm: TRIZ, systematic innovation and technical creativity. USA: Ideation International, 1999 [ISBN: 0964074044].

[13] Altschuller GS, Shapiro RV. About technology of creativity. Questions Psychol 1956;6,37–49 [in Russian].
[14] Arciszewski T. ARIZ-77: a method of innovation design. Des Methods Theor 1988;22(2).
[15] Gassanov A. Birth of an invention: a strategy and tactic for solving inventive problems. Moscow: Interpraks, 1995 [in Russian].
[16] Cavallucci D. Contribution à la conception de nouveaux systèmes mécaniques par intégration méthodologique. Doctoral thesis, Université Louis Pasteur, Strasbourg, France, 1999.
[17] Verduyn D, Alan WU. Integration of QFD, TRIZ, and robust design. In: 2nd Annual Total Product Development Symposium, American Supplier Institute, 1996.
[18] Schulz A, Negele H, Fricke E, Clausing D. Shifting the view in systems development – technology development at the fuzzy front end as a key to success. In: 11th ASME-DETC/DTM Conference, September 12–15, Las Vegas, USA, 1999.
[19] Souchkov V. Knowledge-based support for innovative design. Ph.D. thesis, University of Twente, The Netherlands, 1998.
[20] Cavallucci, Lutz P. How IDM can significantly increase manufacturing systems design. In: 33rd CIRP International Seminar on Manufacturing Systems, June 5–7, Stockholm, Sweden, 2000.
[21] Cavallucci D, Lutz P. Intuitive design method (IDM): a new framework for design method integration. In: 2000 International CIRP Design Seminar, May 16–18, Haifa, Israel, 2000.
[22] Kieffer F. Contribution à l'ingénierie intégrée des systèmes de production: formalisation des mécanismes d'intégration entre modèles et applications sur site industriel. Doctoral thesis, ENS-Cachan, France, 1996.
[23] Nordlund M. An information framework for engineering design based on axiomatic design. Doctoral thesis, Department of Manufacturing Systems, The Royal Institute of Technology (KTH), Stockholm, Sweden, 1996 [ISRN KTH/TSM/R-96/11-SE].
[24] Malmqvist J, Axelsson R, Johansson M. A comparative analysis of the theory of inventive problem solving and the systematic approach of Pahl and Beitz. ASME Design Engineering Technical Conference, August 18–22, Irvine, CA, 1996.
[25] Altshuller GS, Zlotin B, Zusman A, Philatov V. Tools of classical TRIZ. USA: Ideation International, 1999 [ISBN: 1928747027].

Part 3

Tools

10. Synthesis of schematic descriptions in mechanical design
11. An approach to compositional synthesis of mechanical design concepts using computers
12. Synthesis based on function–means trees: Schemebuilder
13. Design processes and context for the support of design synthesis
14. Retrieval using configuration spaces
15. Creative design by analogy
16. Design patterns and creative design
17. FAMING: supporting innovative design using adaptation – a description of the approach, implementation, illustrative example and evaluation
18. Transforming behavioural and physical representations of mechanical designs
19. Automatic synthesis of both the topology and numerical parameters for complex structures using genetic programming

Synthesis of schematic descriptions in mechanical design

10

Karl T. Ulrich and Warren P. Seering

Abstract This article describes a schematic synthesis problem and one of its solution techniques. The problem domain consists of devices that can be described as networks of lumped-parameter, idealised elements in the translational-mechanical, rotational-mechanical, fluid-mechanical, and electrical media. Such devices include speedometers, accelerometers, pneumatic cylinders, and pressure gauges. Design problems in this domain are specified by an input quantity, an output quantity, and the desired relationship between the input and output. The solution technique is based on three steps: (1) generate a candidate design, (2) derive and classify the behaviour of the candidate, (3) based on the derived behaviour and domain knowledge, modify the candidate (if possible) to bring it in line with the specification. The key idea behind this technique is that an abstract characterisation of the essential properties of the candidate design expedites the analysis and modification. The results of this work are aimed at computer tools for preliminary mechanical design.

10.1 Introduction

This article describes a class of schematic synthesis problems and one solution technique in the domain of single-input single-output (SISO) dynamic systems. The computer program implementation of the ideas in this article generates a schematic description of a device in response to a specification of the desired relationship between an input quantity and an output quantity. An example of this kind of synthesis is the generation of a schematic description of a speedometer – in terms of idealised elements like inertias, compliances, and resistances – in response to the specification that the angular velocity of an input and the deflection of an output should be proportionally related. This article presents an introduction to the schematic synthesis problem, an explanation of our choice of problem domain, a description of our synthesis technique, and a discussion of the project results. Our research is aimed at a more fundamental understanding of how designs can be generated from a specification of their behaviour. This understanding is a necessary foundation for better computer tools for design and for enhanced design teaching and practice.

Reproduced from the journal *Research in Engineering Design* (1989)1:3–18 © 1989 Springer-Verlag New York Inc.

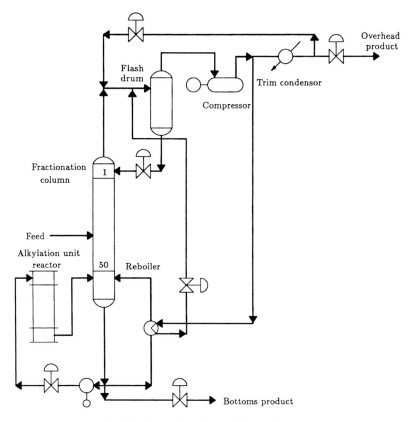

Fig. 10.1 Schematic description of a chemical process.

10.1.1 What is schematic synthesis?

We define schematic descriptions as graphs of functional elements. A design described schematically consists of a specification of its constituent functional elements and their interconnections. The key distinction between schematic descriptions and other types of design description is that schematic descriptions generally contain no information about the design's geometrical and material properties. Generating a schematic description[1] in response to a specification of desired device behaviour is *schematic synthesis*. Researchers and practitioners have developed schematic languages for several engineering domains. Here are two examples of schematic descriptions and associated synthesis problems. Figure 10.1 is a schematic description of a section of a process plant. The functional elements are idealised pumps, heat exchangers, mixers, condensers, *etc*. A synthesis problem in this domain might be: *given a library of equipment elements, generate a schematic description of a heat pump fractionation column for a certain set of specified input and output streams*. Figure 10.2 is a schematic description of an analog circuit. The description

---

[1] We use the terms *schematic description, schematic,* and *description* synonymously in this article.

Synthesis of Schematic Descriptions in Mechanical Design

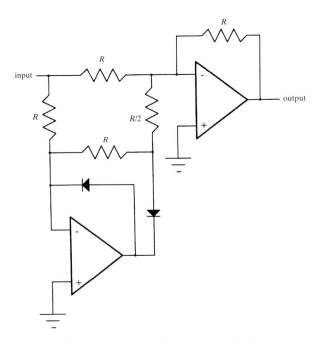

Fig. 10.2 Schematic description of an analog circuit.

consists of idealised elements – resistors, diodes, operational amplifiers – connected together in a graph. A synthesis problem in this domain might be: *generate a circuit that will take a voltage signal input and produce an output voltage corresponding to the absolute value of the input voltage.*

10.1.2 Schematic synthesis of SISO systems

This article presents a schematic synthesis problem and one of its solution techniques in the domain of SISO dynamic systems. Figure 10.3 shows several schematic synthesis problems in the dynamic systems domain along with a sketch version of example solutions. The figure is intended to give a rough sense of what this article is about and of the kinds of problem our solution technique addresses. The first example problem is the design of a pressure gauge. The solution is to connect the pressure source and output spring with a piston–cylinder. The second problem is to design a device to produce a voltage on a resistor that is proportional to an input velocity. The solution is to use a rack-and-pinion connected to a generator. The third problem is to produce a pressure in a fluid capacitance that is proportional to an input angular velocity. The solution is to connect the input through a rotary damper to a rack-and-pinion. The rack is then attached to a piston-cylinder connected to the capacitance. The final problem is to convert a voltage to an angular deflection. The solution is to connect the voltage source with a series resistor to a motor that is attached to the output. The output is also connected to ground with a torsion spring. The schematic synthesis problem in our domain is: *given a description of the input and of the output of a system, what is the schematic description of a system that provides the desired relationship between input and output.*

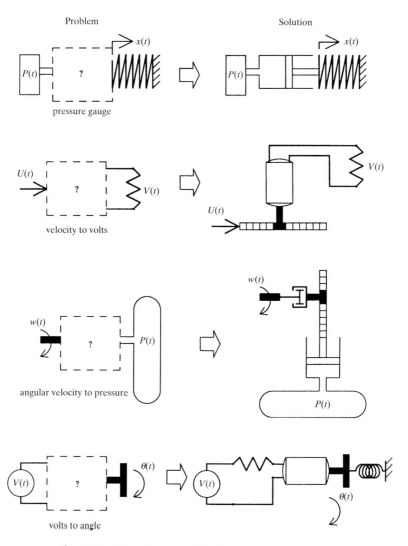

Fig. 10.3 Example problems and solutions in the dynamic systems domain.

10.1.3 Importance of schematic synthesis

Schematic synthesis is an important part of the larger task of generating a physical description of a device from a description of its behaviour. This larger task can be broken into two steps: (1) generate a schematic description of the design in terms of functional elements; then (2) from the schematic description, generate a physical description of a device that approximates the ideal properties of the schematic description. The generation of a schematic description as a first step in design is justified because the strategy reduces the complexity of the problem-solving task, and because the strategy decouples functional and physical design issues. The second step, generating a physical description from the schematic description, is also quite important in mechanical design. Our approach to this problem, discussed thoroughly in Ref.[1], is first to create a very modular physical description by direct substitution of physical elements for functional elements, and then subsequently to look for opportunities to simplify this design by using particular physical features of the design to perform more than one function.

10.1.3.1 Reducing complexity

A schematic description is an abstraction of a design, excluding all physical information except for topology. Working on a design problem first in this abstract problem space is a less complex problem than trying to design directly in the design space described by the physical representation language of the domain. In general, there will be many fewer functional elements in a schematic description language than there will be physical elements in a physical description language. Therefore, the space of possible schematic descriptions is usually considerably smaller than the space of possible physical descriptions. Given these factors, an approach to reducing the complexity of the larger design problem is to work first in the smaller space of schematic descriptions, and then, with the resulting schematic description as a starting point, work in the physical description space. In building computer tools for design synthesis this schematic-physical decomposition is one way of coping with combinatorial complexity.

10.1.3.2 Decoupling functional and physical issues

Focusing the initial problem-solving effort on schematic descriptions decouples the functional and structural aspects of the design problem. Generating a schematic description, without concern for a possible physical implementation, forces the design system to get the ideal device behaviour right first. This strategy is sound, since only devices with correct schematic descriptions will meet the design specifications. Once the schematic description is right, the synthesis of an efficient physical implementation can proceed. This decomposition helps to focus both human and computer problem solving.

10.2 Domain description

This section describes our choice of the lumped-parameter dynamic systems domain, explains the schematic description representation, and defines problem specifications.

10.2.1 SISO dynamic systems

The domain in which we have applied the schematic synthesis concepts consists of devices that can be described as networks of lumped-parameter idealised elements, and whose behaviour can be specified by a relationship between a single input quantity and a single output quantity. We call this domain *SISO dynamic systems* or just *dynamic systems*. Examples of devices in this domain include pressure gauges, speedometers, pneumatic cylinders, and accelerometers. We chose this domain because there is a well-defined set of primitive elements with which to build device descriptions, and because the domain is of engineering interest and importance.

The lumped-parameter elements used in the dynamic systems domain are idealised generalised resistances, capacitances, and inertances, as well as transformers and gyrators. These elements have instances in the fluid-mechanical, translational-mechanical, rotational-mechanical and electrical media. For example, in the electrical medium the elements are idealised resistors, capacitors, and inductors. In the translational-mechanical medium the elements are idealised dampers, springs, and masses. In the rotational-mechanical medium the elements are idealised rotary dampers, torsional springs, and rotary inertias. And in the fluid-mechanical medium the elements are idealised fluid resistances, fluid capacitances, and fluid inertances. In addition to these elements there are elements that can transform quantities in one medium to quantities in another medium – like idealised motors, piston–cylinders, or racks-and-pinions. This domain can be thought of as a kind of generalised analog circuit domain, in that the elements impose constraints on generalised effort and flow variables (corresponding to voltage and current) and their interconnections obey generalised Kirchoff's laws. The devices described in Fig. 10.3 are instances of schematic descriptions that fall into this category. An important characteristic of this domain is that it does not deal with any information about the geometrical or material properties of these devices.

10.2.2 Representing schematic descriptions

The descriptions in Fig. 10.3 are shown with icons that suggest real physical components like motors and racks-and-pinions. These icons are purely mnemonic. We represent these devices as networks of idealized elements.[2] Associated with each element is an effort variable and a flow variable whose product is the power associated with an element. Figure 10.4 shows each of the elements in each of the four media we deal with. Figure 10.5 shows a set of transforming and gyrating elements that convert quantities in one medium to quantities in another medium. Transformers convert efforts to efforts and flows to flows. Gyrators convert efforts to flows and flows to efforts. All of these elements are connected together in series or in parallel to form networks. The networks obey a generalised version of Kirchoff's current and voltage laws. That is, series elements share the same flow quantity and parallel elements share the same effort quantity.

10.2.3 Classifying the behaviour of a schematic description

In order to evaluate a schematic description with respect to a design specification, a system for performing design must be able to derive behaviour from schematic descriptions. In this domain, behaviour consists of a general characterisation of the equations of motion for the schematic. Classifying this behaviour involves first deriving the equations of motion, and second, characterising the equations in terms of their global properties.

---

[2] In performing this research and implementing the results as computer programs we use Paynter's bond-graph language to describe these devices. For clarity of presentation, we have avoided introducing the bond-graph notation here. For those readers interested in a more detailed description of the representation of this problem domain, see Ulrich and Seering [2] and Appendix 10A.2 to this chapter.

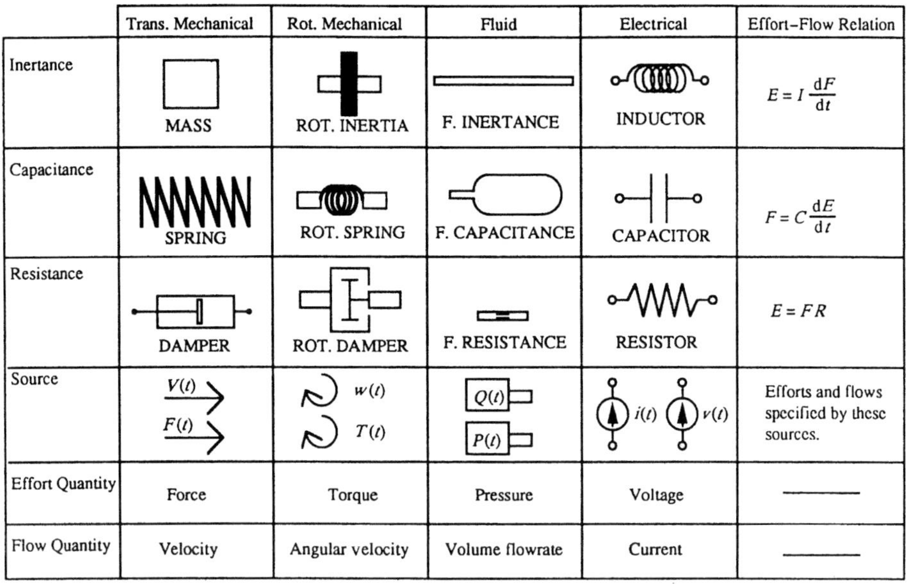

Synthesis of Schematic Descriptions in Mechanical Design

	Trans. Mechanical	Rot. Mechanical	Fluid	Electrical	Effort–Flow Relation
Inertance	MASS	ROT. INERTIA	F. INERTANCE	INDUCTOR	$E = I \dfrac{dF}{dt}$
Capacitance	SPRING	ROT. SPRING	F. CAPACITANCE	CAPACITOR	$F = C \dfrac{dE}{dt}$
Resistance	DAMPER	ROT. DAMPER	F. RESISTANCE	RESISTOR	$E = FR$
Source	$V(t)$, $F(t)$	$w(t)$, $T(t)$	$Q(t)$, $P(t)$	$i(t)$, $v(t)$	Efforts and flows specified by these sources.
Effort Quantity	Force	Torque	Pressure	Voltage	—
Flow Quantity	Velocity	Angular velocity	Volume flowrate	Current	—

Fig. 10.4 Elements used to represent schematic descriptions.

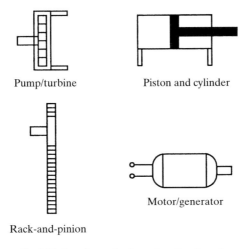

Pump/turbine

Piston and cylinder

Rack-and-pinion

Motor/generator

Fig. 10.5 Example transforming and gyrating elements.

The derivation of equations from the schematic description is a straightforward application of systems theory, and consists of three steps. First, impose the constraints from the generalised Kirchoff's laws by writing a set of network equations relating the effort and flow variables at each junction point between elements. Second, impose the constraints from the element constitutive laws. For each element (capacitance, resistance, inertance, transformer, gyrator, or source), establish the appropriate constraint between effort and flow variables on the element. Third, given the equations generated in steps one and two, eliminate all of the effort and flow variables except for the variables corresponding to the desired input variable and the desired output variables. The result of this procedure is a differential equation relating the input variable and its derivatives to the output variable and its derivatives.

The next step in deriving the behaviour of the schematic description is to characterise its equations of motion. We have developed a characterisation technique aimed at determining the nominal relationship between the input quantity and output quantity. The system is characterised according to its *type number*, an idea borrowed from control theory [3]. The intuition behind this characterisation is that the type number of the system specifies which integral or derivative of the input is proportional to which integral or derivative of the output. A simple proportionality between an effort or flow on the input and an effort or flow on the output corresponds to a type number of 0. For each derivative of the input (or integral of the output) the type number is increased by 1. So, if the first derivative of the input is proportional to the output, then the system is of type 1. If the input is proportional to the first derivative of the output then the system is of type −1, and so on. Four canonical examples illustrate this concept. An accelerometer is a system with a type number of 2, because the deflection of the indicator (the integral of the velocity of the output) is proportional to the input acceleration (the derivative of the velocity of the input). A speedometer is a system with a type number of 1, since the deflection of the output (the integral of the output velocity) is proportional to the input velocity. A piston–cylinder is a system with a type number of 0, since the output force is proportional to the input pressure (*i.e.*, no derivatives or integrals). A mass with a force applied to it, in which the mass velocity is the output, is a system with a type number of −1, because the acceleration of the mass (the derivative of the output velocity) is related to the force on the mass. The formal procedure for determining the type number is given in Appendix 10A.1.

The complete equations for the system are important in refining the particular details of a design, and when choosing design parameters, but only this type number specification is important in getting the gross relation between input and output quantities right.

10.2.4 Specifying a problem

A schematic synthesis problem in the dynamic systems domain is specified by a relationship between an input quantity and an output quantity. This specification includes three parts: (1) specification of the variables of interest, (2) specification of a relationship between variables, and (3) specification of schematic descriptions corresponding to a dynamic model of the input and output environment of the device.

1. The specification of the variables of interest simply involves identifying either an effort or flow variable in the desired medium. For example, if a voltmeter is desired, the input is a voltage (or effort in the electrical medium) and the output is perhaps a velocity (flow in the translational-mechanical medium). Note that the input and output variables must be efforts or flows. Derivatives and integrals of these efforts or flows are specified when the relationship between variables is specified.
2. Specifying a relationship between variables requires the identification of the appropriate derivative or integral of the input and output variables. For example,

if we want a displacement in response to a voltage, we specify that the nominal relationship between the voltage and velocity will be a first derivative (*i. e.*, velocity will correspond to change in voltage, so displacement will correspond to voltage). If we want a displacement proportional to voltage rate-of-change, then we specify that the output velocity will be nominally related to the second derivative of the input voltage. This specification is represented as a *type number*. For example, the first derivative case is type 1, and the second derivative case is type 2. The problem specification can only include nominal, steady-state behaviour of the device. That is, the dynamic response of the device is not part of the specification. The assumption is that, in preliminary design, engineers first want to get the nominal behaviour right, and will then optimise parameters to get a particular frequency response.

3. The third part of the specification is the identification of the dynamic model of the input and output. If, for example, we want to design a voltmeter that measures the voltage of a source whose voltage drops with increasing current, we will specify the input as consisting of an idealised voltage source (one that can provide an arbitrary amount of power) in series with a resistance. And if our voltmeter must actuate a spring-loaded indicator, we would specify the output flow as acting on a translational-mechanical capacitance (a spring). This part of the specification is provided by a lumped-parameter model of the input and output environment of the device. The intuition behind this is that all dynamic systems interact in some way with their inputs and outputs. These interactions partially determine the behaviour of the system. In order to synthesize a system description correctly, one must know what the properties of the input and output interactions are.

In summary, a specification consists of an effort or flow variable in a selected medium, an identification of the derivative or integral relation of interest (indicated with a type number), and a specification of a lumped-parameter model, of the input and output.

10.2.4.1 Example specification

Consider the following complete specification. In designing an aircraft instrument that must actuate a spring-loaded valve with a displacement proportional to rate-of-change in air pressure, we specify the following (shown in Fig. 10.6).

1. The input variable is fluid pressure (effort in the fluid medium). The output variable is translational velocity (flow in the translational-mechanical medium). These variables correspond to the air pressure and the velocity of the valve.
2. The relationship between output velocity and input pressure should be one of second derivative, yielding the desired relationship between *displacement* and *rate-of-pressure*, and so would be specified as a type 2 system.
3. The input schematic description is a pressure source attached in series with a fluid resistance. The output schematic description is a grounded spring.

The synthesis problem now is to find a completed schematic description that will behave as specified. A satisfactory solution would be the description shown at the bottom of Fig. 10.6.

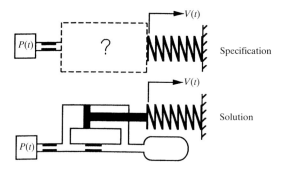

Fig. 10.6 Example specification and solution.

10.3 Solution technique

Our solution to the schematic synthesis problem is based on the following strategy:

1. generate a set of *candidate* schematic descriptions from which all possible solutions can be obtained through augmentations to one of these descriptions;
2. derive and classify the behaviour of these candidate descriptions;
3. based on the behaviour of the candidates, the desired behaviour, and domain knowledge, modify the candidates to meet the specifications.

The following three subsections correspond to these three problem-solving steps.

10.3.1 Generating candidate descriptions

Part of the specification of the schematic synthesis problem is a description of the input and output environment of the desired device. The procedure for generating a set of candidate descriptions is based on the observation that any complete schematic description that performs a transformation from a quantity in the input description to a quantity in the output description must contain either a direct connection between elements in the input and output descriptions, or it must contain a connection between input and output consisting of a sequence of transforming or gyrating elements. We call this connection a *power spine*.

10.3.1.1 Concept of a power spine

A schematic description with a non-trivial relationship between a quantity associated with the input part of the description and the output part of the description must contain some sort of power connection between the two parts. This connection can either be a direct junction between the input and output parts of the schematic, or it can be a connection formed by a sequence of transforming or gyrating elements. Any device that meets the design specifications must have this connection, and the power spine represents the minimal possible connection for a given set of intermediate media. Therefore, any schematic description that will meet the specifications must be derivable from augmentations to this connection between the input and output.

Synthesis of Schematic Descriptions in Mechanical Design

10.3.1.2 Connecting input to output

The synthesis strategy for generating candidate descriptions is to generate a set of schematic descriptions consisting of the input and output descriptions connected by a power spine. These candidate descriptions are generated exhaustively, subject to the constraint that the number of intermediate medium transformations be small (less than three). The size limitation is a constraint imposed by engineering practice – because of cost and reliability, it rarely is a wise design strategy to use more than three media in a SISO dynamic system. *The most important property of the candidate descriptions generated by constructing the power spine is that they are in a certain sense minimal. Minimality in this context means that any schematic containing fewer than three inter-medium transformations and meeting the design specification, must be derivable through the addition of elements to one of the candidates.*

10.3.2 Classifying behaviour

Given a minimal schematic description that will provide some relationship between input and output, the next step is to classify this initial description. It is possible that the candidates meet the specifications; it is also possible that there is a mismatch between the specified behaviour and the actual behaviour of the candidate descriptions, requiring some modifications. The degree of agreement between the behaviour of the candidate descriptions and the specifications is determined by first deriving the equations of motion, and then categorising these equations according to type number. If the type number of the candidate matches the type number of the specification, then coincidentally the candidate meets the design specifications, and the synthesis problem is solved. More likely, a modification to the candidate is required.

At this point the synthesis problem can also be assessed as being impossible. Since increasing the type number of a system requires the addition of elements, and decreasing the type number of a system requires the removal of elements, it is not possible to decrease the type number of a candidate description. This is because no elements may be removed from the candidate. The input and output parts of the description cannot be modified, since they are part of the problem specification; and no elements can be removed from the power spine, since the spine is a minimal connection between the input and output (assuming a given set of inter-medium transformations). Therefore, if the specification of the design calls for a type number that is less than that of the candidate, the synthesis problem has no solution.

10.3.3 Modifying candidate schematic descriptions

The final step is to modify a candidate that does not exhibit the specified behaviour. This process involves three steps: (1) transform the candidate to a compact description; (2) based on explicit domain knowledge, derive and perform a set of modifications; (3) reverse the compacting transformation. The rest of this subsection is divided into parts corresponding to these steps.

10.3.3.1 Transform the candidate design to a compact description

The key concept that allows for correct modification of faulty schematics is the use of a compact representation that highlights only the features of the description that contribute directly to its nominal behaviour. The transformation of a schematic description to its compact representation consists of first converting it to a generalised equivalent description and then applying some simplifying rewrite rules. The rewrite rules eliminate elements that do not influence the type number of the system. The procedure for performing this transformation is as follows.

- The first step converts a description in terms of particular media (electrical, fluid, *etc.*) into a description containing only generalised resistances, capacitances, inertances, and sources. We do not give the details of this transformation here, although it is a straightforward syntactic operation. The transformation basically involves removing all transforming and gyrating elements, and changing specific resistances, capacitances, inertances and sources into their generalised equivalents. The only subtlety in this process involves the removal of gyrators. Since a gyrator converts flows to efforts and efforts to flows, the removal of a gyrator from a schematic requires that the region of the schematic associated with the gyrator be modified to its complementary form. That is, elements connected in parallel are switched to being connected in series and *vice versa*; and capacitive elements are replaced with inertial elements and *vice versa*. Converting a particular description to its generalised equivalent compresses the space of possible schematic descriptions about which the design system must reason.
- The second step is to use rewrite rules to eliminate elements that do not influence the type number of the description. Two example rules are shown in Fig. 10.7 (shown with translational-mechanical elements for clarity). There are eight such rules. Using the rewrite rules to eliminate these non-essential elements completes the transformation of the schematic description into its most compact form.

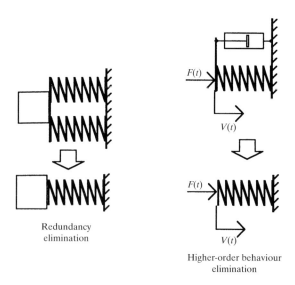

Fig. 10.7 Example simplification rules.

Synthesis of Schematic Descriptions in Mechanical Design

The schematic description now consists of a network of idealised inertances, capacitances, resistances, and sources. This is a minimal characterisation of the candidate schematic description. A system described this way represents an infinite set of graphs that have the same type number.

10.3.3.2 Based on domain knowledge, generate modifications

The power of the compacting transformation in this domain is that once the candidate description is reduced to its compact form the domain knowledge can be expressed concisely. There is one special case and a set of general principles for modifying the design to meet the problem specifications. Here we explain the intuition behind the modification procedure, although the formal explanation is left for Appendix, 10A.

The special case is when the schematic is of type -1, and one desires a type 0 system. There are only two cases in which a schematic can be of type -1. The first is if an inertial element is connected directly to an effort source and the variables of interest are the input effort and the output flow. In this case, the flow will increase linearly in time for a constant input effort (thus a system of type -1). In order to change the type of this system to 0, a resistive element must be added between the inertial element and ground. An example of this case is a mass connected to a force source. The mass accelerates at a constant rate, and therefore the first derivative of the mass velocity is directly related to the force (type -1). If a translational damper is added between the mass and ground, the steady-state relationship between velocity and force is now linear (type 0). The second instance is if a capacitive element is connected directly to a flow source and the variables of interest are the input flow and the output effort. In this case the effort will increase linearly in time. In order to change the type of this system to 0, a resistive element must be added between the flow source and the capacitive element.

In all other cases the type number of the system is exactly the number of *isolated* capacitive or inertial elements in the schematic. The intuition behind an isolated capacitive or inertial element is as follows. Capacitances and inertances are the only lumped-parameter elements that have constitutive relations containing derivatives. That is, the relationship between the effort and the flow on a capacitance or an inertance is an equation between one quantity and the derivative or integral of the other quantity. This is in contrast to resistances and sources whose constitutive relations contain no derivatives or integrals. The only way to change the type number of a system is to add to it elements that add derivatives to the equations of motion; in order words, to add capacitive or inertial elements. However, there are only certain specific conditions under which a capacitive or inertial element can influence the behaviour of the system. For example (we present a translational-mechanical case for clarity, although the procedure would be carried out on the minimal description of the system in terms of generalised elements), imagine a force source connected to ground by a spring (Fig. 10.8). If the force is applied to the spring, it deflects to a position that is determined by the stiffness of the spring and the magnitude of the applied force. Since the integral of the spring velocity is directly related to the spring force, the system is of type 1. Now, imagine that we add another spring to the system in series with the first spring (Fig. 10.8). Although we have added a capacitive element to the design, we have not altered the type number of the system. This new spring does not influence the type number of the system because it is not isolated. If, however, we added a spring that is separated from the original spring by a series

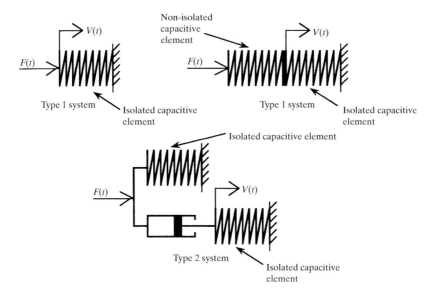

Fig. 10.8 Illustration of the concept of an isolated element.

damper (a resistive element), then the type number of the system increases to 2 (Fig. 10.8). In this second case each of the springs is isolated.

This concept of isolation is defined formally in Appendix 10A of this chapter, but a simplified explanation of the rule for isolation is that the capacitive and inertial elements must be connected into the description in such a way that they are distinct from existing capacitive and inertial elements. In certain cases, this can be facilitated by the addition of resistive elements that serve an isolating function. Given this statement of the domain knowledge, the derivation of modification operators is straightforward. Except in the case of systems of type −1, the difference between the type number of the candidate design and the type number of the specification is equal to the number of new isolated capacitive or inertial elements that must be created in the description. If one isolated element must be created then an isolated inertial or capacitive element could be added, or a non-isolated inertial or capacitive element already in the description also could be isolated by adding a resistive element in an appropriate location. The only constraint that must be enforced when creating new isolated groups is that new elements be added to the power spine between the compact descriptions of the input and output environment. This is because the addition of elements to the input and output regions of the description would require that the input and/or output environment of the device be modified.

10.3.3.3 Reverse compacting transformation

The final synthesis step is to reverse the compacting transformation on the modified schematic in order to arrive at instantiations in specific media. The reverse transformation is straightforward:

Synthesis of Schematic Descriptions in Mechanical Design 167

1. add the elements removed by rewrite operations;
2. change the generalised system back to specific media by reintroducing the transformers and gyrators.

The only decision that must be made in this process is where, with respect to the newly added elements, the gyrators and transformers should be reintroduced. The approach we have taken to this decision is to create a version of the design for each possible location of the elements with respect to the gyrators and transformers. There is a range of possible positions in the various media associated with the system for locations of the newly added elements.

10.4 A complete example

This section is aimed at clarifying the synthesis technique by presenting a concrete example. The example is illustrated by Figs. 10.9 to 10.15. Imagine that an engineer wanted to design a device that would take as an input a current from an electrical circuit and actuate a mass with a displacement proportional to the current. Perhaps the device is to serve as a kind of governor on an electric motor circuit – an increase in motor current will move a mass, causing an increase in the rotary inertia of the system. The problem can be stated in the dynamic systems representation as follows: synthesise a system that will connect the electrical system input description with the mechanical system output description (Fig. 10.9) such that the integral of the flow on the output is proportional to the flow on the input (*i.e.*, a system with a type number of 1). The solution technique proceeds according to the steps developed in Section 10.3.

1. *Construct the power spine.* In this case the power spine is a motor-like element (a gyrator) and a rack-and-pinion-like element (a transformer). The system elements that form the power spine are between the two vertical dotted lines. Just one candidate schematic description is shown (Fig. 10.10), although there will in general be several possible power spines that connect the input graph chunk to the output graph chunk and that consist of fewer than three gyrating or transforming elements.
2. *Derive the behaviour of the candidate schematic description.* Deriving the equations of motion that relate the output flow to the input flow reveals that the candidate schematic description is a type 0 system – the output flow (a velocity) is proportional to the input flow (a current). Since the candidate schematic

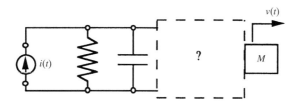

Fig. 10.9 Specification of current meter problem.

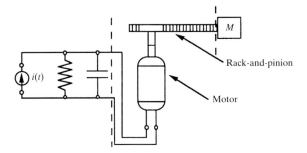

Fig. 10.10 Example of one of the possible power spines.

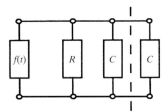

Fig. 10.11 Description converted to generalised elements.

description does not meet the specification, the transformation and modification steps must be carried out. Fortunately, since the type number of the system must be increased (from 0 to 1), the modification is possible.

3. *Transform the candidate schematic description to its compact representation.* The first step is to remove all of the gyrating and transforming elements, replacing particular instances of inertances, capacitances, resistances and sources with generalised inertances, capacitances, resistances, and sources. Note that removing a gyrator requires that all of the *downstream*[3] capacitive elements be changed to inertial elements, and that all of the *downstream* inertial elements be changed to capacitive elements; and that elements that are in parallel be changed to elements that are in series and *vice versa* (Fig. 10.11). The final compacting step is to remove any superfluous elements (execute any rewrite rules that may apply). In this example, one of the capacitive elements closest to the source can be removed without altering the type number of the system (Fig. 10.12).
4. *Derive appropriate modifications.* The design is specified to be a type 1 system – a displacement (integral of output flow) proportional to current (the input flow). Since the candidate schematic description is of type 0, an isolated capacitive or inertial element will have to be added to the power spine. Observing the compacted schematic description reveals that an inertial element can be added along the power spine and will be isolated. Adding a capacitive element, however, will require adding a resistive element as well, since it would otherwise not be

---

[3] *Downstream* in this context is an arbitrary choice of direction within the schematic description. When a gyrator is removed from the design, one chooses a direction within the description as the downstream direction and then converts the elements as explained in this paragraph.

Synthesis of Schematic Descriptions in Mechanical Design

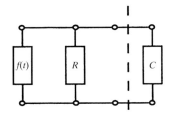

Fig. 10.12 Removal of redundant capacitive element.

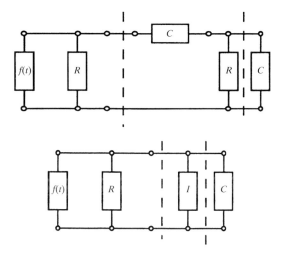

Fig. 10.13 Two possible additions of isolated groups to a minimal description.

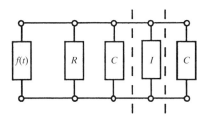

Fig. 10.14 Reintroduction of redundant capacitive element (only one description shown).

isolated from the existing capacitive element. These two new descriptions are shown in Fig. 10.13.

5. *Reverse the compacting transformation.* The next step is to reverse the steps carried out during the compacting transformation (only one description is shown in Fig. 10.14). First the capacitive element is added back to the schematic description. Next the gyrators and transformers are reintroduced. Note that adding the gyrator back into the description reverses the swapping that occurred

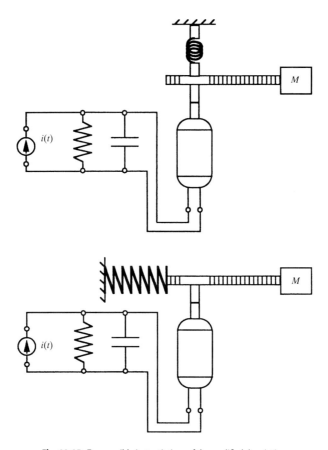

Fig. 10.15 Two possible instantiations of the modified description.

in step 3; that is, the generalised capacitance is converted to a mass, and the generalised inertance is converted to a spring.

Note that there are several possible locations for the newly added element in the schematic description. It could be a torsional spring on the motor output, a translational spring on the rack, or it could be an electrical inductance in the electrical portion of the system. Two of these possibilities are shown in Fig. 10.15.

10.5 Discussion

This chapter has presented a solution to a synthesis problem that people have traditionally solved in an *ad hoc*, heuristic way. This section analyses the importance, completeness, extensibility, and computer implementation of the synthesis technique.

10.5.1 Importance and utility of the technique

When we began thinking about the SISO dynamic systems synthesis problem, we gave several engineers some sample problems to see how they performed the task; most of them found ways to avoid doing the exercise. One or two engineers actually found a solution. No one found more than one solution. These informal observations suggest that the synthesis problem is not simple for people and that synthesis techniques are not widely known. The technique we have developed provides all possible solutions (containing fewer than three inter-medium transformations) to a specified problem. This results in an improvement in the state of the art in this area of synthesis. These techniques are useful not only for computer programs, but also for pencil and paper approaches. The results of this single experience suggest that design in general will benefit from formalising some domains and from finding synthesis techniques in those domains. In many ways human engineers have been successful because they have been able to find *some* solution to a problem. Synthesis techniques for even fairly simple engineering tasks can help engineers find a much greater number of feasible solutions, many of which will be solutions the designer has not thought of.

This work is also important because of its contribution to the larger task of synthesising a physical description of a device from a specification of its behaviour. We propose that performing this larger synthesis task in two steps – first generating a schematic description, and from it a physical description – reduces the problem-solving complexity and focuses the problem-solving effort. In this chapter we have presented a solution to the first half of the larger design problem. We present one solution to the second half in Ref. [1] that is based on the idea of *function sharing* – the use of a single part in a design to implement more than one functional element in the device schematic description.

10.5.2 Completeness of the technique

Our schematic synthesis technique is in a trivial sense incomplete – it will not generate all possible schematic descriptions that meet the problem specifications. The incompleteness results from two factors. First, there is an infinity of schematic descriptions that meet the specifications. A pair of inter-medium transforming or gyrating elements can always be added to a schematic description to create a new schematic description: a back-to-back motor–generator pair added to an electrical device, for example. Second, elements that do not change the nominal behaviour of a schematic description can always be added to a schematic description to create a new schematic description. An example of the second case is the addition of a redundant capacitance to a region of the schematic description that already has a capacitive element. Completeness should, therefore, be defined with respect to the compact representation of the schematic description, after all of the extraneous elements have been removed, and the schematic description has been converted to a generalised form without transformers and gyrators. Under these conditions the technique is complete. The completeness rests on the fact that the type number of a system is determined by the number of isolated capacitive and inertial elements within the power spine of the description. This can be proven by the following argument. A positive type number indicates the number of times the input quantity has been differentiated. There are only two possible ways a quantity can be differentiated in this domain. Either a capacitance differentiates an effort to produce a flow, or an inertance differentiates a flow to produce an effort. Resistances serve to convert efforts to flows and *vice versa*. The only way to differentiate an effort source (there is a similar argument for flow sources) is to connect it to a capacitance (an instance of this case is a force source attached to a spring), or to connect it to a series resistance (to convert the effort to a flow) and then to a parallel inertance (to differentiate the flow into an effort). These two cases correspond to single isolated capacitances and inertances. Since the power spine of a device is a sequence of elements, this argument applies to an arbitrary number of differentiations. For each additional differentiation (increase in the type number) an isolated capacitance or inertance must be added. Because the synthesis technique produces all possible ways of modifying the candidate schematic description in order to increase the number of isolated capacitances and inertances in the system, it produces a complete set of compact schematic descriptions.

10.5.3 Extensibility

The ideas in this chapter have been directed at a very specific synthesis problem. This subsection deals with the problem of extending the ideas within the dynamic systems domain and to other domains.

10.5.3.1 Extension within dynamic systems domain

The devices that can be synthesised with the technique described in this chapter are limited in several ways: they are SISO; there is no feedback or amplification; and they can not be specified to have particular dynamic response characteristics. Extension of the solution technique to erase these limitations requires either an extension of the representation language, an extension of the synthesis procedures, or both.

To deal with feedback or amplification, we would have to introduce the use of active elements – elements with external power sources. Along with the active elements, signal flows would be required. This extended representation would also require extensions to the synthesis procedures. Specifically, the connection property of the domain that leads to the concept of a power spine is still valid, although the kinds of connections that are allowed are increased to include signal-only connections. Under these conditions the synthesis procedure may require the consideration as special cases each of several classes of devices – those with feedback, those with hidden sources, those without any signal flows.

If the problem were extended to address multiple-input multiple-output systems, the synthesis procedure would become more complex. The power spine concept no longer is as strong as in the SISO case. There are now many possible ways for the multiple inputs and outputs to be connected together other than by a single sequence of idealised transforming and gyrating elements. Prabhu and Taylor have explored some of the issues in the multiple input and output problem [4].

The problem addressed in this chapter deals only with the nominal behaviour of a dynamic system – we have not considered dynamic issues like bandwidth. To some extent these issues can be addressed by optimising parameter values numerically. In other cases, filtering elements or damping must be added to the schematic descriptions to achieve the specific numerical specifications desired. The synthesis procedure in this chapter, however, is a necessary first step towards meeting more detailed numerical design specifications. Any schematic description that meets a detailed numerical specification must also meet the specification of the nominal behaviour of the device. The work in this chapter is directly applicable to this more specific synthesis problem.

10.5.3.2 Extension to other domains

The three major segments of the solution to the dynamic systems synthesis problem are: development/selection of a schematic language, a procedure for generating candidate schematic descriptions, and a procedure for categorising and altering schematic descriptions falling into different classes of behaviour.

The particular schematic language we use in the dynamic systems domain cannot be extended to any other domain, although the properties of the lumped-parameter systems language that make it a good choice for the dynamic systems problem can be used as guidelines for selecting schematic languages in other domains.

The power spine idea in the dynamic systems domain has incarnations in other domains as well. In the domain of material-handling systems (i.e., design of conveyers) there must be a connection between input and output. In the domain of engineering structures (beams, brackets, supports, struts) there must be a connection between input and output. In fact, in the structures case, perhaps the candidate schematic descriptions could be maximal connections – a structure that occupies all of the possible space between the input and output points of the structure. Then, any valid schematic description might be derivable from material removal operations on this candidate schematic description. The classification and modification procedures in the dynamic systems domain are problem specific; however, the general idea is applicable in many other domains. Imagine for example, a synthesis problem in the shaft and bearing system domain. A minimal characterisation of the system in terms of the types of interaction between adjacent elements could be derived in order to compute the degrees of freedom of the device, or to determine whether or not the device could be assembled.

10.5.4 Computer implementation

In this chapter, we have focused almost exclusively on the principles and the procedures for performing the schematic synthesis task. We have deliberately avoided a discussion of any particular computer program that implements these ideas. Our belief is that programs in this context serve two purposes: first, they may be tools for solidifying and testing hypotheses about how to perform the synthesis task; second, they may be embodiments of the final procedures, useful as tools or demonstrations. We have written a computer program that implements a solution to the schematic synthesis problem described in this chapter. The program was written to fulfil the first purpose – to clarify the thinking involved in devising a solution technique – and to that extent was quite useful. The final version of the program does not exactly match the solution technique presented in this chapter. The actual program deviates from the theory we present in one major way: it relies on a set of debugging rules to modify the candidate schematic descriptions rather than performing the compacting transformation and subsequent modifications. An example debugging operator is: *if the output is a flow variable, the input is a flow variable, and differentiation is desired, add a series resistance and capacitance to ground.* These rules were derived by analysing human design reasoning in debugging deficient schematic descriptions, and in fact correspond to the same operations as are derived by the technique presented in this chapter. The debugging rules are, however, much more complicated than the procedure using the concept of a compact representation and isolated inertances and capacitances. It was not until we began exploring the completeness of the debugging procedure that we realised that there was a more compact way of characterising the schematic descriptions and that the domain knowledge could be expressed more concisely.

10.6 Related work

A more detailed presentation of this work, along with a procedure for generating physical descriptions from the schematic descriptions, can be found in Ref. [2]. The key idea behind generating a physical description from a schematic description is to first use a standard physical component for each functional element in the schematic description, and then subsequently to look for opportunities to simplify the resulting device. Doyle [5] describes work aimed at generating explanations of device behaviour. In particular, the input to the explanation problem is a time history of events or device states. The output is a network of causal connections between the events in the observation. Each connection corresponds to a physical phenomenon or mechanism. This causal network can be thought of as a sort of schematic description of the device. Doyle works with a very broad class of devices, including toasters, bicycle coaster brakes, and tyre gauges. To deal with this wide variety of devices, he has developed a very rich representation for physical phenomena, including many kinds of transduction and transport. The key contribution of this work is the exploration of the power of various types of constraint in controlling the combinatorial complexity of the explanation problem. Prabhu and Taylor [4] derive some mathematical properties of multi-input, multi-output devices characterised by power flows between inputs and outputs. Prabhu and Taylor use a bondgraph representation for these devices, and concentrate on synthesising the *junction-structure* or network of transducing elements that can link the inputs with the outputs while dividing the power flows according to specification. Because the bond graph is a formal mathematical representation, certain results such as soundness and completeness can be proven. Kannapan and Marshek [6] developed an instance of a schematic language for mechanical design. The language includes primitives shafts, bearings, handles, supports, *etc*. Pieces of devices described with this language are combined together to form new designs. Williams [7] is one of the very few pieces of work that deals with the synthesis of a design configuration. Williams explores the domain of digital circuit design at the FET level. His program is based upon using qualitative analysis of the behaviour of a preliminary circuit design to guide design modifications. Williams develops a time-based dependency scheme for tracking the influence of one event in time on another event. Roylance [8] is probably the first effort at synthesising analog circuit configurations computationally. In this master's thesis, Roylance describes a design-rule-based system that designs simple *RLC* circuits. The design rules allow the system to backward chain from an equation specifying the desired circuit behaviour. Ressler's thesis [9] describes a computational procedure for designing operational amplifiers. The procedure is based upon a hierarchical circuit grammar. Amplifiers are viewed as consisting of three stages. Stage one may be implemented as a differential pair, a current cancellation configuration or a super beta circuit. A differential pair may be viewed as a load and emitter coupler pair. A load can be resistive, a current mirror, a simple pair, or a Darlington pair. There are similar expansions for the other amplifier stages. The design procedure is to implement the amplifier with the simplest possible pieces, then analyse the resulting circuit at each stage in the instantiation process. If the circuit will not meet the specifications, then a more complex option is chosen. Rieger and Grinberg [10] was one of the first attempts to describe the causal structure of a mechanical device. This article presents a language for describing devives, and develops a model in this language for a thermostat. The functional elements of the model are terms like: continuous and one-shot enablement, state coupling, state equivalence, state antagonism, threshold, and rate confluence. This language was designed to allow descriptions of complex, highly non-linear device behaviour that cannot be described with differential equations. Rieger and Grinberg used these descriptions to simulate the behaviour of the devices by assigning a computational behaviour of each functional element in the device description.

Acknowledgements

This article describes work performed at the Artificial Intelligence Laboratory of the Massachusetts Institute of Technology. Support for the laboratory's research is provided in part by the Defense Advanced Research Projects Agency of the United States Department of Defense under the Office of Naval Research contract N00014-85-K-0124, and the National Science Foundation under grant DMC 8618776. The presentation and substance of the research described in this article were improved in response to helpful comments by Randall Davis, David Gossard, Patrick Winston, and the reviewers.

Appendix 10A

10A.1 Determining the type number from the system equations

The procedure for determining the type number of a system given a differential equation relating an input and output quantity is as follows.

1. Rearrange the equation so that the terms consisting of derivatives of the input are on the right-hand side (RHS), and the terms consisting of derivatives of the output are on the left-hand side (LHS).
2. Perform a Laplace transformation on the expression leaving an equation between polynomials in the Laplace operator s and the input and output quantity. (In the case of equations of this form, the Laplace transform is performed by simply replacing s^n for d^n/dt^n.) Multiply both sides of the equation by s raised to some power to eliminate any s terms in the denominators of any fractions. Factor out as many s terms as possible from both sides of the equation. If an s term can be factored out of both sides of the equation, simplify the equation by dividing both sides by the lower-order of the two factored-out s terms.
3. We call the order of the remaining factored-out s term the *type number* of the system. If the s term is on the RHS, then the type number is positive. If the s term is on the LHS, then the type number is negative.

As an example, consider the following equation between an input pressure and an output velocity (corresponding to a piston–cylinder connected to a mass, a spring and a damper):

$$M\frac{d^2}{dt^2}V + B\frac{d}{dt}V + KV = A\frac{d}{dt}P.$$

Performing the Laplace transform leaves:

$$MVs^2 + BVs + KV = APs.$$

Since there is one s that can be factored out of the RHS, the system has a type number of 1, meaning that the input pressure changes with the integral of the output velocity (position). One important property of this derivation procedure is that it is algorithmic and can be automated.

10A.2 An explanation of isolated groups using bond graphs[4]

Although we have used mnemonic icons to show the schematic descriptions in this chapter, the representation we have used in this project is bond graphs [11,12]. In the bond graph representation, our descriptions are simply sequences of 2-ports and zero- or one-junctions with attached 1-ports that connect a source element to an n-port corresponding to the output quantity of interest. In this context, a power spine is a sequence of transformers and/or gyrators connecting a graph chunk that represents the input environment of the device to a graph chunk that represents the output environment of the device. Given this representation, the concept of an isolated group can be defined as a syntactic pattern within the graph sequence. In this section, we give the procedure for modifying a candidate graph once it has been simplified to contain only generalised resistances, capacitances, inertances, and sources. This procedure corresponds to the informal description given in Section 10.3.3.2.

1. In the case of a bond graph of type −1 with an effort source as input, adding a 1—R (a new one-junction with an attached R) along the power spine will make it a type 0 bondgraph. In the case of a bond graph of type −1 with a flow source as input, adding a 0—R along the power spine will make it a type 0 bond graph.
2. In all other cases, the type number of a graph is exactly the number of *isolated* 1—C or 0—I groups along the power spine of the graph. An explanation of isolation is given below.

The power spine of a graph connects the input n-port to the output n-port. The spine can be thought of as a string of bond graph groups of the form $(x—A, y—B, z—C)$ where x, y and z are n-ports and $A, B,$ and C are 1-ports. The bond graph groups 1—C and 0—I cause differentiation of the input if they are isolated. This differentiation is what determines the type number of a system. An isolated 1—C or 0—I group is one whose neighbours (on either side of it in the power spine string) are *isolating groups*.

---

[4] This section presumes familiarity with bond graph notation.

1. An isolating group for a 1—C is a 0—R, a 0—I, or a 0—SE; a 1—I or 1—R occurring at the last 1-port in the bond graph sequence; or combinations of 1—Rs or 1—Is and any other isolating groups (SE indicates effort source, SF indicates flow source). This last case – a combination of 1—Rs or 1—Is and any isolating group – is also an isolating group because a 1—R or a 1—I next to a 1—C does not influence the type number of a graph.
2. The isolating groups for a 0—I are either 1—Rs, 1—Cs, or 1—SFs; 0—Cs or 0—Rs occurring as the last 1-port in a bond graph sequence; or combinations of 0—Cs or 0—Rs and any other isolating group.

Given this statement of the domain knowledge, the derivation of modification operators is trivial. Except in the case of systems of type –1, the difference between the type number of the candidate design and the type number of the specification is equal to the number of new isolated 1—Cs or 0—Is that must be created in the graph. If one isolated group must be created then an isolated 0—I or a 1—C could be added, or a non-isolated 0—I or 1—C already in the graph could be isolated by adding an isolating group.

References

[1] Ulrich KT, Seering WP. Function sharing in mechanical design. In: Proceedings of the Seventh National Conference on Artificial Intelligence (AAAI-88), St. Paul, MN, August 1988.
[2] Ulrich KT, Seering WP. Computation and pre-parametric design. MIT Artificial Intelligence Laboratory Technical Report 1043, October 1988.
[3] Ogata K. *Modern control engineering.* Prentice-Hall, 1970; 284.
[4] Prabhu D, Taylor DL. Some issues in the generation of topology of systems with constant power-flow input–output requirements. In: Proceedings of the 1988 ASME Design Automation Conference, Kissimmee, FL, September 1988.
[5] Doyle RJ. Hypothesizing device mechanisms: opening up the black box. Ph.D. thesis, Massachusetts Institute of Technology Department of Electrical Engineering and Computer Science, 1988.
[6] Kannapan S, Marshek KM. Design synthetic reasoning: a program for research. Mechanical Systems and Design Technical Report 202, University of Texas at Austin, Department of Mechanical Engineering.
[7] Williams B. Principled design based on topologies of interaction. Ph.D. thesis, Massachusetts Institute of Technology Department of Electrical Engineering and Computer Science, 1988.
[8] Roylance G. A simple model of circuit design. Massachusetts Institute of Technology Artificial Intelligence Laboratory Technical Report 703, 1983.
[9] Ressler AL. A circuit grammar for operational amplifier design. Massachusetts Institute of Technology Artificial Intelligence Laboratory Technical Report 807, January 1984.
[10] Rieger C, Grinberg M. The declarative representation and procedural simulation of causality in physical mechanisms. In: Proceedings of the Fifth International Joint Conference on Artificial Intelligence, vol. 1, 1977; 250.
[11] Paynter HM. *Analysis and design of engineering systems.* Cambridge (MA): MIT Press, 1961.
[12] Rosenberg RC, Karnopp DC. *Introduction to physical system dynamics.* McGraw-Hill, 1983.

An approach to compositional synthesis of mechanical design concepts using computers

11

Amaresh Chakrabarti, Patrick Langdon, Yieng-Chieh Liu and Thomas P. Bligh

Abstract Usually, there are many solutions to a design problem; therefore, there is scope for producing improved designs if one could explore a large solution space widely. An approach would be to use the computer to synthesise a wide variety of concepts for a given problem, and allow designers to explore these before developing the most promising ones. Adopting a research approach based on developing basic representations, knowledge base and reasoning procedures adequate for synthesising concepts of existing devices and mechanisms, a computer program for synthesising solutions to a class of mechanical design problems has been developed. For a given design problem, the program can produce an exhaustive set of solution concepts, in terms of their topological, spatial and generic physical configurations, which can then be explored by designers. In order to aid designers in the exploration process, an approach for clustering and browsing these concepts has been developed. The program has been tested for its ability to: (1) generate existing as well as innovative ideas of solutions at the various levels; (2) cluster these ideas using measures of similarity used by designers and the difference this makes to the quality of solution space exploration.

11.1 Objective

The objective of this paper is to establish the importance of the conceptual design stage in the design process, identify relevant areas of research within this stage, propose a method to do research in these areas, present an overview of an approach developed to support concept generation and exploration, present some results of testing this approach, and indicate further research directions, especially for further supporting exploration of concepts generated using this approach.

Conceptual design is that stage of the design process where concepts of solutions are developed to meet the functional requirements of the design problem [1]. Conceptual design, being an early stage of design, is characterised by information that is often imprecise, inadequate and unreliable. Apart from the observation that the recognition and generation of functional requirements and the generation of solutions are highly coupled in this stage [2,3], there is little understanding as to how this is done, and consequently little support is available. Notwithstanding these difficulties, it is at this early stage that substantial costs are committed. As pointed out by Berliner and Brimson [4]: on average, as high as 80% of the cost of a product over its total life cycle is committed by the conceptual stage of the design process.

The challenge is how to support designers so as to increase their chances of producing the best possible concepts. The key to answering this question is that there is often not one, but a multitude of possible alternative solutions to a given design problem. A comparative analysis of the styles of work of designers revealed that designers who explored, in a balanced way, a wider range of ideas have been more successful in developing solutions of better quality than those who explored fewer solutions [5,6]. Therefore, it could be argued that if a designer can be supported and encouraged to generate and explore a wider range of solution alternatives using a wider range of evaluation criteria, this should increase their chances of producing better designs. The hypothesis that being exposed to a wider range of solutions increases the chances of generating solutions of greater novelty and quality is supported, in a different context, by the work of Heylighen and Verstijnen [7].

Usually, however, only a few design alternatives are considered, and using only a few evaluation criteria. There are several possible reasons. Often designers are not aware of the existence of potential solutions in a different domain. This is particularly true of novice designers, whose repertoire of designs is limited owing to lack of experience. Designers can be biased towards using certain kinds of solutions, perhaps because they have used them before. Evaluation criteria often emerge from the observation of key positive and negative features of the design alternatives at hand. This means that the more and widely varied the considered solution alternatives are, the wider and richer the criteria for their evaluation should be. However, if none of these difficulties existed, designers would still not be able to consider more than a few alternatives without being supported by an enhanced information processing capability. This is because, as each design is detailed, the information generated around it grows quickly, making it quite impossible to explore more than a few alternatives. This is where computers, with their potential for information processing, could make a difference.

One possible route is to devise a computational framework where the designer are presented with a wider range of ideas than is possible at present, and could be supported to evaluate and modify these before homing in on the most promising ones for further development.

The goal of this project is to develop such a computational framework; consequently, there are two central objectives: (i) to support the generation of a wider range of ideas than is possible at present, and (ii) to support exploration (browsing and visualisation) leading to modification and evaluation of these ideas so as to fulfil the emerging functional requirements. The results presented in this paper will deal mainly with the first objective, but also with the second. The approach is evaluated using several case studies and experiments, the results of which are used to identify key issues for further research. Section 11.2 describes the research approach, Section 11.3 explains, using a simple example, the synthesis approach developed, Section 11.4 summarises the research results and issues identified, and Section 11.5 details further developments undertaken to resolve these issues. Section 11.6 draws conclusions and identifies further work.

11.2 Research approach

The research approach for realising the first objective has been based on the assumption that new concepts can be generated by combining, and/or adapting, existing ele-

Compositional Synthesis of Mechanical Design Concepts Using Computers

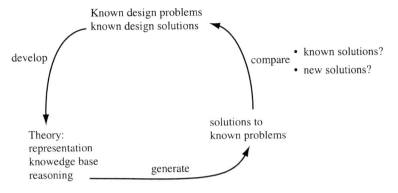

Fig. 11.1 The research approach.

ments so as to satisfy new functionality. The reason is that a wide variety of designs do seem to have common elements, and, therefore, if these elements could be distilled, and combined in different ways, this should generate new designs. We start our research by analysing a range of known design problems and their known design solutions (see steps in Fig. 11.1) to identify commonalties among them. What things are common across the given set of design problems, and what things are common across these design solutions? Are there some general ways of representing them? Can we identify the common elements across these designs which could form a collective knowledge base? Can we then develop reasoning procedures that would use this knowledge and representation to generate solutions to these known problems? If this could be done, we could then compare the outcome of the procedures with the solutions already known to us. At the very least, do they generate the solutions we already know? But, more importantly, do they generate solutions that are new? If this can be done, this should help provide designers with a wider range of ideas than currently available.

In terms of the overall design research methodology presented in Fig. 11.2 [8], the above approach is only the first step towards validating the usefulness of a design method (termed a prescription, in the methodology). Each design method is developed with the intention of making a specific difference to the world: its impact at a high level of the business process must be evaluated using appropriate criteria, e.g., the resulting improvement in market share of the company. It is possible for a method to make this impact at the high level by directly influencing some intermediate factors, called influences (such as development of novel, high-quality products), that have an influence on the high-level factors represented by the criteria chosen. The development of a method should, therefore, be preceded by studies that identify which important factors of influence can be targeted by the intended method to bring about the intended change in criteria (Description I), and followed by studies (Description II) to evaluate whether this change is indeed brought about, i.e., in our case, whether the approach developed can generate a wider range of ideas than the designer, and whether offering this to designers can influence them to generate more novel products than otherwise.

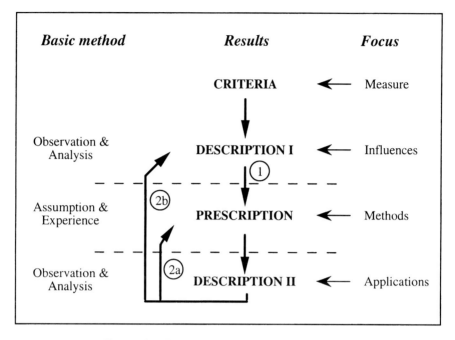

Fig. 11.2 Overall research methodology (from Blessing et al. [8]).

11.3 Synthesis approach: representation, reasoning and example

In this section, each of the steps outlined in Fig. 11.1 will be illustrated.

11.3.1 Develop theory from known design problems and solutions

We start off with a set of existing devices and mechanisms, which includes devices such as door latches, paper punches, earth-moving equipment, *etc.*, and identify common features among these devices. In terms of their functions, each of them requires some input, and produces some output. The number of inputs or outputs can be more than one (*e.g.*, the two outputs of a paper punch). Each of these inputs and outputs has some characteristics that may change with time. For instance, as the handle of the door latch is pressed down, it moves down, causing the output point (the wedge) to retract into the casing. If this is continued, at some point the handle stops moving down, and with it the retraction motion of the wedge also stops. If we now let the handle go, it moves in the opposite direction to the original position, and simultaneously the wedge moves back to its original position. Now, this is a temporal sequence of activities defined by changes in the characteristics of the inputs and outputs of the device with time. The characteristics of the inputs and outputs, in this case, include their kinds (*e.g.*, force, torque, linear motion or angular motion), the directions of their action (*e.g.*, up, down or sideways), their magnitudes, and their positions. For further details, see Chakrabarti and Bligh [9,10].

Let us, for the discussion that follows, not consider the temporal characteristics of the inputs and outputs, and focus on them at an instant of time. What parameters are sufficient to describe the functionality of each of the above problems? We could have an adequate representation of the function of a design in terms of a set of inputs and outputs, each of which has a kind, a direction (positive or negative directions along the i, j, or k axes), a magnitude, and a position.

Now the question is whether there is anything common between the structures of devices and mechanisms. The first thing common is that each design is composed of a number of elements, which are connected in certain ways. The second thing is that some of these elements function in a similar way, although their embodiment might be quite different. For instance, the handle of a door latch is such that when pushed on the input end, it rotates at the other. In other words, it has a single input and output, one of which is a force (or translation) and the other is a torque (or rotation). Although visually different, functionally this element is similar to the top part of a paper punch, which, when pushed down on its input edge, produces a rotation at its pivot end. If rotation is taken as a pseudo-vector (using the right-hand rule for instance, where curled fingers point in the direction of the rotation and the thumb points in the vector direction), the spatial configurations of the input and output of this element have a definite relationship: they are orthogonal and non-intersecting to each other. Interestingly, the crank of a scotch-yoke mechanism is very similar, where the input and output kinds are just reversed. Another interesting feature is that the directions of the inputs and outputs (though always being orthogonal and non-intersecting to each other) depend on the direction in which the element is laid out in space. We could observe similar functional similarities between the wedge assembly (which can move back and forth), the two punches, and the output element of the scotch-yoke mechanism. Also, their inputs and outputs are of the same kind (translation) and parallel to each other. What we also notice is that the way these elements are connected to form the device allows them to transfer their inputs and outputs. In a common door latch, for instance, the output rotation from the handle is transferred to the input of a cam, which produces a translation at its output. This translation is taken at the top portion of a wedge assembly, which is transferred to its output to produce the desired effect. In other words, if the function of each element is known in terms of its inputs and outputs, and if they are appropriately connected so that the inputs and outputs match at the connections, then, they together provide a causal account of the internal functioning of the device.[1] Figure 11.3 presents a representation of the design problems and solutions presently investigated in this research. A design problem is described by its functions, where a function is represented by a number of inputs and outputs, each having a kind, a direction and a magnitude. Design solutions are described as combinations of a set of functional elements (such as a lever, of which a door-latch handle is an example embodiment). Each such element is defined as one of five basic element types,[2] or combinations

---

[1] Even though we have chosen to illustrate these by elements having inputs and outputs with mechanical characteristics, this approach is not restricted only to such elements. Generally, an element could have inputs and outputs with any characteristics.
[2] These element types are basic in the sense that these are the only possible distinct spatial relationships that an input vector I, an output vector O, and the length vector L, which denotes the spatial separation between I and O, could have (they can be coaxial, parallel, intersecting with I being coaxial with L, intersecting with O being coaxial with L, and skew). These would give rise to more elements when the kind of input and output are specified. For more details, see Chakrabarti and Bligh [10].

Problem representation

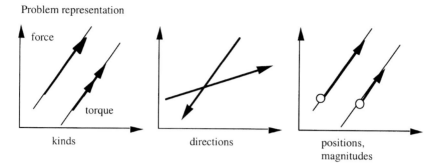

Solution representation, showing the five basic elements

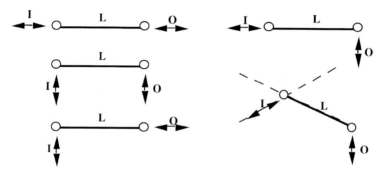

Fig. 11.3 A representation for design problems and solutions.

thereof. These elements are distinguished by the spatial relationships between their inputs, outputs and their spatial separation.

Using this representation, for example, we could now present the function of a common door latch as a vertical downwards force input to be transformed into a horizontal leftwards translational output, as shown in Fig. 11.4a. The door-latch solution could be represented using the functional elements as (see Fig. 11.4b) a combination of a lever taking the input, transferring its output (a torque) to a cam, which, in turn, produces a translational output to be transferred by two tie rods to the desired output point.

Assuming that we have a set of known functional elements that were distilled from a host of existing designs, the next question is how we can use these to generate these existing designs and, perhaps, other new ones. As an illustration of how this could be done, a single-input single-output (SISO) synthesis algorithm, a simpler version of the reasoning procedure developed that can exhaustively generate multi-

---

[3] In graph theory [11], a graph is a mathematical structure that consists of two kinds of set that may be interpreted geometrically as nodes (or vertices) and as arcs (or edges) whose end points are nodes. A path is a sequence of arcs such that the initial node of the succeeding arc is the final node of the preceding arc. It is directed if each of its edges can be traversed in one direction only. A path is simple if, in going along the path from its initial to its final vertex, one reaches no vertex more than once. It is unlabelled if its vertices and edges have not been labelled. See Reinschke [11] for further details.

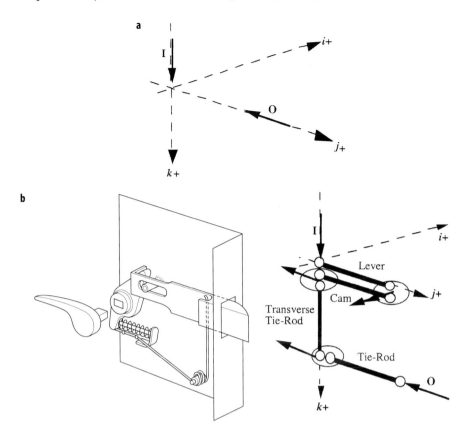

Fig. 11.4 a Door-latch function; **b** one existing concept and its representation.

ple-input multiple-output designs, is presented in Fig. 11.5. The general idea is that, having arbitrarily chosen the maximum allowable number of elements to be used in a design solution, the procedure generates each possible distinct unlabelled simple directed path.[3] For instance, if the maximum allowable number of elements is two, the two possible paths are: with a single element connecting the input to the output, or with two elements in series. Each arc in such a graph is an element, and each node is an input–output connection. If we now label each of the nodes in each path using the information about the functional requirements and the kinds of I/O that can be handled by the database of functional elements, we get a number of distinct paths with labelled nodes. For instance, if the system input should be a force **F**, and the system output a torque **T**, and if the elements in the database can only have forces and torques as inputs or outputs, the only two possible paths with labelled nodes are **F–F–T** and **F–T–T** (which means that this function could be achieved either by a force to force followed by a force to torque transformation, or by a force to torque followed by a torque to torque transformation). If we now label each arc with each possible alternative database elements capable of doing the transformation indicated by the labelled nodes, the alternative solutions are found. For instance, in the **F–F–T** path,

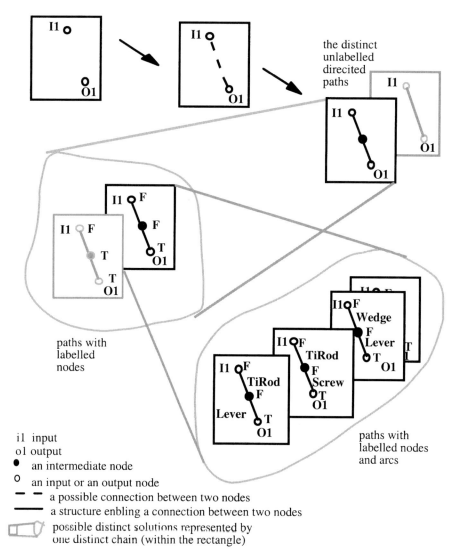

Fig. 11.5 The SISO synthesis algorithm.

if the F–F transformation could be done in two ways, *i.e.*, using a wedge or using a tie rod, and the F–T transformation could be done in two alternative ways, *i.e.*, by a screw or by a lever, then the overall function could be achieved by their possible combinations (see Fig. 11.5). For further details, see Refs [9,10,12,13].

11.3.2 Generate solutions to known problems and compare with existing designs

The next question is: what is the result of applying the above theory (representation, knowledge base and reasoning) to the problems with which we started? As an illus-

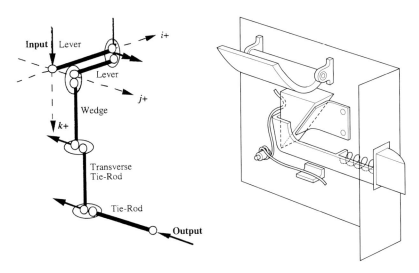

Fig. 11.6 One of the solutions generated by the program for the door-latch problem.

tration, solutions to the door-latch problem are generated. When we give the SISO door-latch function described before as an input to the program based on the above theory, does it generate at least the existing designs? Further, does it generate other novel, exciting ones?

This is what we did for a range of design problems, including those used to develop the theory. As an illustration of the results, using a maximum of five elements per solution and for a given database of six elements, the program generated 193 different designs to the door-latch problem,[4] so as to have a tie rod at the output end of each. It generated a number of alternative (stick-diagram-like) spatial layouts for each of these solutions as well. One layout of one such design is shown in Fig. 11.6. The left-hand diagram is computer generated, whereas the right-hand diagram is one of its possible embodiments. It should be noted that each such spatial layout could, in principle, be developed into a number of alternative embodiments; that shown is just one, and not necessarily the best one at that. As we see, in this embodiment, when the handle is pressed down it presses down the wedge, which then moves down as well as to the left, thereby pushing the latch assembly leftwards. The constituent elements are similar to those extracted from existing designs such as door latches and paper punches, but the combination is a different one, leading to a substantially different concept. This is the assumption we mentioned in Section 11.2, that new designs can be generated by combining or adapting existing designs.

---

[4] The exhaustive set of solutions, which can be synthesised by combining these six elements (a lever to transform a force into a torque, a lever to transform a torque into a force, a cam to transform a torque into a force, a wedge to transform a force into a force, a transverse tie-rod to transform a force into a force, and an axial tie-rod to transform a force into a force) so as to use at most five elements in each solution so generated, would contain 687 distinct solutions. This can be calculated using Equation 5 given in Appendix A of Chakrabarti and Bligh [10]. There would be 193 of these solutions that would have a tie rod as their output element.

11.4 Evaluation of the synthesis approach

The synthesis approach described in Section 11.3 was originally implemented using the Common-LISP language on a LispWorks® [14] package, which is a LISP-based environment for developing knowledge-based tools. The program is called FuncSION, which is an acronym for **Func**tional Synthesiser for Input Output Networks.

FuncSION was evaluated in two ways. One was the use of an in-house project, called the Mobile Arm Support (MAS) project, carried out at the Cambridge University Engineering Department, as a case study. The other test was the evaluation of the concepts generated by FuncSION, by three experienced designers, for their novelty and usefulness.

11.4.1 MAS project case studies

The MAS project [15] was to design a means for enhancing the mobility of muscular dystrophy (MD) sufferers. The project ran in two phases for over 3 years, which led to the development of two prototypes. The MAS was designed by two designers who worked closely with each other. The designers were assisted by a brainstorming session involving eight people, which gave them an initial pool of ideas. They explored these ideas, and eventually came up with three concept variants, from which one was selected for embodiment. All these ideas are documented by Bauert [16]; see Fig. 11.7 for some examples. As a retrospective study, an input–output requirement, which describes the intended instantaneous function of the arm support, was given to FuncSION for it to generate solutions and their spatial configurations. A total of 162 solutions were generated by FuncSION. See Fig. 11.8 for a list of the database of elements used by FuncSION for generating these solutions. The solutions generated were compared with the ideas generated by the designers. The statistics of how designers' ideas with sketches relate to the solutions generated by FuncSION are as follows. Of the 43 ideas generated by designers, 18 were incomplete, abstract, infeasible, incomprehensible or based on a different domain of knowledge, and therefore could not have been generated by FuncSION. The 162 solutions that FuncSION generated included 22 of the remaining 27 ideas generated by designers, and a large number of novel ideas not generated by designers [17].

11.4.2 Hands-on experiments by experienced designers

Three experienced designers were asked to evaluate the solutions generated by FuncSION for aspects of their originality, feasibility, variants and redundancy. They were also asked to comment on the ease of using FuncSION, and to make suggestions.

The designers in the above experiments found that FuncSION, in general, generates a range of interesting solutions, and often comes up with surprisingly clever ideas and provides insight. However, it also generates a large number of redundant solutions, and this makes it difficult or frustrating to evaluate and explore the ideas in any depth. They had some difficulty in visualising possible embodiments of the solutions with the present representation, and could visualise only when these solutions were shown at a lower level of abstraction. On the whole, the above experiments suggested that the designers had a common understanding of when solutions were considered similar to each other: they felt that if two solutions were different only by translational elements (*e.g.*, a tie rod, or a shaft), then they were similar. If this

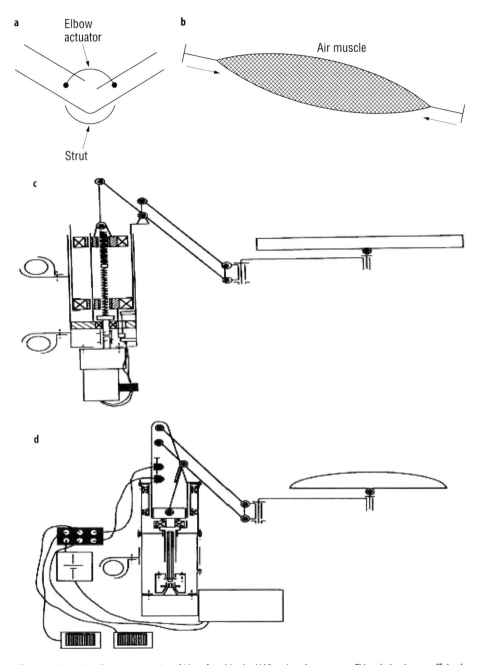

Fig. 11.7 Examples of various categories of ideas found in the MAS project documents. **a** This solution is not sufficiently detailed to be discussed within the realm of FuncSION. **b** This solution cannot be generated at present by FuncSION as it is in a different domain. **c** This solution cannot be generated by FuncSION as it contains a cycle in its network. **d** This solution can be generated by FuncSION.

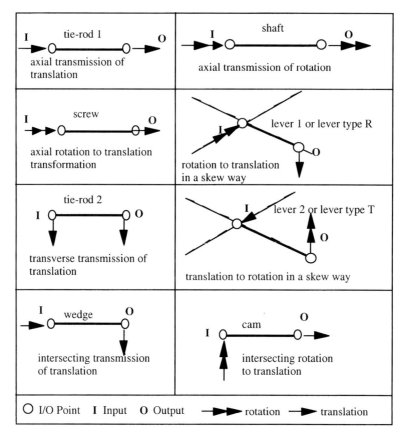

Fig. 11.8 The entire set of elements used by FuncSION in synthesising concepts.

criterion, *i.e.*, of what similar solutions mean, was applied to group the solutions that FuncSION suggested in the MAS cases described before, these would fall into 17 different groups of solutions of an average size of about ten. Of these, only four groups were considered at any length at all by the designers in the MAS project, and just a single instance was detected in the documents for two of the 13 other groups. This indicated the potential of FuncSION for suggesting different ideas and idea types. It is important to note that the above groups were the result of solving the MAS problem using a database of only five elements. If this database were to be increased to seven for instance, the number of distinct clusters would be as high as 29, of which only six would then have been considered at all by the designers, and only four of these in any depth.

11.5 Further developments

Based on the experience gained from the above case studies and experiments, further research focused on two areas of deficiency of FuncSION.

- *Too many solutions.* One problem associated with the synthesis approach embodied in FuncSION was that it had the potential to generate too many solutions to be meaningfully explored by designers. A strategy was needed to generate or present these solutions in a way that allowed browsing with reasonable effort. A related issue when many solutions are generated is that the program had to be efficient in order to be able to generate them all. Three strategies were undertaken: (1) to use various measures to improve the efficiency of the synthesis procedure; (2) to use additional constraints from the specification, not used in the original version of FuncSION, during the synthesis process in order to eliminate generation of solutions that did not meet these constraints; (3) to group the solutions generated into clusters of similar solutions, and allow easy review of them.
- *Too difficult to interpret and visualise.* Two issues concerning the interpretation and the visualisation of the synthesis results were considered vital by the designers. The first was that the representation of elements was too abstract to visualise their potential embodiments. The second was that the static representation for functional elements and conceptual solutions made it hard for the designers to imagine their likely temporal behaviour. Thus a means of visualising solutions and their elements was necessary. Two strategies were undertaken: (1) addition of a further, more detailed yet generic level of representation for the solutions, and embodiment of solutions generated earlier at the topological and spatial levels, at this level; (2) a three-dimensional visualisation of the design problem and all the solutions in all the clusters.

11.5.1 Resolving the first problem: managing the number of solutions generated

11.5.1.1 Improving efficiency of the synthesis procedure

The efficiency of compositional synthesis depends both on the size of the database of building blocks used, and the characteristics of the algorithm that uses this database. One way to make a synthesis process efficient is to ensure that the database of elements used contains only those elements that will contribute to the generation of solutions to the current problem; this can be done by pre-processing the database to its optimum, useful size before using it in synthesis [18]. The effect of doing this, as shown in the above work, is substantial. For instance, for a database with 44 elements, a tenfold reduction of memory required is achieved by pre-processing the database before using unidirectional search [19] for synthesis, when the number of building blocks per solution synthesised is nine.

Another way to make a compositional synthesis process more efficient is to use bidirectional search rather than unidirectional search [20]. If a design problem is expressed as a function with an input and an output, unidirectional search can be seen as concatenating elements from the database to form chains whose input requirement matches that of the intended function, and checking to see if their output matched the output required. In contrast, bidirectional search can be seen as trying to form two chains, one with the input of the intended function and the other with the intended output, and checking to see if the I/Os at their other ends match. Use of bidirectional search instead of unidirectional search, for instance for solutions having nine elements synthesised using a database of 53 elements leads to a 60-fold reduction in the memory required, and a similar reduction in computation time. Together, these two measures can have more substantial benefits.

11.5.1.2 Using additional constraints

The initial version of FuncSION did not take into account constraints on the relative position or magnitude of the inputs and outputs of the intended function. An example of a position constraint, in the case of a door latch, would be that the door handle that takes the input must remain on the left, on top and in front of the wedge at the output. Concepts that do not obey this constraint will have to be eliminated during synthesis. An example of a magnitude constraint, in the case of the latch, would be that the user would only need to move the handle by a small amount before the door is disengaged from the frame, allowing it to open with small movement. In this case, a magnification of movements may be necessary between the input and output. Current developments include use of position constraints for pruning the solution space, and rules for checking the ability of a design to satisfy a given magnitude constraint [21].

The other sets of constraints that can be employed do not necessarily have a concrete, generic theoretical grounding, but are known often to be useful. We call these elimination heuristics. Some of these are domain specific; for instance, a designer may specify that all concepts must start with a handle, or end with a wedge, or that a specific element should never be used with another, or that it should be used only with another element. The heuristics can also be domain neutral: that only the minimum number of elements necessary to form a concept to be synthesised first; more "complex" concepts are to be generated only if the designer is not satisfied with these. These heuristics are now implemented and can be used at the discretion of the designer to prune a solution space [22,23].

11.5.1.3 Grouping solutions using similarity

Two overall strategies for clustering are adopted. One is based on concepts of similarity, the knowledge of which is neutral to any specific domain of application. For instance, it may be useful to cluster all concepts that have the same number of elements, or force a solution set into a given number of clusters, or into the number of the clusters that give the most compact cluster average. It may also be based on the morphological similarity, e.g., having the same elements, or similar relationships. A clustering technique has been developed that is used to plot each concept in a Euclidean space, where the axes are formed by each possible element-pair that can be formed using the elements of the database; the similarity between two concepts is measured by calculating the Euclidean distance between these concepts in this space [24]. Once these similarities are calculated, one of many standard clustering algorithms can be used to cluster them, and concepts that best represent each cluster can be determined. The idea is that designers can browse a clustered solution space and get a fair overview by only reviewing one of a few concepts from each cluster, rather than having to go through each single concept, thereby reducing the difficulty of meaningfully exploring a large solution space [25]. The cognitive validity of the clustering technique has already been tested by comparing its outcomes with that of designers, and an average 75% match is found [24]. It was also tested to see whether designers found it easier to obtain an overview of a solution space with the help of clustering; the results indicated that clustering-assisted designers solved a design problem much faster than those not assisted [26].

The second kind of heuristics that can be used to cluster concepts is based on domain knowledge. For instance, solutions can be clustered into the same group if they are different only in transmitters (*e.g.*, shafts or tie rods, see Ref. [22]), or those that use the same elements, *etc.* [23]. A further clustering of solutions can be done at the topological level by treating as similar those elements that are based on the same principle but differing in embodiment. For instance, tie rod1 and tie rod2 in Fig. 11.8 can be treated as similar elements at the topological level, since they both use pure translation as their motion principle. Currently, all these heuristics have been implemented and can be used at the designer's discretion. Using these heuristics together led to an average of a sevenfold reduction in the solution spaces tested [23].

11.5.2 Resolving the second problem: strategies for aiding visualisation

11.5.2.1 Embodiment at the generic physical level

In order to enhance visualisation of the solutions generated by FuncSION, a further, more detailed yet generic level of representation for the solutions has now been developed [27]. The three levels of solution representation – topological, spatial and physical – are illustrated in Fig. 11.9 (taken from Ref. [27]). The figure represents a spatial variant of the door-latch concept shown in Fig. 11.6.

At the topological level, the concept is described by a chain of elements, each of which has a name, an input kind and an output kind that it accepts. For instance, a tie rod takes a force and associated motion as input and transmits the same as output; see Fig. 11.9a. At this level, a solution has information about the kinds of inputs and outputs and the connectivity among the elements that constitute the solution.

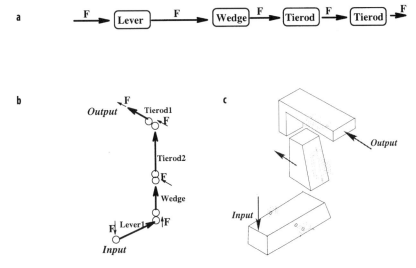

Fig. 11.9 Three levels of representation of a solution (from Ref. [27]). **a** Topological solution; **b** spatial configuration of the solution; **c** physical solution.

At the spatial level, a distinction is made between the spatial variants of an element. For instance, there are two variations of a tie rod at this level: tie rod1 transmits force and motion along its nominal axis of orientation in space, whereas tie rod2 does this perpendicular to its nominal orientation (see Fig. 11.9b). At this level, a solution has information on the relative position and orientation of the inputs and outputs, and the relative position and orientation of the elements.

At the newly added level – called the generic physical level – a solution has further details about its elements and interfaces. These include the generic shape of each element, the physical nature of its interfaces, and spatial constraints imposed on the solution. For instance, the combination of tie rod1 and tie rod2 shown in Fig. 11.9c has a generic shape made of rectangular chunks of material, has a contact interface with the wedge, and is constrained to move only along a horizontal axis. A synthesis algorithm developed to operate at this level can transform a spatial solution, using compositional synthesis, into a variety of designs as potential alternative embodiments of the spatial solution [21].

The addition of this physical level and associated synthesis not only provides better visualisation of the concepts generated by FuncSION, but also enhances its potential for providing a wide variety of solutions. The latter has been tested by comparing the physical embodiments proposed by the software with those available in a number of product catalogues, *e.g.*, patents of corkscrew designs. As well as "predicting" most of these designs at the topological and spatial levels, the software can predict more than 80% of all the embodiments available, as well as novel, interesting ones, which further stresses its potential as a stimulant of creativity [28].

11.5.2.2 Three-dimensional representation of the solution space

In order to represent the space of solutions generated, a three-dimensional user interface is being implemented. This has three parts [24].

- A requirement specification tool is being implemented as a three-dimensional graphic representing the spatial relationships between the inputs and outputs.
- A browsing and clustering tool has been implemented (see Fig. 11.10) that utilises a two-dimensional "star" representation of clusters and is intended to aid the understanding of the grouping of functional solutions by similarity. A two-dimensional "sun-and-planet" graphic represents a cluster calculated by the algorithm, where the representative central member (called medoid) and the average distance of cluster members (planets) from that medoid are shown.
- In conjunction with this, a three-dimensional visualisation for displaying an overview of the solution space and spatial layouts of individual solutions is under development. This will display a three-dimensional representation of an individual solution, allow the manipulation of a three-dimensional viewpoint of these, and allow the underlying data to be used for further development and simulation.

11.6 Conclusions and further work

One of the crucial reasons why supporting generation and exploration of mechanical design concepts has been so hard is the essentially coupled geometrical nature of these designs. Abstraction poses significant challenges, owing to the potential loss of

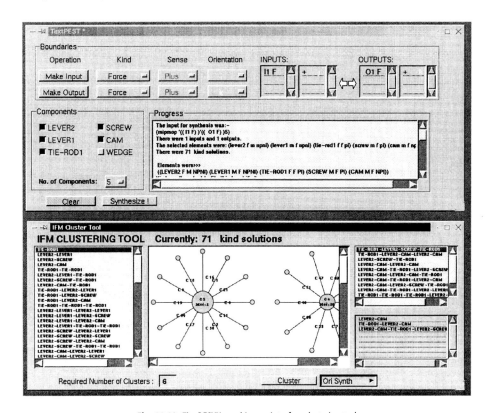

Fig. 11.10 The DESYN graphic user interface clustering tool.

geometric information that may be essential for the functioning of a mechanical device. We have tried to deal with this by synthesis at multiple levels of abstraction, where synthesis makes use of, at a given level, only those parts of the specification that are appropriate at that level. The other problem with compositional synthesis in general is that of managing complexity: a large number of potential solutions are generated, and reviewing and exploring them all is hard. This translates, in the context of our approach, to the following issue: how can designers be supported in exploring solutions without compromising their range? We feel that a major part of complexity arises from the conflict between the level of abstraction that allows high explorability and that which can generate a wide range of solutions [22]. Our approach to resolve this has been by:

- providing range by generating solutions at a high level of abstraction, while allowing visualisation at lower levels;
- pruning solutions as early as possible at each level, and clustering remaining solutions using cognitive notions of similarity, so that designers can get overview without having to explore each individual solution;
- allowing designers to navigate the solution space, so that new ideas are triggered even if the solutions proposed may not be complete or sufficient in solving the entire problem.

To summarise, the functional synthesis approach presented in this paper has, while retaining the underlying philosophy of using compositional synthesis of building blocks, evolved in two directions since its original development [10]. One is the addition of a further, more detailed level of representation and associated synthesis that enhance visualisation; the other is the development of a user interface and associated methods that make synthesis more efficient and exploration more manageable. Each has been implemented and tested using a number of case studies and experiments. On the whole, besides being able to generate a variety of ideas for the cases studied, which included most of the ideas generated by the designers, the system also consistently suggested a wide variety of other feasible and novel ideas that the designers did not generate. Although more validation is required to be conclusive, this provides substantial evidence that a designer can be assisted in the generative aspects of design using compositional synthesis approaches.

Future work is pursued in three areas: completion of the development of the three-dimensional user interface, application of the approach to other domains [29], and integration of FuncSION with the detailed embodiment software being developed in the Cambridge EDC [30].

Acknowledgements

The work presented in this chapter has been funded by the EPSRC, UK, Nehru Trust for Cambridge University, India, Cambridge Commonwealth Trust, UK, and National Ministry of Defence, Taiwan.

References

[1] Pahl G, Beitz W. Engineering design: a systematic approach. 2nd ed. London Berlin: Springer, 1996.
[2] Ullman DG. The evolution of function during design. In: Workshop on Function and Function-to-Form Evolution, Engineering Department, University of Cambridge, 1992; 117–36.
[3] Nidamarthi S, Chakrabarti A, Bligh TP. the significance of co-evolving requirements and solutions in the design process, Proceedings of the International Conference on Engineering Design (ICED97), Tampere, Finland, vol. 1, 1997; 227–30.
[4] Berliner C, Brimson JA, editors. Cost management for today's advanced manufacturing: the CAM-I conceptual design. Boston: Harvard Business School Press, 1988.
[5] Fricke G. Experimental investigation of individual processes in engineering design. In: Cross N, Doorst K, Roozenburg N, editors. Research in design thinking. Delft University Press: 1992; 105–9.
[6] Ehrlenspiel K, Dylla ND. Experimental investigation of the design process. In: Proceedings of the International Conference on Engineering Design, Harrogate, UK, vol. 1, 1989; 77–95.
[7] Heylighen A, Verstijnen IM. Exploring case-based design in architectural education. In: Proceedings of 6th International Conference on AI in Design (AID00), Worcester, USA, 2000; 413–32.
[8] Blessing LTM, Chakrabarti A, Wallace KM. A design research methodology. In: Proceedings of the International Conference on Engineering Design (ICED95), Praha, 1995; 50–5.
[9] Chakrabarti A. Designing by functions. Ph.D. thesis, University of Cambridge, UK, 1991.
[10] Chakrabarti A, Bligh TP. Functional synthesis of solution-concepts in mechanical conceptual design. Part I: knowledge representation. Int J Res Eng Des 1994;6(3):127–41.
[11] Reinschke KJ. Multivariable control: a graph theoretic approach. Thoma M, Wyner A, editors. Lecture notes in control and information sciences, vol. 108. Delft: Springer, 1988.
[12] Chakrabarti A, Bligh TP. An approach to functional synthesis of solutions in mechanical conceptual design, part II: kind synthesis. Res Eng Des 1996;8(1):52–6.

[13] Chakrabarti A, Bligh TP. An approach to functional synthesis of solutions in mechanical conceptual design, part III: spatial synthesis. Res Eng Des 1996;8(2):116-24.
[14] LispWorks reference manual. Harlequin Limited, UK, 1991.
[15] Chakrabarti A, Abel C. The mobile arm support project: evolution of the management, design, tools, and documentation processes. In: Proceedings of the 16th Annual International Conference of the IEEE Engineering in Medicine and Biology Society Conference, Baltimore, MD, USA, vol. 1, 1994; 486-7.
[16] Bauert F. The mobile arm support phase I: design, manufacture, testing, software tools. Technical report no. CUED/C-EDC/TR 13, Cambridge University, UK, 1993.
[17] Chakrabarti A, Bligh TP. An approach to functional synthesis of design concepts: theory, application, and emerging research issues. Artif Intell Eng Des Anal Manuf 1996;10:313-31.
[18] Chakrabarti A. Increasing efficiency of compositional synthesis by improving the database of its building blocks. Artif Intell Eng Des Anal Manuf 2000;14:403-14.
[19] Pohl I. Bi-directional search. In: Meltzer B, Mitchie D, editors. Machine Intelligence 6. Edinburgh University Press: 1971; 127-40.
[20] Chakrabarti A. Improving efficiency procedures for compositional synthesis by using bi-directional search. Artif Intell Eng Des Anal Manuf 2001;15:67-80.
[21] Liu Y-C. A methodology for the generation of concepts in mechanical design. Ph.D. thesis, Department of Engineering, University of Cambridge, 2000.
[22] Chakrabarti A, Tang MX. Genenerating conceptual solutions on FuncSION: evolution of a functional synthesiser. In: Gero JS, Sudweeks F, editors. Proceedings of Artificial Intelligence in Design '96 (Proceedings of the Fourth International Conference on AI in Design, Stanford, USA). Kluwer Academic, 1996; 603-22.
[23] Liu Y-C, Chakrabarti A, Bligh TP. Using engineering heuristics for efficient exploration of design solution space. Des Stud J 2001;submitted.
[24] Langdon PM, Chakrabarti A. Browsing a large solution space in breadth and depth. In: Proceedings of the International Conference on Engineering Design (ICED99), Munich, vol. 3, 1999; 1865-8.
[25] Chakrabarti A. A framework for browsing a solution space in depth and breadth. Technical report CUED/C-EDC/TR62, Department of Engineering, University of Cambridge, January, 1998.
[26] Langdon PM, Chakrabarti A. Improving access to design solution spaces using visualisation and data reduction techniques. In: ICED01 Conference; Design Research-Theories, Methodologies, and Product Modelling, Glasgow, 2001; 379-86.
[27] Liu Y-C, Chakrabarti A, Bligh TP. A computational framework for concept generation and exploration in mechanical design: further developments of FuncSION. In: Proceedings of the Sixth International Conference on AI in Design (AID00), Worcester, July 2000; 499-519.
[28] Liu Y-C, Chakrabarti A, Bligh TP. Developing physical embodiments of concepts in mechanical design. Res Eng Des 2001;submitted.
[29] Chakrabarti A, Johnson AL, Kiriyama T. An approach to automated synthesis of solution principles for micro-sensor designs. In: Proceedings of the International Conference on Engineering Design (ICED97), Tampere, vol. 2, 1997; 125-8.
[30] Bracewell RHB, Johnson AL. From embodiment generation to virtual prototyping. In: Proceedings of the International Conference On Engineering Design (ICED99), Munich, vol. 2, 1999; 685-90.

Synthesis based on function–means trees: Schemebuilder

12

Rob Bracewell

Abstract This chapter describes the development of the Schemebuilder computer-supported methodology for the conceptual design of mechatronic and other multidisciplinary systems. This employed function–means trees, implemented using an artificial-intelligence-derived context management system, hierarchical schematic diagrams, and bond graph theory. Together, these allowed the generation of complete design *schemes* by creating, and forming combinations of, alternative partial design solutions. Quantitative dynamic simulation models were automatically created for the evaluation of the generated schemes. An illustrative example shows how the step-by-step application of Schemebuilder's bond-graph-based rules were used to synthesise, from first principles, a novel and eventually commercially successful concept for a remote telechiric manipulator.

12.1 Background

Schemebuilder is a computer-supported design methodology that was originally aimed at improving support for the early stages of mechatronic[1] systems design, but developed into a "virtual prototyping" design system of wider applicability. It was one of four original projects of Lancaster Engineering Design Centre (EDC) [2], set up in 1990 with support from the UK Science and Engineering Research Council, under the directorship of Michael French. On his retirement in 1992, the EDC directorship passed to John Sharpe and Schemebuilder was developed into a broad unifying theme of the EDC's research [3]. This chapter will not attempt to summarise the whole of that work, but instead will focus on the core topological synthesis and design context management concepts at its heart.

The Schemebuilder project started with a clearly identified need to address, but with few preconceived ideas as to what theoretical direction it should take. The following sections describe the sources of inspiration, the development and the integration of the key concepts of the design synthesis process that emerged.

---

[1] Mechatronics may be considered as the integration, at all levels and throughout the design process, of mechanical engineering with electronics and computing technology to give both functional interaction and spatial integration in components, modules, products and systems [1]. Common examples of products resulting from a mechatronic approach to design are camcorders and auto-focus cameras.

12.2 Key concepts of Schemebuilder

12.2.1 Hierarchical schematic diagrams

The multidisciplinary nature of mechatronics can create considerable complexity, even in relatively simple products. The most fundamental tool available to designers in dealing with complexity is hierarchical decomposition. As applied to the design process, Ringstad [4] suggests that its uses include creating manageable design tasks, and enabling parallel development, evaluation and combination of subsystem solutions. Furthermore, for many design tasks, it has been shown to be advantageous to perform hierarchical decomposition on the basis of function. Existing functional decomposition notations of particular relevance to mechatronics design are the function structures of Pahl and Beitz [5], and the data flow diagrams of structured systems analysis and design methods, such as the Yourdon structured method [6]. The former are intended for the conceptual design of mechanical systems, whereas the latter are commonly used in designing embedded control software. There are obvious similarities between the two, the chief difference being that in the former, flows of material, energy or signal may be defined between functions, whereas in the latter only signal flows are considered. Thus some hybrid of these two forms of hierarchical schematic diagram seemed a promising starting point for a method capable of addressing both mechanical and software aspects of mechatronics design. In electronics too, schematic diagrams have long been the basic representation used to support function-based design, with an increasing trend towards hierarchical schematics, to manage complexity and promote reuse of functional solutions. This reinforced the feeling that hierarchical schematics could form a suitable basis for an effective, integrated mechatronic design methodology.

12.2.2 Scheme generation by combination of alternative subsolutions

Schemebuilder took its name from French's view of the conceptual design stage as a process of generating and evaluating candidate schemes, the most promising of which should be selected for further embodiment [7]. He defined a scheme as an:

> outline of a solution to a design problem, carried to a point where the means of performing each major function has been fixed, as have the spatial and structural relationships of the principal components. A scheme should be sufficiently worked out in detail for it to be possible to supply approximate costs, weights, and overall dimensions, and the feasibility should have been assured as far as circumstances allow. A scheme should be relatively explicit about special features or components but need not go into much detail.

In addition to function structures, another feature of the Pahl and Beitz method [5] that strongly influenced the development of Schemebuilder was the use of tables of options (sometime called morphological tables) to generate schemes by enumerating combinations of alternative subsolutions. Hundal [8] had reported implementing such a facility in a prototype design support system based on Pahl and Beitz. However, there are two main drawbacks with his approach. Firstly, scheme generation using two-dimensional tables of options becomes very unwieldy when the function structure is hierarchical. Secondly, it only supports alternative solutions for performing a function, not alternative ways in which a function might be decomposed into subfunctions. These drawbacks were avoided by adopting Andreasen's elegant function–means tree concept [9].

12.2.3 Function–means trees

In essence, a function–means tree is a Pahl and Beitz style function hierarchy extended to allow the representation of both alternative decompositions and solutions for its functions. It does this by enforcing a strict alternation between function and means nodes in successive levels of the hierarchy, where a means can represent either a solution or a decomposition. Thus a function is a type of exclusive OR logic element – a valid scheme must contain exactly one of its child means. Conversely, a means is an AND element – a valid scheme containing it must also contain means for all of its child functions. Using this logic, a simple algorithmic procedure can enumerate all of the valid, complete, distinct schemes that can be generated from a particular function–means tree. The extended function logic diagrams of Sturges *et al.* [10] are closely related, but, instead of using two alternating types of node, the AND/OR logic is realised by two different types of arc.

As various authors have noted, it is useful to augment the basic function–means tree with information about the configuration of the functions and solution elements associated with each means. For example, Hansen [11] suggests that this information may be attached graphically, employing various levels of detail from simple concept sketches to detailed layout drawings. Alternatively, particularly if software support is available, it is attractive to combine the function–means tree with a hierarchical schematic editor. In this, every means in the tree has a separate, associated configuration view, showing the flows of energy, material and signal between subfunctions and physical entities contained by the means, and also those flows "pulled down" from the parent level of the tree. This was the approach adopted in Schemebuilder, and also by Malmqvist and Schachinger in their function–means tree implementation [12].

12.2.4 Artificial intelligence (AI) support for design context decomposition and recombination

The main research method applied in the Schemebuilder project was rapid software prototyping. The arguments in favour of this approach were perfectly summed up by Welch and Dixon [13]:

> Writing computer programs is an effective research method, because they require that the representations and reasoning processes be made explicit, which often leads to discovering holes in proposed models. Moreover, a computer program serves as an experimental tool. The data, search strategies, *etc.*, can be changed and the results noted. The outcome is not a computer program, but more importantly an understanding of how and why the program (and therefore model) works the way it does.

However, the effectiveness of this approach is heavily dependent on the suitability of the programming languages, software components and tools applied. Research projects such as the Edinburgh Designer System [14] had recently demonstrated the possibilities offered to design researchers by the toolbox of techniques emerging from AI research. With this in mind, the Lisp-based KEE environment was selected as arguably the most powerful and comprehensive AI and object-oriented software development system then available [2]. KEE proved a good choice, which provided another important piece of the Schemebuilder methodology jigsaw. This was the application of assumption truth maintenance system (ATMS)-based context management, in the form of the recently added KEEworlds system [15].

KEEworlds is an advanced AI facility that allows the creation of directed acyclic graphs (DAGs) of contexts or *worlds*, in which the facts known in a particular world comprise those inherited from its ancestors, plus and minus additions and deletions made locally. A graphical display was provided that automatically laid out the DAG of worlds for easy visualisation. Another type of node that could be added to the DAG of worlds was termed an *exclusion set*. The purpose of an exclusion set was to declare to the context management system that any world with two or more of its children as ancestors was to be regarded as inconsistent.

It was soon apparent that KEEworlds was very well matched to the task of providing computational support for function–means synthesis of schemes. Each means or scheme might be represented by a world, and each function by an exclusion set. The multiple parents of a scheme world would be the tree leaf means of which the scheme was composed. The in-built exclusion set logic ensured that a scheme containing two or more means inadvertently attempting to perform the same function would automatically be marked as invalid. Figure 12.1 is an annotated screen dump of the resulting Schemebuilder function–means synthesis implementation, showing the partial tree and generated schemes for a device for controlled intravenous drug infusion. This software was demonstrated "live" at the ICED'93 conference in The

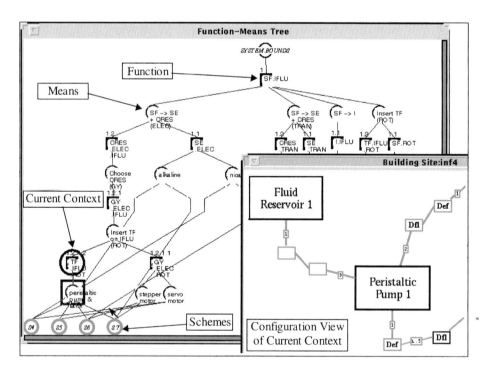

Fig. 12.1 Screen dump of the Schemebuilder function–means tree synthesis system implemented using KEEworlds.

Hague, on an early Sun workstation laptop. All valid complete scheme contexts were generated automatically by forming combinations of the manually created alternative means in the tree. The ATMS-based design context management was used in conjunction with object-oriented representations of functions, components, ports and links, to implement hierarchical schematic editing of means configurations, in a "building site" window. This is also shown in Fig. 12.1, displaying the configuration of the highlighted means labelled as the current context. This is performing a power transformation function between the rotational and incompressible fluid domains. The means consists of two components, a fluid reservoir and a peristaltic pump connected to rotational and fluid links that have been pulled down from the parent level. The component topologies in the generated schemes were automatically inferred by inheriting and combining configuration information from all of the scheme's ancestor means in the tree. An example generated scheme topology is shown in the upper window of Fig. 12.2.

Apart from flows of signal, energy and material, other types of design context information can be efficiently managed using this ATMS/worlds approach. For example, generalised reference connections of energetic components, such as electrical earth wires, hydraulic tank lines, and mechanical mountings, are awkward to manage using pure function-based schematics. However, using worlds, a global reference can be defined for each energy type at the top level context, and inherited so that component reference ports are connected to it automatically unless overridden in a lower level context, *e.g.*, where components are mounted on a floating subassembly rather than direct to a main chassis. Similarly, environmental information may be specified at the top level design context and inherited everywhere except where special conditions apply, *e.g.*, inside a hermetically sealed enclosure, in which context it is overridden. Requirements too, may be specified at the top level and

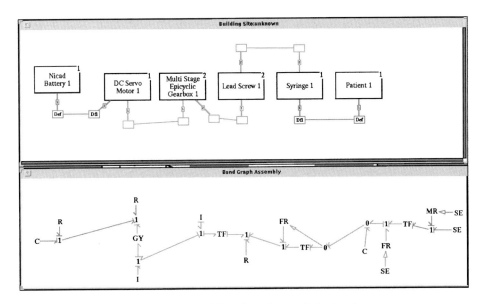

Fig. 12.2 A generated drug infuser scheme and its bond graph.

flowed down to components that meet them. Weight budgets or critical parameter allocations, such as transformer ratios and amplifier gains, can be divided among and inherited by subfunction contexts [16], as in the allocation arithmetic concept of Sturges et al. [10].

It is noteworthy that the downward direction of propagation of configuration and contextual information in the Schemebuilder function–means tree is opposite to that in Hansen's concept [11]. In his decomposition procedure, after selecting a means, the next step of is to *"Adjust the configurations on all levels above the current means level, by adding this new insight"*. This is valid in Hansen's case because he is not using the tree to support morphological scheme generation, his decomposition procedure requiring the immediate selection of a single "best" means from those generated for any function. The selected means is the only one to be elaborated further; thus the tree only generates one complete design solution. The disadvantage of this approach is that, as the Pahl and Beitz method acknowledges [5], it is often difficult to evaluate subsolutions or means without combining them to form complete concept variants or schemes.

A contemporary of Schemebuilder was Bañares-Alcántara and Lababidi's KBDS system; this explored closely related ideas, but in support of chemical and process-engineering design [17]. It also used ATMS to support the decomposition of design problems, together with the generation and valid recombination of partial solutions. Instead of using KEEworlds, KBDS was written in CLOS and interfaced externally to de Kleer's original C language ATMS implementation [18]. Several years earlier, the Edinburgh Designer System (EDS) used ATMS to control inferencing in hierarchies of mechanical design task contexts [14].

12.2.5 Computer support for simulation and evaluation of schemes

The obvious potential of Schemebuilder's morphological scheme generation system to create large numbers of schemes made it imperative to have efficient facilities for their evaluation and filtering. Otherwise, the best scheme was liable to be lost in a sea of mediocre ones. Because the schemes were synthesised as topologies of known components, it was desirable to interface with a modelling and simulation system to allow automatic generation of behavioural models as a basis for evaluation. Direct application of mainstream block-oriented simulators, such as Simulink, was apparently ruled out by the general dissimilarity of topology between the structure of energy transforming systems and their equivalent block diagram model [19]. However, two promising approaches to *composable simulation* existed in the form of qualitative process theory [20], and bond graphs [21,22]. Both were being actively investigated by various research groups to support modelling of, and reasoning about device function and behaviour [13,23–26]. Schemebuilder adopted bond graphs, as these offered the possibility of high-fidelity quantitative simulation, coupled with their potential for supporting useful symbolic reasoning about the model's causal structure [27,28]. Figure 12.2 shows the bond graph modelling facility implemented. The upper window shows a scheme for the drug infuser generated from the function–means tree in Fig. 12.1. The lower window shows the bond graph automatically composed from that scheme. Each component could be represented by a series of models of increasing complexity, the level index of which can be seen in the top right corner of each component block. As in the MAX system [29], the appropriate level of model was the choice of the designer and could be changed at any time.

12.2.6 Bond-graph-based functional synthesis

Schemebuilder in its early stages was envisaged as recording the development of a function–means tree, and generating schemes from it [30], rather than actively guiding the functional decomposition and embodiment process. This changed radically in late 1992 on Michael French's succession as Director of Lancaster EDC by John Sharpe. A practising designer of considerable experience, Sharpe's approach to design had years previously been profoundly influenced by learning about bond graph theory from its originator, Henry Paynter. A decade before its widespread discovery by the design research community [25,26], Sharpe had developed the concept of bond graph synthesis and been routinely using it, in manual form, as a practical design tool. It also formed an important topic of his engineering design methodology course at the University of London. Sharpe's earliest (1978) paper on the bond graph synthesis of force-feedback telechiric manipulators [31] clearly describes the thinking involved. However, this being a short paper written primarily to report a particular design rather than as a piece of generic design research, the whole synthesis process is not spelled out in detail. Hence Section 12.3 will follow it through step by step to demonstrate the application of the bond-graph-based synthesis rules later implemented in Schemebuilder [32].

12.3 Design synthesis example: telechiric hand

This example, though simple, shows how bond graph synthesis generated the working principle for a totally new class of robotic manipulator with many unique properties and excellent dynamic behaviour. Following the construction of a number of two and three degrees of freedom systems over a period of 15 years, Sharpe's research student, K.V. Siva, was instrumental in applying the synthesised design to a commercial seven-axis robotic manipulator used in the nuclear industry.

The starting point was a simple, one degree of freedom telechiric "hand". This may be regarded as having the required behaviour of a pair of pliers, but where the two halves are to be split and made remote from one another. Thus the function may be expressed as a transformer (TF) in which both ports are in the mechanical translation energy domain. The desired causality is that the operator's hand determines the flow, and the resistance to crushing of the work-piece determines the effort required, which is reflected back to the operator. Figure 12.3 shows the scheme that was synthesised from this required function [31].

Figure 12.4 shows the synthesis process, with the sequence of applied rules in the left-hand column and the transformed bond graph function structure in the right-hand column. For a full introduction to bond graph concepts and notation, see Ref. [22].

- Rule *a* replaces one of the two mechanical translation bonds with a TF. Rotation is selected in a free choice of energy domain of the bond between the two TFs.
- Rule *b* replaces the rotational bond with two gyrators (GYs) in series. Again, there is a free choice of energy domain for the bond between them. Electrical is selected.
- Rule *c* replaces the electrical bond with modulated sources of electrical effort and flow (MSE, MSF) and two signal flows in opposite directions. The flow i of the MSE is sensed and used to modulate the MSF, while the effort v of the MSF

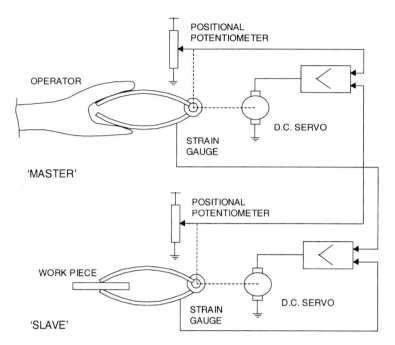

Fig. 12.3 Experimental telechiric hand (after Sharpe [31]).

is sensed and used to modulate the MSE. Thus the electrical bond has been replaced by its equivalent effort and flow signals, the directions being determined by the causality of the bond. This may be favourable if the two halves of the manipulator are widely separated, since no power then needs to be transmitted between them.

- Rule *d* states that instead of sensing the effort variable on one side of a GY, it is equivalent to sense the flow variable on the other side, or *vice versa*. Applying this twice results in the MSE being modulated by the angular velocity ω on the operator side, while the MSF is modulated by the torque τ on the work-piece side. Exploiting this choice of which variables to sense and which to estimate is an important principle of mechatronics design [33].
- Rules *e* and *f* apply the principle of closed-loop control to synthesise an MSF and MSE respectively from a source of effort SE, a modulated resistor MR, a flow or effort sensor and an unspecified controller block CONT.
- Rule *d* is applied twice again, followed by two applications of rule *g*, which states that instead of sensing the effort variable on one side of a TF, it equally may be sensed on the other side. The same is true for the flow variable. Thus, on the operator side the force felt by the operator *f* may be sensed rather than the electrical current *i* of the GY. On the work-piece side, the angular velocity ω may be sensed rather than the voltage *v*, and the crushing resistance force *f* rather than the GY torque τ.
- Rule *h* states that, since the controller blocks CONT are unspecified, the first block inside on any input could be an integration with respect to time INT, so that can be pulled outside CONT as a separate function. Applying this twice,

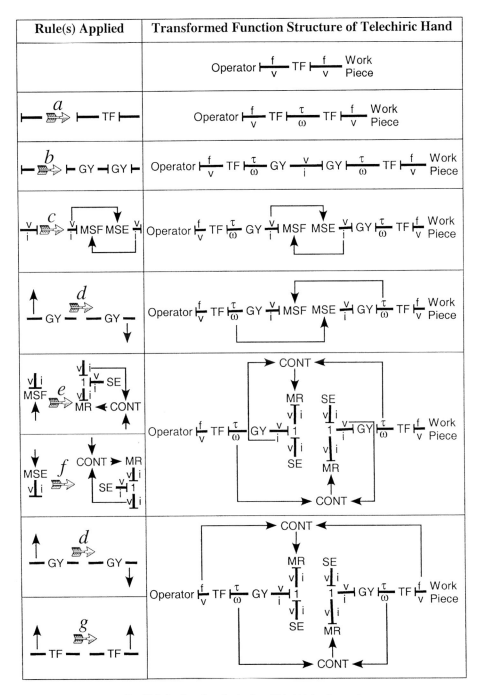

Fig. 12.4 Bond graph synthesis of novel telechiric hand concept.

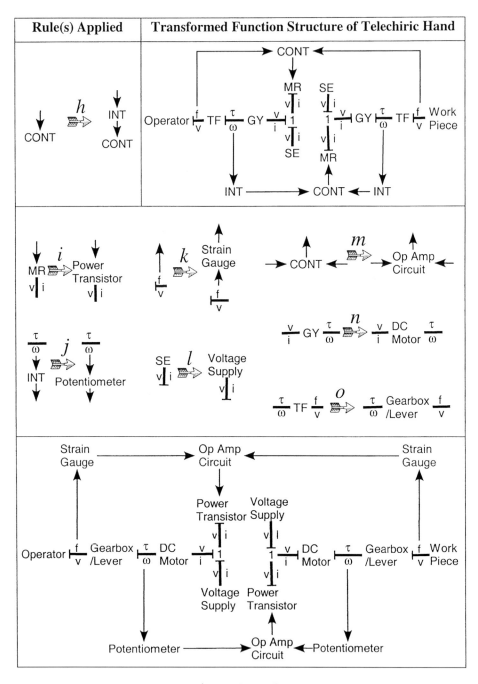

Fig. 12.4 (continued)

both GY angular velocities are integrated before passing them as inputs to the lower CONT. In other words, angular-position sensors can be used instead of angular-velocity sensors. Principles for selecting and designing such controllers were later added to Schemebuilder by Counsell [33].

This is the final state of the function structure, from which each of the functions may be directly embodied by the components shown in rules *i* to *o*, giving the complete scheme shown, which is identical to that reported in Ref. [31].

Of course, if this process were to be combined with the development of a function–means tree, changing the rules selected, their order of application, and the free choices of new intermediate energy domains would have led to a host of alternative schemes. For example, on application of rule *a*, the choice of incompressible fluid as the intermediate energy domain would have suggested a hydraulic-brake-like device. The tree can be used to record the entire design space explored, with the rationale behind the choices taken and rejected.

12.4 Implementation of function–means-based synthesis

The next step in the research was to integrate the bond-graph-based functional synthesis techniques, described in Section 12.3, with the existing function–means tree, context management and scheme generation system. The decision was taken at this point to discontinue the use of KEE and instead adopt the CLIPS object/rule-based software development tool [34]. Among other factors in that decision was the growing conviction that, to be effective, function–means development would need to support both function sharing and reuse, which might best be achieved by transforming the tree into a particular form of DAG [35]. The closed and inflexible nature of KEE-worlds would have made this impossible, but the open-source CLIPS, which also has a truth maintenance system, was suitable for the task.

The belief was that the most effective approach to computer-based design tools is for the computer to provide decision support and allow the human designer to supply the judgement. Thus the designer controlled the generation process, choosing the working context to focus on, and selecting which rule to apply from those applicable [35]. Context-sensitive hypermedia guidance was provided in the form of component data sheets, application notes and principles of good design, and an audit trail created of the designer's actions [36].

Like the function decomposition hierarchies for hydraulic circuit synthesis of Kota and Lee [37], the knowledge as to how functions can be decomposed or performed by known components was stored in an abstraction hierarchy, the functional embodiment knowledge-base. However, the two approaches were rather different. In Kota and Lee's case it was feasible for them to predefine function–means decompositions of their limited number of hydraulic domain-specific functions for direct reuse in a new design. To deal with Schemebuilder's coverage of a wide range of functions, a different approach was required. Its hierarchy was a partly bond-graph-based function taxonomy, with energy and material flows and signal carrier types progressively defined towards the leaves of the tree. This allowed the storage of functional decomposition rules at a high level of abstraction, with energy types undefined, then inherited into all appropriate energy domains. Figure 12.5 shows how rules stored in the functional embodiment knowledge base were applied in developing the function–means tree for the conceptual design of a pilotless aircraft fin

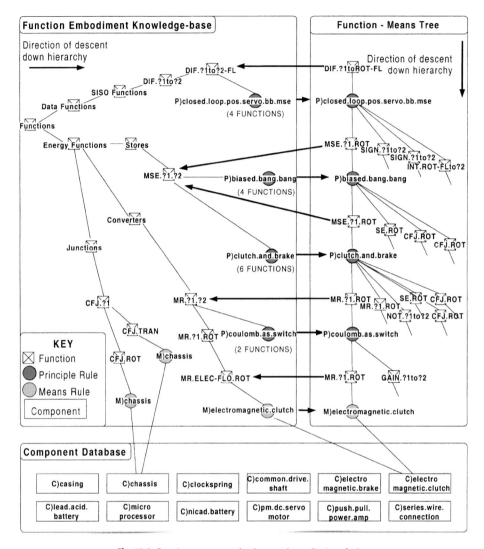

Fig. 12.5 Function–means tree development by application of rules.

actuator (described more completely in Ref. [35]). Successive rules defined independently of energy type are instantiated into the rotational domain. Another important feature was the ability of a single component type, such as "chassis", to appear in different means, embodying different functions at different points in the tree. This allowed components to be applied in alternative and potentially unusual ways.

Unfortunately, after this system had been demonstrated by graphical knowledge modelling and manual testing of the synthesis logic, a combination of circumstances led to the departure of the entire original research team before its software implementation could be completed. Development did continue, however, leading to the release of a package that has been used in mechatronics design teaching and for industry-linked projects [38]. Functional synthesis and scheme simulation were fully implemented, as was a new knowledge base of mechatronic and servo controller design principles [33]. However, lack of resources prevented the implementation of design context management and morphological scheme generation. Function–means trees reverted to something similar to Hansen's concept [11], where only one scheme is generated at a time.

The author, meanwhile, continues to research the function–means approach to design synthesis, context management, rationale capture and reuse [39].

References

[1] Buur J. A theoretical approach to mechatronics design. Ph.D. thesis, Institute for Engineering Design, Technical University of Denmark, 1990.
[2] Bracewell RH, Chaplin RV, Bradley DA. Schemebuilder and layout-computer based tools to aid in the design of mechatronic systems. In: IMechE Conference on Mechatronics – The Integration of Engineering Design, Dundee, 1992; 1–7.
[3] Bracewell RH, Sharpe JEE. Functional descriptions used in computer support for qualitative scheme generation – Schemebuilder. Representing functionality in design [special issue] Artif Intell Eng Des Anal Manuf 1996;10(4):333–46.
[4] Ringstad P. A comparison of two approaches for functional decomposition-the function/means tree and the axiomatic approach. In: 11th International Conference on Engineering Design (ICED 97), Tampere, Finland, 1997.
[5] Pahl G, Beitz W. Engineering design. The Design Council, 1984.
[6] Yourdon E. Modern structured analysis. Prentice-Hall, 1989.
[7] French MJ. Conceptual design for engineers. London: Design Council, 1985.
[8] Hundal MS. A systematic method for developing function structures, solutions & concept variants. Mech Mach Theory 1990;25(3):243–56.
[9] Andreasen MM. Syntesemetoder på systemgrundlag. Ph.D. thesis, Lund Technical University, Lund, Sweden, 1980.
[10] Sturges RH, O'Shaughnessy K, Reed RG. A systematic approach to conceptual design. Concurr Eng Res Appl 1993;1:93–105.
[11] Hansen CT. An approach to simultaneous synthesis and optimization of composite mechanical systems. J Eng Des 1995;6(3):249–66.
[12] Malmqvist J, Schachinger P. Towards an implementation of the chromosome model-focusing the design specification. In: 11th International Conference on Engineering Design (ICED 97), Tampere, Finland, 1997.
[13] Welch RV, Dixon JR. Guiding conceptual design through behavioural reasoning. Res Eng Des 1994;6(3):169–88.
[14] Logan B, Millington K, Smithers T. Being economical with the truth: assumption-based context management in the Edinburgh designer system.: In First International Conference on Artificial Intelligence in Design, Edinburgh, UK, 1991.
[15] Filman RE. Reasoning with worlds and truth maintenance in a knowledge-based programming environment, Commun ACM 1988;31(4):382–401.
[16] Yan XT, Sharpe JEE. Unified dynamic mixed mode simulation of mechatronic product design schemes. In: International Workshop on Engineering Design, Lancaster University, 1994; 259–80.
[17] Bañares-Alcántara R, Lababidi HMS. Design support systems for process engineering – II: KBDS: an experimental prototype. Comput Chem Eng 1995;19:279–301.
[18] De Kleer J. An assumption-based TMS. Artif Intell 1986;28:127–62.
[19] The Modelica Association. Modelica – a unified object-oriented language for physical systems modeling: tutorial accessed 18 April 2001, from http://www.Modelica.org/current/ModelicaTutorial14.pdf, v1.4, 2000.
[20] Forbus KD. Qualitative process theory. Artif Intell 1984;24(3):85–168.
[21] Paynter HM. Analysis and design of engineering systems. Cambridge (USA): MIT Press, 1961.

[22] Karnopp DC, Margolis DL, Rosenberg RC. System dynamics: a unified approach. Chichester: Wiley, 1990.
[23] Iwasaki Y, Vescovi M, Fikes RE, Chandrasekaran B. Causal functional representation language with behavior-based semantics. Appl Artif Intell 1995;9:5–31.
[24] Tomiyama T, Umeda Y, Ishii M. Knowledge systematization for a knowledge intensive engineering framework. In: IFIP WG 5.2 Workshop on Knowledge Intensive CAD, Espoo, Finland, 1995; 55–80.
[25] Ulrich KT, Seering WP. Synthesis of schematic descriptions in mechanical design. Res. Eng Des 1989;1(1):3–18.
[26] Finger S, Rinderle JR. A transformational approach to mechanical design using a bond graph grammar. In: Design Theory and Methodology – DTM'89, DE-vol. 17, 1989; 107–15.
[27] Bracewell RH, Bradley DA, Chaplin RV, Langdon PM, Sharpe JEE. Schemebuilder: a design aid for the conceptual stages of product design. In: 9th International Conference on Engineering Design, The Hague, 1993; 1311–8.
[28] Yan XT, Sharpe JEE. Reasoning and truth maintenance of causal structures in interdisciplinary product modelling and simulation. In: International Workshop on Engineering Design, Lancaster University – Ambleside Campus. London: Springer, 1995; 405–25.
[29] De Vries TJA, Breedveld PC, Meindertsma P. Polymorphic modelling of engineering systems. In: International Conference on Bond Graph Modelling ICBGM'93, La Jolla, CA. The Society for Computer Simulation: 1993; 17–22.
[30] Bradley DA, Bracewell RH, Chaplin RV. Engineering design & mechatronics-the Schemebuilder project. Res Eng Des 1993;4(4):241–8.
[31] Sharpe JEE. Application of bond graphs to the synthesis and analysis of telechirs and robots. In: 3rd Symposium on Theory and Practice of Robots and Manipulators, Third International CISM-IFFToMM Symposium, Udine, Italy. Amsterdam: Elsevier, 1978; 217–27.
[32] Bracewell RH, Sharpe JEE. The use of bond graph reasoning for the design of interdisciplinary schemes. In: International Conference on Bond Graph Modeling and Simulation (ICBGM'95), Las Vegas, vol. 27(1). SCS Publishing, 1995; 116–21.
[33] Counsell JM. Design principles for mechatronic systems. In: Computer Aided Conceptual Design'97: Lancaster International Workshop on Engineering Design, Lancaster University, 1997; 17–26.
[34] Giarratano J, Riley G. Expert systems: principles and programming. Boston (MA): PWS Publishing, 1994.
[35] Bracewell RH, Sharpe JEE. Computer-aided methodology for the development of conceptual schemes for mixed energy-transforming and real-time information systems. In: International Workshop on Engineering Design, Lancaster University, 1994; 79–94.
[36] Langdon PM, Cheung HF. A hypertext documentation system for Schemebuilder: the 'mechatronics' book. In: Joint Hungarian–British Mechatronics Conference, Budapest, 1994.
[37] Kota S, Lee CL. A functional framework for hydraulic systems design using abstraction/decomposition hierarchies. In: ASME International Computers in Engineering Conference, Boston, MA, 1990; 327–40.
[38] Porter I. Schemebuilder mechatronics. In Engineering Design Conference '98: Design Reuse, Brunel University, 1998; 561–8.
[39] Bracewell RH, Wallace KM. Designing a representation to support function–means based synthesis of mechanical design solutions. Accepted for International Conference on Engineering Design, ICED 01 Glasgow, August 21–23, 2001.

Design processes and context for the support of design synthesis   13

Ralf-Stefan Lossack

Abstract Artefacts are man-made objects that, to an increasing degree, have a determining influence on our world alongside natural systems. We live in geometrically formed houses made of stone and concrete, glass and metal, and by making use of heating, air-conditioning and lighting we can create living conditions that make us independent of external conditions. If one considers the constitutive role played by engineering, it is amazing how little research has actually taken place in the area of designing, of synthesising artefacts. This chapter discusses the problem of synthesising artefacts by proposing an integrated concept of supporting designing. The concept tries to find answers on how to describe indeterministic design processes, knowledge about design objects and how to model them in a context. The concept proposed is the result of ongoing research, and most of it is a compilation of research results from about the last 8 years. The chapter first gives an introduction and overview of the design process to state the assumptions of the conceptual framework in which the whole concept fits in. Then the concept of solution patterns is described, which build the basis of representing artefact and process knowledge. Design working spaces are the means to support synthesis and represent contextual information. Finally, a design system that was used to implement the concept is presented.

13.1 Introduction and overview of the design process

Proposals for methodological approaches to design were first made in the 19th century. One of the first was made by Reuleaux and Moll [1], with a process plan for the synthesis of kinetic machines. During the 1940s the studies of Wögerbauer [2] had a great impact, and Zwicky's morphological chart [3] has frequently been quoted and used for an algorithmic approach. Since the 1960s, scientists have made diverse attempts in the area of mechanical design to investigate the design process in a scientific manner and work out prerequisites for making it possible to view the design process from a more theoretical viewpoint, thus building a basis for design synthesis. These efforts have led to the design methodologies that are available today and provide a rich source of methods and rules that describe a systematic design process in the form of activity models [4–12]. These activity models subdivide and structure the design process into individual, systematic *steps* with defined *work results*.

Regarding a more stringent semantic definition, terminology and its use is marked by a rather close correlation to application areas, such as mechanical products, as well as by a variety of terms and an accompanying lack of clarity and standardisation. Definitions in the area of design methodology in the German language can be found in Hansen [13], Müller [14], Rodenacker [15], Roth [5], Pahl and Beitz [10] as well as in Hubka [7,16,17], Grabowski and Rude [18] and Ehrlenspiel [19]. An

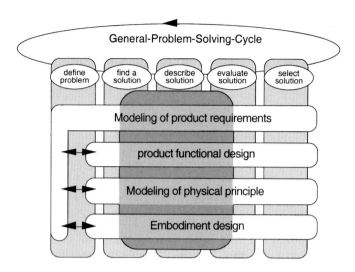

Fig. 13.1 Design process.

agreement in terminology was documented in the VDI-Guidelines [11,12,20], which build a basis for this article.

According to these guidelines, and from the viewpoint of a more computational model, the process of design can be subdivided into different stages (Fig. 13.1).

Each of the design levels depicted in Fig. 13.1 has different semantics and depends on the purpose one would like to achieve. The explication of the terminology will be described with a design example of a robot gripper (Fig. 13.2). The working principle of the robot gripper is as follows. A force is generated by a pneumatic energy source. The resulting force is then transmitted through a piston rod to a wedge, splitting the force into two resulting forces, which are applied, to the gripper jaws. The applied pressure causes a movement of the piston rod towards the jaws. As a consequence the wedge causes a turning motion of the jaws, which results in a gripping force as an output.

In this example there are four abstraction levels used to describe the design process. The requirement, functional, physical principle and embodiment level.

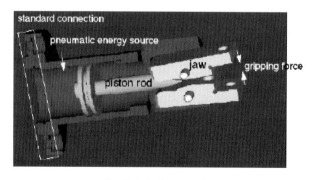

Fig. 13.2 Product example.

Design Processes and Context for the Support of Design Synthesis

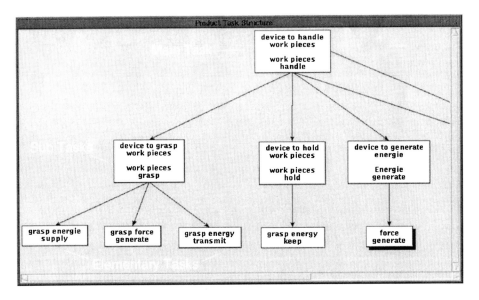

Fig. 13.3 A task structure of the robot gripper.

Requirement level. This abstraction level serves for the computational projection of the results won by the clarification of the design task. Here the *preconditions* of the design, the *to-be properties* of the future product, are specified. Requirements are the starting point[1] in design and represent the formulated *needs* of the customer and other sources, such as the product planning department.

Functional level. Functions describe the product behaviour at a non-physical level. There are several types of function. One type is the *task structure*. The task structure is defined by a *noun* and a limited set of *task verbs*. Nouns and task verbs establish basic elements. The relationships between tasks build a hierarchy. An example of a task structure of the robot gripper is shown in Fig. 13.3.

Here the task structure consists of a *main* task, the *subtasks* and the *elementary subtasks*. In this example a designer develops a *device* that is able to *handle work pieces*.[2] The main task is decomposed into *subtasks*. This level is finished when the designer has reached *elementary subtasks*.

Each elementary subtask is mapped onto an input–output function structure [21]. Figure 13.4 (1) shows an example of the function structure of the robot gripper. Here the energy type *pressure* (input) is *changed* into the energy type *force*. After that the force is *channelled*, *distributed* and *amplified* (output). In this example the function structure with its inputs and outputs, such as *pressure* as input or *change* as function verb, are the basic elements in this level.

Physical principle level. At this abstraction level the physical solution is described covering all information of the physical interrelationships of an artefact. In this sense the term physical means the description of the product behaviour based on physical

---

[1] A starting point does not imply that requirements have to be complete.
[2] Verb (noun).

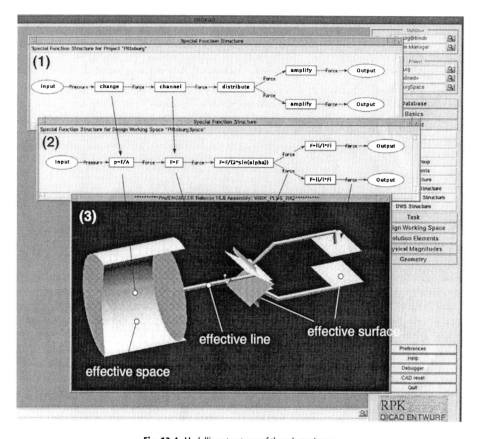

Fig. 13.4 Modelling structures of the robot gripper.

laws. The behaviour is established by physical effects, which are described by a *mathematical equation* and geometrical information, such as *effective lines, effective surfaces* and *effective spaces*. Figure 13.4 depicts the established physical principle structure (2) and the derived effective geometry structure (3) of the robot gripper.

Embodiment modelling layer. The embodiment is the most concrete of the abstraction levels (not depicted). In this layer the design process is completed by the geometrical specification of the product, such as points, lines, surfaces and design features, parts and assemblies.

A generalised approach supporting (indeterministic) design processes is formulated in a universal design theory (UDT) [22]. Here a set of modelling layers is assumed,[3] between which the transitions are defined by so-called solution patterns and thus support design synthesis. Solution patterns represent knowledge of an artefact, as well as of the design process itself.

---

[3] Not strictly four, such as the four modelling layers in the robot gripper example.

13.2 Solution patterns

Solution patterns are used to implement *problem-oriented design knowledge*. They establish concrete relations between elements in the modelling layers, *e.g.*, relationships between functions and physical principles. The knowledge is subdivided into object and process patterns describing artefact and process knowledge.

13.2.1 Artefact and process knowledge

Product knowledge or knowledge about artefacts is explicit knowledge on its use and its structure. Product knowledge consists of explicit properties of an artefact at different levels of abstraction. Product knowledge is described by means of object patterns.

> **Object pattern.** An object pattern O is an application-neutral description of a solution that can be adapted to certain types of problem. An object pattern directly supports the problem-solving process within design by a parameterised solution S together with a solution supposition P and a solution context $C: O = \{P,S,C\}$. P, S and C are sets of properties defined in the design layers. Object patterns explicitly and declaratively represent design knowledge in basic units.

An example of an elementary object pattern that supports a transition between the function and the physical principle layer might consist of the function P = "change pressure into force" and the physical principle S = "pressure effect", represented by the law $p = F/A$, where P is the proposition or supposition, S is the solution, p is pressure, F is force and A is area (see also the example in Fig. 13.4 (1) and (2)); and there are no constraints defined, $C = \{\}$; then the object pattern O is: O = {change pressure into force, $p = F/A$,}. Deduction is usually used as the reasoning process, because the knowledge is represented in the premises P and C and the conclusion S. Object patterns can be composed of other object patterns, thus building higher aggregations, such as machine elements, gearboxes, *etc*.

Higher aggregations are usually used in catalogue design and do not necessarily support creative or new design solutions. Therefore, a distinction is made between complete and incomplete patterns. The completeness/incompleteness depends on the degree of their granularity. Complete patterns contain knowledge of all abstraction levels; incomplete patterns contain knowledge of only some levels. For example, given n abstraction levels an incomplete pattern represents knowledge for the transition between $\{1, K, n - 1\}$ abstraction levels (Fig. 13.5). A complete object pattern, on the other hand, gives a solution for all abstraction levels. For example, a machine element or a gearbox can be interpreted as a complete object pattern because it includes all specifications on the function, physics, geometry, costs, *etc*.; a physical principle, such as a "lever effect", with a given function, such as "to amplify force", is an incomplete object pattern because it only provides knowledge of two abstraction levels, function and physical principle level.

Figure 13.5 shows a schematic example for complete and incomplete object patterns. There are n abstraction levels, each of which contains nodes. Each node represents a set of properties of a design description and each node is linked with other nodes by arrows. An arrow represents a reasoning process, which is defined as a mapping between nodes (via solution patterns). Object patterns are those knowledge units that map one node onto another, thus supporting a reasoning between one set

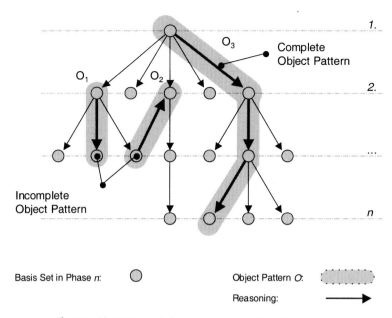

Fig. 13.5 Schematic example for complete and incomplete object patterns.

of design properties of an individual design stage and another set of design properties of another design stage. Incomplete object patterns, such as O_1 and O_2, only map nodes between some stages, whereas complete object patterns, such as O_3, describe the knowledge of all stages $(1, \ldots, n)$. Complete object patterns are also known under the notion of design prototypes [23,24].

Object patterns have a well defined structure in the design process, and in this sense they are unique, but they belong to the class of case-based reasoning approaches, *e.g.*, design prototypes [23], design patterns [25], *etc.*

What happens if there is no knowledge available, if there is no object pattern? Usually a good designer finds a way to reach his goal, to meet the requirements of an intended product. The designer usually has some strategies in mind, solution directions at hand, that might lead to a successful design. This kind of knowledge is intrinsically procedural and not declaratively organised, as the object patterns are. Such kinds of procedural pattern are called *process patterns*.

Process patterns describe knowledge of how to reach a goal starting from a given problem or task. Process patterns are used to implement object patterns following a certain design strategy or if there is no object pattern available.

Process patterns are units of knowledge that contain all the information necessary to transfer a design object from a solution state SS_i into a following solution state SS_{i+1}. This implies knowledge of all activities required for a given task, a design object description in the solution state SS_i and SS_{i+1} as well as the corresponding solution context. Process patterns explicitly and procedurally represent design knowledge in basic units.

Process patterns. A process pattern P is an application-neutral description of a non-empty set of modelling commands M_C together with a solution state SS and a solution context C; $P = \{M_C, SS, C\}$.

Example: M_C can represent design strategies, such as "design along⟨function flow⟩", or solution directions, such as "abstract⟨physical principle⟩", "decompose⟨function⟩", or be a very concrete set of commands, such as "insert⟨line⟩", "insert⟨effective surface⟩", *etc*. Process patterns also represent design activities, such as *evaluation of* and *find*, of the general problem-solving cycle in Fig. 13.1. Implemented in a computer-aided design (CAD) system, a process pattern returns M_C if there is a given solution state *SS* and a design context *C*. For a detailed set of design examples of M_C please refer to Ref. [26].

A specific design domain, such as computer science or mechanical engineering, uses specific domain knowledge, which means that a design object of a specific domain requires a corresponding design process. For example, a process pattern that leads to a *decomposition* of design patterns in computer science requires different knowledge than a *decomposition* of a gearbox in mechanical engineering. The process *to decompose* is the same in each domain; only the solution state and respective design properties differ. Therefore, object patterns and process patterns always belong together and complement each other; a generalised view of object and process patterns supporting a design solution is, therefore, called a *solution pattern* (Fig. 13.6).

Figure 13.6 shows the organisation of solution patterns in unified modelling language (UML) notation. According to this, a solution pattern is an *aggregation* of object and process patterns, each of which has premises and solutions. A solution pattern, *e.g.*, requires *P* and *C* in order to return the solution *S*, and a process pattern requires *SS* and *C* in the premises and returns M_C.

There is no solution; no design and no synthesis are possible without context. Therefore, modelling and representing context are respectively very important to support solution-finding processes and synthesis in design in general, but these issues are far from being solved [27,28]. Context can be seen as the conditions or circumstances under which something exists or is understandable. For example, if a word is out of context, then without the surrounding words it is not fully understandable. A proposal for defining contexts in mechanical engineering has been introduced using the concept of design working spaces [29]. Similar approaches, although for different aspects, have been followed elsewhere [30–35].

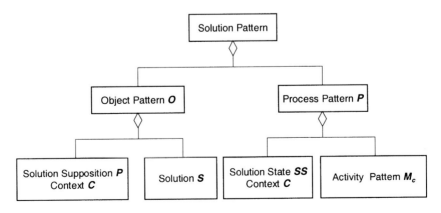

Fig. 13.6 Solution, object and process patterns.

13.3 Design working space

The concept of design working spaces makes use of system theory and defines a context as a system. Each context is described by properties formally defined by the elements of the design layers. The definition of a design working space provides the framework to support solution-finding processes, together with object and process patterns. Design working spaces are specified as follows.

1. A design working space consists of a *set of elements* and a *set of relationships* between the elements.
2. *Elements* of a design working space are properties of the design layers, such as product requirements, functions, physical principles. etc. *Relationships* between the elements are defined by magnitudes, such as energy, information, matter or force, torque, *etc.*
3. Every design working space can be subdivided into *independent* (*disjunct*) *subspaces*. If elements of different subspaces are grouped, then these subspaces are called overlapping design working spaces.
4. Every design working space and every subspace is defined by a system boundary; e.g., in the function layer the *system boundary* is defined by functional inputs and outputs; at the geometrical layer the boundary is specified by its envelope, giving the maximum geometric space for a design task.
5. A *system boundary* of a design working space has *one* or *more inputs* and *outputs*. If a design working space has no inputs/outputs then it is called a *closed* design-working-space, otherwise it is an *open* design-working-space.

An example of a design working space with the infolded effective geometry of the robot gripper is depicted in Fig. 13.7.

Figure 13.7 (1) depicts three design working spaces, each of which fulfils a certain function. The effective geometry in (2) is the content of the middle design working space. Its context C is described by the set of forces F and the set of geometric border patches G (free form surface (1)): $C = \{F,G\}$.[4] We can also say that the solution, the effective geometry in (2), is constrained by C. Having said this, a computational basis for *synthesis in design* can be supported by the concept of design working spaces together with object and process patterns. All information necessary to fulfil a process $P = \{M_C, SS, C\}$ and an object $O = \{P, S, C\}$ pattern is available.

The formalisation of object and process patterns was mainly the result of studies into a UDT. Five years ago, at IFIP [26] and KIC [29] the conceptual architecture of a design system that is able to support design synthesis by solution patterns and design working spaces was presented (Fig. 13.8).

The architecture consists of four components. The object model, task solution, process model and process control component. The object model component is responsible for the design description, such as requirements, functions, *etc.*, and can be interpreted as the product model of a design system. The task solution component describes modelling commands M_C, or sets of M_C, on a level that is usually known as an elementary CAD system function call, although this should not be con-

---

[4] Functional information, information about needs and requirements, *etc.* are also attached, but these are not shown. Consequently, C should be denoted $C = \{F,G,\ldots\}$.

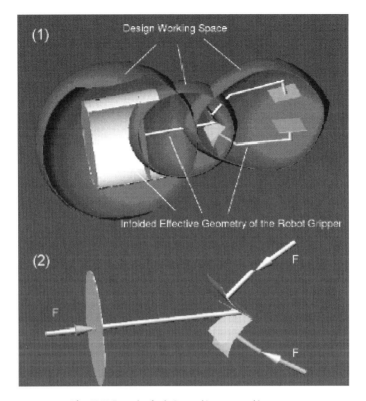

Fig. 13.7 Example of a design working space and its content.

fused with commercial CAD systems because commercial CAD only supports geometric information.

The process model component describes the dynamic model of the design process, specifying possible transitions between different design states of semantic units. It can be interpreted as the model that is responsible for computing *possible* paths of individual design steps.

The process control component is responsible for the design path leading from the requirement specification of an intended product to the product itself, thus supporting the whole design process. This component evaluates the state of the current design and computes on the basis of the process model the next design step that should be performed in order to transform a solution state SS_i^c into a following state SS_{i+1}^c. SSc denotes the state of a solution in a certain context. For a detailed description of the components, with examples, please refer to Refs [26,29].

The main assumption is that design can be modelled in a sequence of states. A state is a certain state of a product model at a certain point in time. In terms of designing, a state can be interpreted as a snapshot of a product model at a certain point in time. Usually, the whole state of a product model does not make sense because a designer has a certain focus on a design, a certain aspect of an interesting area of that design.

In this sense, a design state represents a certain state of a certain context. Design working spaces represent a context with a certain state and, therefore, design working

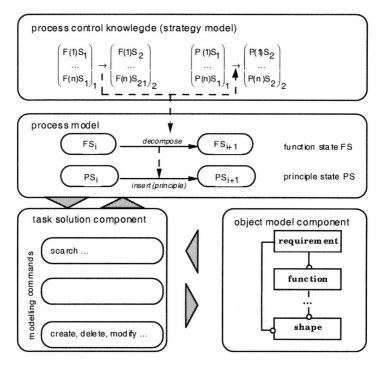

Fig. 13.8 Architecture of a design system to support design processes [26,29].

spaces are used to organise design processes and design knowledge semantically. Not only can this concept be used to support new designs, it can also be used for evaluating them.

Evaluation is closely related to having a view on something. In design, it is typical for a designer to be permanently changing the scope of the design in order to evaluate the results, which depend on a certain context [5,36]. An example of a gearbox design is shown in Fig. 13.9. In this example, a designer's viewpoint changes by clipping and zooming the drawing area. On the one hand, the designer has to keep in mind the overall gearing box in order to meet the requirements regarding size, function, general arrangement, spatial compatibility, *etc.*; on the other hand, the designer has to detail the bearing, otherwise the whole mechanism would not work. Here a design working space can provide a computational means for structuring the scope (context) of the design in order (half-automatically) to evaluate a design according to certain criteria. Criteria cover all aspects, including the product tasks, product functions, principles and effective geometry, and are not limited to the embodiment stage, as Fig. 13.9 might suggest. For an explanation of how context, scope and the means of design working spaces can support evaluation processes, the term context is used synonymously with the problem that has to be solved or evaluated.

Figure 13.10 gives an example of how to apply design working spaces to define a complex problem and then decompose it into subproblems until these subproblems can be evaluated. Figure 13.10 depicts an overall problem that is supposed as being too complex and, therefore, it cannot be evaluated. The idea is to identify and to mark

Design Processes and Context for the Support of Design Synthesis 223

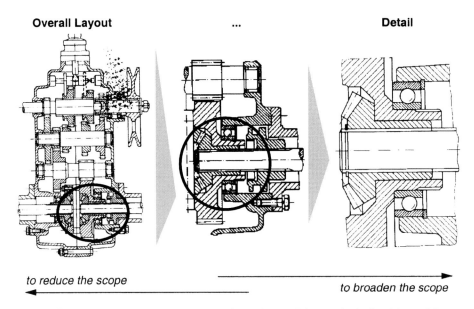

Fig. 13.9 Broadening and reducing the scope of a gear design with the example of technical drawing [5].

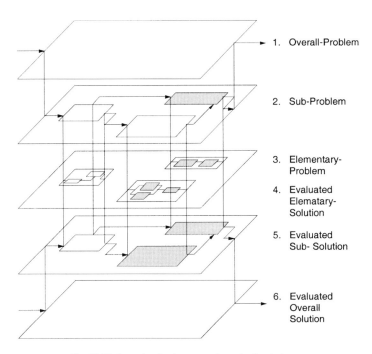

Fig. 13.10 General evaluation process for evaluating designs.

off the critical subspaces.[5] The overall problem has to be decomposed into subproblems and, consequently, subspaces. This step is repeated until an elementary problem is found. The elementary problem is evaluated and solved. Then the subproblem is evaluated on a higher level by evaluating the relations (constraints) between the solved elementary problems while regarding those elementary problems solved as black boxes. These steps are performed repeatedly until the overall solution is evaluated.

The concept of using design working spaces to support evaluation processes can be shown with examples of the design principles of Pahl and Beitz [10]. Detailed descriptions with examples of coded implementations of the knowledge base have been published [37].

13.4 The DIICAD Entwurf design system

DIICAD (dialog-oriented intelligent CAD) Entwurf (which stands for design-working-space modelling) is a design system with which the described concepts of solution patterns and design working spaces have been implemented. The system is divided into three parts: the *graphical user interface*, the *design modeller* and the *object-oriented database* (Fig. 13.11). The graphical user interface (depicted in Figs. 13.3 and 13.4), the design modeller and the application programming interface (API) and the geometric modeller are written in itcl;[6] the knowledge base has been written in CLIPS.[7] For the geometric engine, Pro/ENGINEER is used and controlled by the design modeller with ProDEVELOP (the programmable interface to Pro/ENGINEER).

The design modeller consists of two parts (Fig. 13.11): the *design modeller*, and the *knowledge-based system*. The design modeller consists of five modules, each of which represents a model of a certain stage in the design process, such as requirements modelling, function modelling, *etc.* (Figs. 13.3 and 13.4). The knowledge base is logically divided into a *knowledge acquisition* and an *interpretation* module.

13.5 Conclusion

Alongside natural systems, artefacts have a determining influence on our lives, and it is amazing how little research has taken place in the area of designing, of synthesising artefacts. This paper discussed the problem of synthesising artefacts by proposing an integrated concept of supporting designing. The concept has been implemented in the software system DIICAD, with which the validity of the proposed concept could be demonstrated. Owing to the complexity of the problem of designing and the large amount of knowledge (solution patterns) that is necessary to come up with reasonable design solutions, and owing to the poor performance of research prototypes in general, the whole design system has mainly been used for research

---

[5] A subspace is a critical subspace if the complexity of an overall problem can be reduced.
[6] itcl provides object-oriented extensions to Tcl, much as C++ provides object-oriented extensions to C.
[7] CLIPS is an expert system tool developed by the Software Technology Branch (STB), NASA/Lyndon B. Johnson Space Center. CLIPS is designed for writing expert systems.

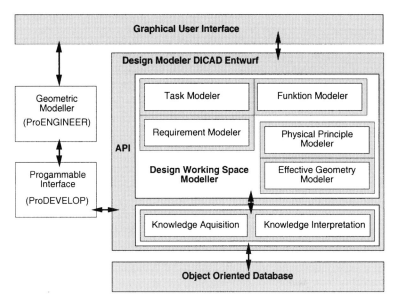

Fig. 13.11 Architecture of the DIICAD Entwurf design system.

and teaching. Parts of the system, such as the requirements and function modeller, could be refined and extended and are successfully used in industry.

13.6 Future work

The concept of a design working space as a means to represent contextual information, object patterns to represent artefact knowledge, and a layered model of designing with the requirements, functional layers, *etc.* have been demonstrated as very useful to support design processes. In the future an important research issue is how to represent design processes themselves. We believe that there exist elementary processes that can be composed to complex processes that represent design steps to synthesising an artefact. We denote them as process patterns. The representation of a process pattern is derived from the experience of object patterns. Design, or rather synthesis, is an indeterministic process and cannot be represented statically, *e.g.*, as a workflow. We believe that the concept of process patterns, together with design working spaces representing contextual information, is a further step in representing an indeterministic process and thus representing synthesis.

References

[1] Reuleaux F, Moll C. Konstruktionslehre für den Maschinenbau. Braunschweig: Vieweg, 1854.
[2] Wögerbauer H. Die Technik des Konstruierens. 2nd ed. Berlin München: Oldenburg, 1943.
[3] Zwicky F. Entdecken, Erfinden, Forschen im Morphologischen Weltbild. München, Zürich: Droemer-Knaur, 1966–1971.

[4] Matousek R. Konstruktionslehre des allgemeinen Maschinenbaus. Berlin Heidelberg New York: Springer, 1957 [reprint 1974].
[5] Roth K. Konstruieren mit Konstruktionskatalogen, Band 1. Bd 1: Konstruktionslehre. 2nd ed. Berlin Heidelberg New York: Springer, 1994.
[6] Hubka V. Theorie der Maschinensysteme. Berlin: Springer, 1973.
[7] Hubka V. Theorie der Konstruktionsprozesse. Berlin: Springer, 1976.
[8] Koller R. Eine algorithmisch-physikalisch orientierte Konstruktionsmethodik. VDI-Z 1973;115:147–52, 309–17, 1078–85.
[9] Pahl G. Die Arbeitsschritte beim Konstruieren. Konstruktion 1972;24:149–53.
[10] Pahl G. Beitz W. Konstruktionslehre. Methoden und Anwendung. 4th ed. Berlin, Heidelberg: Springer, 1997.
[11] VDI-Richtlinie 2221. Methodik zum Entwickeln und Konstruieren technischer Systeme und Produkte. Berlin: Beuth-Verlag, 1986 and 1992.
[12] VDI-Richtlinie 2222. Blatt 1: Konzipieren technischer Produkte. Düsseldorf: VDI Verlag, 1973 [2nd ed. Methodisches Entwickeln von Lösungsprinzipien. VDI-Verlag, 1995].
[13] Hansen F. Konstruktionswissenschaft-Grundlagen und Methoden. München Wien: Hanser, 1974.
[14] Müller J. Arbeitsmethoden der Techniksysteme. Berlin Heidelberg: Springer, 1990.
[15] Rodenacker WG. Methodisches Konstruieren. 4th ed. Berlin Heidelberg: Springer, 1991.
[16] Hubka V. Theorie der Maschinensysteme. Berlin Heidelberg New York: Springer, 1984.
[17] Hubka V, Eder WE. Einführung in die Konstruktionswissenschaft. Berlin Heidelberg: Springer, 1992.
[18] Grabowski H, Rude S. Methodisches Entwerfen auf der Basis zukünftiger CAD-Systeme VDI-Fachtagung "Rechnerunterstützte Produktentwicklung – Integration von Konstruktionsmethodik und Rechnereinsatz", Bad Soden 01.–02.03.1990. VDI-Berichte Nr. 812. Düsseldorf: VDI Verlag, 1990; 203–26.
[19] Ehrlenspiel K. Integrierte Produktentwicklung. München Wien: Hanser, 1995.
[20] VDI-Richtlinie 2223. Methodisches Entwerfen technischer Produkte. Düsseldorf: VDI Verlag, 1999 [Gründruck].
[21] Grabowski H, Lossack R-S, Kunze H. Functional modeling: representation of dynamic aspects in function structures. In: Kumar A, Russell I, editors. proceedings of the 12th International Florida AI Research Society Conference. AAAI Press, 1999; 399–404.
[22] Lossack R-S. A design process model "A concept of modeling object and process knowledge". In: Clarkson J, Blessing L, editors. Design Project Support Using Process Models, AID'00, Pre-Conference Workshop, Worcester Polytechnic Institute, MA, USA, 2000.
[23] Gero J. Design prototypes: a knowledge representation schema for design. AI Mag 1990;11(4).
[24] Qian L, Gero J. Function–behaviour–structure paths and their role in analogy-based design. In: Artificial intelligence for engineering design, analyses and manufacturing. Cambridge University Press: 1996.
[25] Gamma E, Helm R, Johnson R, Vlissides J. Design patterns. Reading (MA):Addison Wesley, 1995.
[26] Grabowski H, Lossack R-S, Weis C. Supporting the design process by an integrated knowledge based design system. In: IFIP WG 5.2 Workshop on Formal Design Methods for CAD, Mexico City, 1995.
[27] Bouquet P, Serafini LP, Brézillon P, Benerecetti M, Castellani F. Modeling and using context. In: Second International and Interdisciplinary Conference, CONTEXT'99, Trento, Italy. Springer, 1999.
[28] Workshop on Reasoning in Context for AI Applications, Orlando, FL, USA, 1999.
[29] Grabowski H, Lossack R-S, Weis C. A design process model based on design working spaces. In: Tomiyama T, Mäntylä M, Finger S, editors. Knowledge Intensive CAD-1, 26–29 September, 1995.
[30] Medland AJ, Jones A. Design for product assembly by controlling space ownership and associativity. In: Proceedings of the 6th Annual Conference on Design Engineering, Birmingham, UK. North Holland, 1992.
[31] Sittas E. 3D design reference framework. Comput Aided Des 1991;23:380–4.
[32] Chakrabarti A, Bligh TP. An approach to functional synthesis of mechanical design concepts: theory, applications, and emerging research issues. Artif Intell Eng Des Anal Manuf 1996;10(4).
[33] Linner S. Konzept einer integrierten Produktentwicklung. Dissertation. Springer: 1995.
[34] Grabowski H, Rude S, Schmidt M. Entwerfen in Konstruktionsräumen zur Unterstützung der Teamarbeit. In: Scheer AW, editor. Simultane Produktentwicklung, Forschungsbericht 4. München: Hochschulgruppe Arbeits- und Betriebsorganisation HAB e.V., gmft-Verlag, 1992; 123–59.

[35] Kramer S. Virtuelle Raüme zur Unterstu"tzung der featurebasierten Produktgestaltung Carl Hanser Verlag Mu"nchen Wien, 1994.
[36] Rutz A. Konstruieren als gedanklicher Prozess. Dissertation. München: TU.
[37] Grabowski H, Lossack R-S. Knowledge based design of complex products by the concept of design working spaces. In: Tomiyama T, Mäntylä M, Finger S, editors. Knowledge Intensive CAD II. IFIP WG 5.2, Pittsburg, USA. Chapman and Hall, 1996.

Retrieval using configuration spaces 14

Tamotsu Murakami

Abstract When designing mechanisms, making use of examples in past designs and handbooks should lead to cost reduction by promoting the sharing of parts and subassemblies among products, as well as reductions in time and effort. At present, however, the process of surveying design examples is left almost entirely to human designers, and little computerised aid has been developed. This chapter describes a computerised method of retrieving mechanism designs from a library by specifying a required behaviour using qualitative configuration space as a retrieval index.

14.1 Introduction

To design a mechanical product, designers usually identify specific kinematic motions to realise a required behaviour of the product, and then determine the specific shape and structure of a mechanism to realise the motion. In the field of machine design, there are basic and general mechanisms widely used in many kinds of machine design, such as the cam and follower, spur gear, internal gear, and ratchet wheel and pawl, as well as specific mechanism design examples in past products and handbooks, *e.g.*, see Artobolevsky [1]. In either case, theoretical/experimental knowledge for synthesis/analysis has already been obtained concerning those mechanisms; so, in many cases, it is worth considering the utilisation of those mechanism cases to realise an intended behaviour. Making use of past mechanism designs should lead not only to reductions in time and effort, but also to cost reduction by promoting the sharing of parts and subassemblies among products. At present, however, the process of surveying mechanism examples in handbooks and past designs is left almost entirely to human designers, and little computerised aid has been developed. In this chapter, a computerised method of storing mechanism cases in a library and retrieving them by specifying required behaviour is described.

14.2 Mechanism library

Figure 14.1 shows the outline of a mechanism retrieval system. The retrieval program searches through the mechanism library for mechanisms that meet kinematic behavioural conditions specified by designers. The designers should be able to obtain answers to their design problems by utilising the retrieved mechanisms without modification, or to come up with a new design by using them as hints.

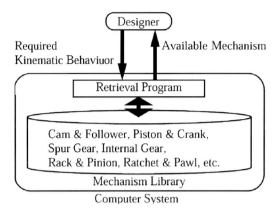

Fig. 14.1 Mechanism retrieval scheme.

14.2.1 Mechanism and configuration space

To enable a computer system to retrieve mechanisms based on behavioural conditions, we adopt configuration space to describe behavioural characteristics of mechanisms. A configuration space is a mathematical space defined in terms of position/orientation parameters of objects in an actual physical space, and is used to solve engineering problems such as spatial planning [2] and mechanical device analysis [3].

For the mechanism retrieval method described in this chapter, a two-dimensional configuration space is defined in terms of two motion parameters of mechanical components in a mechanism [4]. Figure 14.2a–f shows examples of mechanisms and their configuration spaces, and Fig. 14.3 shows configuration space description elements (corresponding to Fig. 14.2a). A point in the space represents a combination of motion parameter values, *i.e.*, a state of the mechanism. An unshaded region is called a free region and represents a set of possible (*i.e.*, without object overlap) states of the mechanism, whereas a shaded region is a blocked region and represents a set of impossible states. A solid line segment represents a boundary between free and blocked regions. A chain double-dashed segment represents (upper and lower) range limits of a single parameter. Symbols such as $p1, p2, q1$, and $q2$ are constant values of motion parameters representing characteristic positions/orientations of mechanism components. Regions of zero width are graphically represented simply as lines in this chapter (Fig. 14.3b).

Note that the mechanism retrieval method described in this chapter is intended to be useful in the conceptual design phase, in which the fundamental direction of the design is decided and possible candidate mechanisms are selected. In conceptual design, behavioural conditions specified by designers should often be simple and qualitative, rather than complicated and quantitatively detailed. Therefore, configuration space descriptions in this chapter are simplified and qualitative in the following senses:

- characteristic constant values are not considered, merely the relationships between values are considered;

Retrieval Using Configuration Spaces

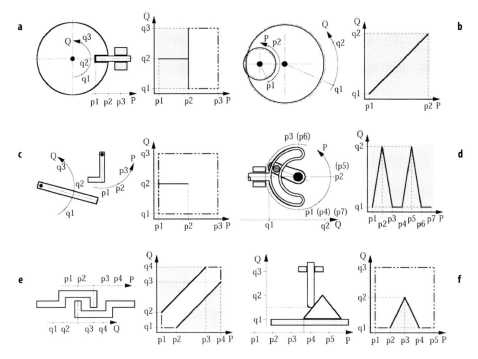

Fig. 14.2 Mechanism and qualitative configuration spaces: **a** notched wheel and slider; **b** internal gear; **c** arm and stopper; **d** crank and cross slider; **e** joint with gap; **f** sliding slope and slider.

Fig. 14.3 Qualitative configuration space description: **a** description elements; **b** zero-width regions represented as lines.

- region boundary and range limit segments, which can actually be curves, are approximated as straight segments.

14.2.2 Kinematic behaviour and configuration space

Since the kinematic behaviour of a mechanism can be viewed as a sequence of transitions between states of the mechanism, a locus of point movements in a configuration space represents the kinematic behaviour. Just as shapes of objects restrict the possible kinematic behaviour of the mechanism in a physical space, shapes of free and blocked regions restrict possible locus patterns in the configuration space. Therefore, we can judge if some mechanism can realise a specific kinematic behaviour in the physical space by checking whether or not a corresponding locus pattern can be drawn in the configuration space of the mechanism. For this reason, qualitative configuration space descriptions are stored in the mechanism library and used as indexes to retrieve mechanisms based on behavioural conditions, as described later.

14.2.3 Additional behavioural information description

Some necessary information about the kinematic behaviour of mechanisms is omitted in the qualitative configuration space (free and blocked regions) representation, and different mechanisms may have the same qualitative configuration space.

For example, an internal gear (Fig. 14.2b) has two rotation parameters, P and Q, whereas a rack and pinion (Fig. 14.4a) has a rotation parameter P and translation parameter Q. Despite this difference in motion type, however, these two mechanisms have the same simplified and qualitative configuration space, as shown in Fig. 14.2b. Similarly, in an internal gear (Fig. 14.2b), one of P or Q can be the input motion to the mechanism and the other the output. In a worm and wheel (Fig. 14.4b), however, motion can be transmitted only when worm P is the input and wheel Q is the output. No information about such possible motion transmission directions is contained in their qualitative configuration space (Fig. 14.2b).

These examples suggest that a qualitative configuration space alone is not sufficient as a mechanism index. These differences can be handled by introducing some additional information to qualitative configuration spaces, such as by labelling parameters by motion types "rotation", "translation", and "rotation/translation" (either rotation or translation), and by motion transmission directions "input", "output", and "input/output" (either input or output).

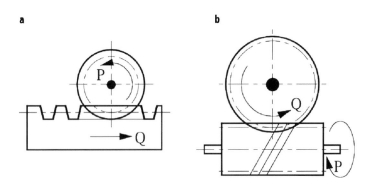

Fig. 14.4 Motion type and transmission direction: **a** rack and pinion; **b** worm and wheel.

14.3 Required behaviour as retrieval key

14.3.1 Required behaviour description

With the mechanism library explained in Section 14.2, designers can retrieve mechanisms to realise the required kinematic behaviour specified in the following way. Since the definition of the coordinates used to specify the required kinematic behaviour by a designer may be different from that of qualitative configuration spaces stored in a library, the designer uses parameters U and V for the required behaviour, as shown in Fig. 14.5, whereas the qualitative configuration spaces in the library use P and Q, as shown in Fig. 14.2.

14.3.1.1 Timing charts of input/output motions

In this chapter, the kinematic behaviour of a mechanism is defined as the generation of the output motion from the given input motion. Therefore, the required kinematic behaviour can be described by the input motion given to a mechanism and the output motion that we want the mechanism to generate. In this approach, the input and output motions are described by timing charts. Consider the problem of designing a portion of a shutter release mechanism of a disposable camera as an example (Fig. 14.5a). When taking a picture with the camera, the user presses the shutter button. This action must cause the exposure lever to move in a specific manner (and hit the shutter blade to open it quickly to expose the film, though this part of the behaviour is not considered here). By assigning parameters U and V to the shutter button and the exposure lever, and using T for time, designers can qualitatively describe the required kinematic behaviour of the mechanism in the form of the two timing charts in Fig. 14.5b (T–U chart for the given input motion and T–V for the intended output).

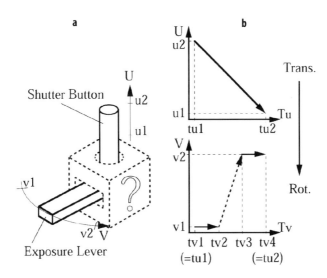

Fig. 14.5 Shutter release mechanism retrieval: **a** shutter release mechanism; **b** required behaviour.

14.3.1.2 Types of input/output motion

Although we often need to specify motion types (*e.g.*, rotation, translation) for the input and output motions of a mechanism in order to describe the design problem accurately, motion types are not represented in timing charts T-U and T-V in Fig. 14.5b. Therefore, we introduce motion-type labels "translation", "rotation", and "*" (translation/rotation) which can be assigned to motion parameters U and V. As for the design problem in Fig. 14.5, we specify that the input motion is translation and the output motion is rotation, as in Fig. 14.5b.

14.3.1.3 Motion speed dependence

In some mechanisms, the speed of the output motion is required to depend on that of the input motion. (In a sewing machine, for example, it is necessary that the speed of the motion with which a needle moves up and down depends on the speed with which the fabric is fed.) In contrast, in the shutter release design in Fig. 14.5, the speed of the lever motion during time interval $tv2$-$tv3$ must be independent of that of the button because the film exposure time should not be influenced by the way in which the user is pressing the shutter button. In order to specify whether the speed of the output motion should depend on that of the input motion or not, we introduce the motion speed dependence labels "dependent" and "independent" which can be assigned to time intervals. In this chapter, output motion, whose speed is specified as independent, is indicated by a dashed arrow, as in Fig. 14.5b.

14.3.2 Required locus pattern generation

By projecting the timing charts T-U and T-V described by designers into two-dimensional U-V space, we can generate a required U-V locus pattern representing the required kinematic behaviour (Fig. 14.6). We can judge whether or not some mechanism can realise a specific kinematic behaviour in the physical space by comparing the U-V space described by designers and the P-Q space in the library and checking whether or not a corresponding locus pattern can be drawn in the qualitative configuration space of the mechanism in the manner described in Section 14.4.

14.4 Locus pattern and configuration space matching

After a required behaviour is represented as a locus pattern, a computerised procedure should examine each qualitative configuration space in a library and attempt to draw the locus pattern in it based on some rules reflecting physical laws, in order to realise kinematic behaviours in a physical space. If the locus pattern can be drawn in some qualitative configuration spaces, then the corresponding mechanisms can probably realise the kinematic behaviour in question in a physical space.

By observing the way mechanisms behave in physical space, and by considering the relationship between mechanism behaviour and locus generation in a qualitative configuration space, we find at least the following three types of locus.

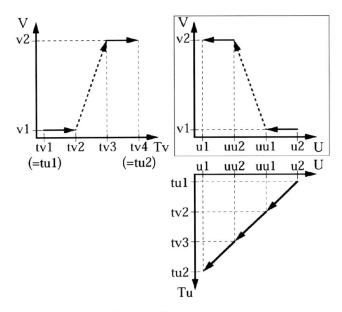

Fig. 14.6 Required locus pattern generation.

14.4.1 Locus along region boundary

14.4.1.1 Motion by object contact

In a physical space, contact between objects restricts motions and results in a specific kinematic behaviour (Fig. 14.7a). In a qualitative configuration space, this corresponds to a case where a locus is drawn along boundaries between free and blocked regions (Fig. 14.7b). Therefore, we can judge that a mechanism can realise the required behaviour by object contact if the required locus pattern can be drawn along region boundaries in its qualitative configuration space (retrieval index).

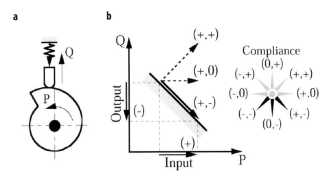

Fig. 14.7 Locus by object contact: **a** object contact; **b** locus along region boundary.

Since we define the kinematic behaviour of a mechanism as generating an output motion from a given input motion, the direction of a locus along a region boundary can be decomposed into input and output components. In Fig. 14.7b, for example, a qualitative direction (dP, dQ) of the locus is (+, −), where the input component is "+" and the output component is "−". This represents the fact that object contact in this mechanism generates output motion in the negative direction of motion parameter Q when input motion is given in the positive direction of motion parameter P.

14.4.1.2 Compliance

Here we consider compliance (realised by gravity or a spring, for example) to ensure object contact in a physical space when the objects are kinematically underconstrained. In a qualitative configuration space, a compliance appears as an effect in which a point drifts in a specific direction as long as it does not conflict with restraint by region boundaries and range limits. Qualitatively, three cases of compliance application for one motion parameter are possible: positive (a point drifts in the positive direction of the parameter), negative, and null (*i.e.*, no compliance is applied) compliance. For a two-dimensional space, nine combinational cases of compliance application are qualitatively possible, as represented by eight arrows and one dot in Fig. 14.7b.

Apparently, a compliance direction in a qualitative configuration space can be decomposed into input and output components in the same way as a locus direction, and we assume that locus generation along a region boundary requires the following conditions on compliance direction.

(1) The input motion is given to a mechanism forcibly and it can override compliance. Therefore, the input component of the locus direction drawn is determined only by the input motion and does not require any condition on compliance direction.

(2) On the other hand, the output component of locus direction is not always determined by only the input motion and thus must be generated by point movement along a region boundary in a qualitative configuration space. This means that the output component of the required locus direction can be generated only when there is a compliance in a specific direction whose output component supports the necessary point movement. In Fig. 14.7b for example, the required locus (a solid arrow) in the direction (+, −) can be drawn only when there is a compliance whose direction has "−" as the output component. Otherwise, the locus to be generated might be in other directions (as depicted by dashed arrows).

Accordingly, considering (1) and (2), generating the required locus along the region boundary in Fig. 14.7b requires the existence of compliance in one of the directions shown by the three black arrows.

14.4.2 Locus along range limit

In a physical space, motion of an object may be limited within a certain range by limiters (Fig. 14.8a). In a qualitative configuration space, this corresponds to a case where a locus is drawn along range limits (Fig. 14.8b). If a range limit in a qualitative configuration space is used to draw a locus, then the range limit is "activated" and a limiter needs to be prepared at the corresponding position in physical space. For a range limit that is not used, no limiter is necessary.

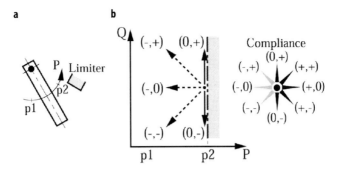

Fig. 14.8 Locus by limiter: **a** limiter; **b** locus along range limit.

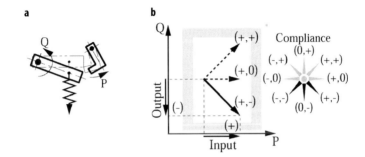

Fig. 14.9 Locus by leaping motion: **a** leaping motion example; **b** locus through free region.

14.4.3 Locus through free region

Although locus generation along boundaries and range limits is crucial in characterising the kinematic behaviour of mechanisms, there can be an auxiliary case of locus generation. When a contact between objects is lost in a physical space, a compliance can drive the underconstrained object to another position/orientation where a new contact arises, as shown in Fig. 14.9a. In a qualitative configuration space, this corresponds to a case where a locus is drawn through a free region, not necessarily along a boundary or range limit (Fig. 14.9b). Thus free regions can be used to complete a required locus, along with boundaries and range limits. Since such a "leaping" motion of an object is terminated when a new kinematic constraint appears, the end point of the corresponding "leap" locus segment must be on a boundary or range limit. Therefore, we can judge that a mechanism can realise the required behaviour by using "leaping" motion if the required locus pattern can be drawn through free regions in its qualitative configuration space. In this chapter, the direction of compliance is assumed to be the same throughout a qualitative configuration space, meaning that a locus segment through a free region must be straight.

14.4.4 Generation of entire locus from segments

The three types of locus generation and conditions on compliance directions described above concern a single segment that is a portion of an entire locus pattern.

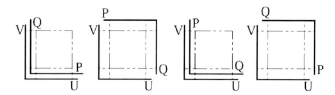

Fig. 14.10 Library (P–Q) and designer (U–V) space matching.

If a specific locus pattern can be drawn in a qualitative configuration space by a sequence of these three types of locus segment, the corresponding mechanism should realise the kinematic behaviour represented by the pattern. In this case, all the conditions on compliance directions are determined by intersecting every condition concerning every locus segment and obtaining the common compliance directions. For example, if the entire locus consists of the three locus segments in Figs. 14.7b, 14.8b, and 14.9b, the possible compliance directions for drawing the entire locus are (0, –) and (+, –). If the intersection has no element, it means that no compliance direction can support the generation of all locus segments simultaneously and, therefore, the generation of the entire locus is impossible.

In a pattern-matching process, two spaces U–V and P–Q should be compared in four ways (combinations by exchanging P and Q axes and flipping P–Q space along P and Q axes), as shown in Fig. 14.10.

14.4.5 Check additional conditions on motion

To check the other conditions, such as motion type and motion transmission direction, labels attached by a designer to parameters U and V are compared with labels attached to parameters P and Q of the qualitative configuration space stored in a library. The motion speed dependence condition is checked based on the following assumptions.

- Output motion speed depends on input motion speed when the locus segment is drawn along a region boundary, neither of whose input or output component is "0" (Fig. 14.7). (If either the input or output component is "0", the boundary segment in Fig. 14.7 is vertical or horizontal, and input motion speed cannot influence output motion speed along this segment.)
- Output motion speed can be independent of input motion speed when the locus segment is drawn along a region boundary or a range limit, either one of whose input or output components is "0" (Fig. 14.8), or through a free region (Fig. 14.9).

14.5 Implementation and execution examples

The mechanism library and the behaviour-based retrieval method described above is implemented as a computer program written in Prolog, since the method is basically a search problem based on pattern matching.

14.5.1 Mechanism library

In the mechanism library implemented, a mechanism is defined by its name, qualitative configuration space representation, parameter specifications (motion types

Retrieval Using Configuration Spaces

and possible motion transmission directions for parameters P and Q), and bitmap file name, which is used to draw a figure of the mechanism. In the qualitative configuration space representation, region boundaries, range limits, and free/blocked regions are defined as a kind of two-dimensional boundary representation used in the field of geometric modelling. Currently, 21 mechanisms, including six in Fig. 14.2, are stored in the library.

14.5.2 Specifying required behaviour

We specify a required behaviour by describing timing charts T–U and T–V, which represent the kinematic behaviour, in the form of a text file. Besides the basic arrow patterns for the charts, the additional information to be specified is motion direction (which of U or V is the input and the output), motion type (translation/rotation) for U and V, and motion relation (whether the output motion speed should be dependent on or independent of the input motion speed during specific time intervals).

14.5.3 Example 1: mechanism for shutter release

With the system implemented, we attempted to retrieve possible mechanisms for shutter release behaviour (Fig. 14.5). Figure 14.11 shows the screen display of the retrieval system. The system read the required behaviour file, displayed the timing charts graphically (the left window), and generated the required locus pattern shown

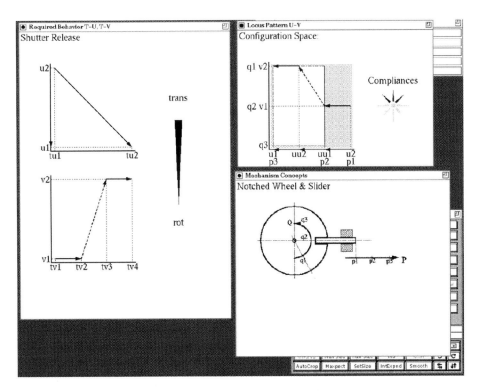

Fig. 14.11 Screen of system implemented.

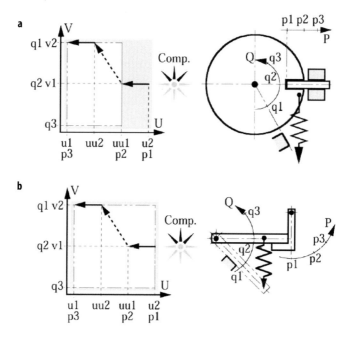

Fig. 14.12 Retrieved solution examples: **a** retrieved solution using notched-wheel and slider; **b** rejected solution using arm and stopper.

in Fig. 14.6. Then the system examined each qualitative configuration space in the library and checked whether or not it can generate the required locus pattern and satisfy the conditions of motion type, motion transmission direction, and motion speed dependence. Once a possible mechanism was retrieved, its image and qualitative configuration space were displayed graphically (two right windows in Fig. 14.11).

Figure 14.12a shows the answer retrieved, and indicates that the qualitative configuration space in Fig. 14.2a can generate the locus pattern. The three black arrows mean that compliance in one of these directions is necessary to draw the locus. Since a part of the locus is drawn along a range limit at $Q = q1$, we find that a limiter is necessary at $Q = q1$ in a physical space. These results indicate that the mechanism "notched wheel and slider" in Fig. 14.2a can possibly be used, provided the following:

- the slider and the wheel are used respectively as the shutter button and the exposure lever in Fig. 14.5a;
- a compliance, by a spring for example, is realised for the wheel in the negative direction of parameter Q;
- a limiter is prepared at $Q = q1$.

Although another mechanism, "arm and stopper" in Fig. 14.2c, can generate the same locus pattern as in Fig. 14.12b, it was rejected because it does not satisfy the motion type condition that the input be translation and the output be rotation.

Retrieval Using Configuration Spaces

The retrieval time required to obtain all possible candidates was about 3 s on SUN SPARCstation10.

14.5.4 Example 2: mechanism in sewing machine

Similarly, we tried retrieving a mechanism for a sewing machine. The given input to the mechanism is the feeding of fabric, and the intended output is the reciprocation of the needle. The speed of the needle motion must depend on the fabric feed speed so that the pitch of the stitches is constant. The required behaviour was specified by the timing charts in Fig. 14.13a, where U is the input and V is the output. The system derived the required locus pattern shown in Fig. 14.13b and searched the library for mechanisms to generate it.

Figure 14.13c, d shows examples of retrieved candidates corresponding to the mechanisms in Fig. 14.2d, f respectively. Although Fig. 14.13e indicates a similar locus pattern to that in Fig. 14.13d, it was rejected because it contains a leaping motion (dashed arrow in the figure), which means that the speed of needle motion is independent of fabric feed speed and thus the condition of motion speed dependence is not satisfied. The retrieval time required to obtain all possible candidates was about 5 s.

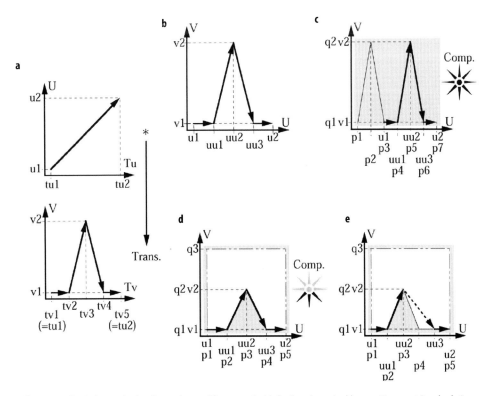

Fig. 14.13 Retrieving mechanism for sewing machine: **a** required behaviour; **b** required locus pattern; **c** retrieved solution using crank and cross slider; **d** retrieved solution using sliding slope and slider; **e** rejected solution using sliding slope and slider.

In the field of machine design, the same kinematic behaviour can generally be realised by many mechanisms. Thus it is important to examine as many possibilities as possible during the conceptual design stage to achieve a better solution. Using our experimental program, we found that different mechanisms in the library may realise similar kinematic behaviours (*e.g.*, Figs. 14.12a, b, and 14.13c, d) and that a mechanism can be used in different ways (*e.g.*, Fig. 14.13d, e). These findings show the effectiveness of this approach.

14.6 Conclusions and discussions

This chapter described a computerised method of retrieving mechanisms from a library, by specifying a required kinematic behaviour using qualitative configuration space as a retrieval index.

The limitations of the method are simplified and qualitative representation (no detailed or quantitative conditions), single mechanism as an answer (no mechanism combination/composition), and retrieval based on "exact match" of required behaviour and mechanism in library. Studies on eliminating these limitations are necessary to support actual and practical mechanism design problems.

Related studies using configuration spaces involve a method of generating contours of two contacting objects based on a configuration space as a desired behaviour description [5]. Simplification and abstraction of configuration spaces and some methods of comparing and classifying mechanisms have been discussed [6]. In order to analyse the mechanism behaviour, a computational method of generating a region diagram from the configuration space in order to describe the mechanism operating modes [7] and an approach to generate place vocabulary from the configuration space to describe the mechanism behaviour qualitatively [8,9] have been proposed.

In related studies on kinematic design aid based on input and output motion specification, various types of vector transformer realised by existing mechanisms were used to generate a design solution based on given translations between input and output motions represented as temporal functions [10]. There is also a study on synthesising a chain of mechanisms using qualitative configuration spaces represented as constraint equations [11].

Although a diagrammatic approach, as described in this chapter, is useful for kinematic consideration of mechanical design, it cannot support mechanical design in a more general manner. Another necessary technology is design case retrieval based on the comparison of physical phenomena. In a mechanical design, we need to consider some physical phenomena represented by a set of physical quantities. When designing a hydraulic cylinder, for example, we should consider and calculate quantities such as the "force" to be generated, "pressure" of fluid and "area" of the cylinder. The design of an electric motor, on the other hand, should contain "torque", "electric current" and "voltage". Since a physical quantity is represented by its magnitude and unit (*e.g.*, see Gruber and Olsen [12]), a set of physical quantities characterising a design case can be plotted as a cluster in a mathematical space defined by nine axes corresponding to nine fundamental SI quantities: length, mass, time, electric current, thermodynamic temperature, amount of substance, luminous intensity, plane angle and solid angle. By defining the distance between and similarity of two (clusters of) physical quantities, we can objectively define the relevance of two design cases from the viewpoint of physical phenomena. I call this nine-dimensional space "quantity dimension space", and am currently working on design case retrieval based on this space [13].

References

[1] Artobolevsky I. Mechanisms in modern engineering design, vols 1 & 2. Japan: Gendai-kogaku-sha, 1979; 416 [Japanese translation].
[2] Lozano-Perez T. Spatial planning: a configuration space approach. IEEE Trans Comput 1983;C-32(2):108–20.
[3] Joskowicz L. Shape and function in mechanical devices. In: Proceedings of the Sixth National Conference on Artificial Intelligence, 1987; 611–15.
[4] Murakami T, Gossard DC. Mechanism concept retrieval by behavioral specification using configuration space. In: ASME Design Theory and Methodology – DTM'92, vol. DE-42, 1992; 343–50.
[5] Joskowicz L, Addanki S. From kinematics to shape: an approach to innovative design. In: Proceedings of the Seventh National Conference on Artificial Intelligence (AAAI-88), vol. 1, 1988; 347–52.
[6] Joskowicz L. Mechanism comparison and classification for design. Res Eng Des 1990;1(3–4):149–66.
[7] Joskowicz L, Sacks E. Computational kinematics. Artif Intell 1991;51:381–416.
[8] Forbus K, Nielsen P, Faltings B. Qualitative spatial reasoning: the CLOCK project. Artif Intell 1991;51:417–71.
[9] Faltings B. A symbolic approach to qualitative kinematics. Artif Intell 1992;56:139–70.
[10] Chakrabarti A, Bligh TP. An approach to functional synthesis of solutions in mechanical conceptual design. Part I: introduction and knowledge representation. Res Eng Des 1994;6(3):127–41.
[11] Subramanian D, Wang CS. Kinematic synthesis with configuration space. Res Eng Des 1995;7(3):193–213.
[12] Gruber TR, Olsen GR. An ontology for engineering mathematics. In: Fourth International Conference on Principles of Knowledge Representation and Reasoning. USA: Morgan Kaufmann, 1994.
[13] Murakami T, Shimamura, J, Nakajima N. Design case relevance in quantity dimension space for case-based design aid. In: Thirteenth International Conference on Engineering Design (ICED01), Design Management – Process and Information Issues, 2001; 51–8.

Creative design by analogy 15

Lena Qian

Abstract Design synthesis is considered as a search for possible candidates that satisfy initial design requirements, but how to navigate a designer to find creative design alternatives becomes very important to defining the boundary of the design searching space. One technique, among others, to produce creative designs in a computer system is to use analogy. The role of analogy is to change the space of possible designs. Most current design systems that utilise analogy produce designs from analogous designs in the same domain. It is claimed that using a process that radically changes the space of possible designs is likely to produce more creative designs than one that only marginally modifies that space. The process of designing based on analogies with other designs from different and distant domains is presented in this paper as a computational approach to creative design using the concept of designing by exploration, since this process radically changes the space of possible designs. The processes required for such an approach to designing are developed. Examples of their use are given.

15.1 Introduction

Analogy is a particularly useful approach to solving an unfamiliar problem without adequate or directly applicable knowledge. Through analogy, relations between the new problem and experience or knowledge about the relevant phenomena can be found. This experience or knowledge can be transferred to a new situation to enable understanding or solving of a new problem. Analogy is not only an effective problem-solving approach, it is also a good exploration resource. Because analogy does not require the two situations to have a direct resemblance, it leaves space for imagination and creativity.

Unlike similarities, such as mere-appearance and literal similarity, analogy maps relational structures, and especially higher-order relations. In analogy-based design (ABD), similarity in higher-order relations such as function and behaviour is the concern. Whereas functions specify what a design does, behaviours describe how a design achieves its functions. The aim of ABD is to obtain new ideas of the structure and associated operations from an existing design with similar functions and/or similar behaviours.

A number of knowledge-based design systems use analogy techniques to produce creative results. Software engineering design, mechanical device design, architectural and structural designs, among many other design fields, have used analogy [1–3]. As design is a problem-solving process, transformational analogy and derivational analogy are widely used in engineering designs [4–6]. A design description is transformed with or without a design plan to produce a creative product. Structure-mapping theory is a feature of most existing design systems in which causal and other kinds of higher-order relations are considered in mapping during the analogical reasoning process [2,7].

However, a limitation of most design systems is that analogy is applied only within the same design domain. Although some rules are domain independent, issues of representation, computation and implementation are not addressed for between-domain analogies. ABD is potentially more creative if the source and target domains are different. The more remote the source and target domains, the more difficulties there will be for naive designers, and the more exploration there will be; therefore, the greater is the likelihood of a creative design being produced.

In this chapter, an ABD system is described. The computational model includes a knowledge base and consists of three subprocesses: (1) analogy retrieval; (2) analogical elaboration, including mapping, transference and transformation; and (3) inference evaluation. An example using between-domain knowledge to produce a creative design is demonstrated.

15.2 Knowledge representation for design retrieval based on analogy

To set up a platform for ABD, the primary consideration is the knowledge representation for design artefacts, upon which the reasoning process can be carried out efficiently. As this research covers analogous design in different design domains, a unified representation with necessary and sufficient knowledge of a design artefact is proposed. Finally, the notion of a design prototype as it satisfies these representation requirements is described.

ABD requires design knowledge represented with: (1) three distinctive categories – function, behaviour and structure; (2) qualitative causal relation of how design structure achieves a function through relevant behaviour; and (3) generalised design knowledge. Whereas the structure specifies what elements the design is composed of, what the attributes of the elements are and how they are related, behaviour manifests functionality and state changes of the structure decide the structure's behaviour. The function tells what the design does and is often used for specifying design requirements. Function, behaviour, and structure are characterised as design variables and play different roles in understanding the design. The qualitative causal relation provides the deep knowledge of the design. It also provides a platform for analogical reasoning. Generalised knowledge provides sufficient information of the design and avoids using unnecessary data to carry out the reasoning task.

15.2.1 Structure

The design goal is achieved by conscious ordering and planning of design elements. A design element is either a physical or logical entity and the relationship between elements is physically connected or logically linked. Each element can be defined with many attributes. The elements, their attributes and their relationships form a design description and are considered as shallow, or superficial, knowledge.

Design artefacts are comprised of physical and logical elements. However, some designs may include an internal process as part of the design, whereas others involve an external process that operates on the design. The existence of this internal process affects both the design description and the operations involved in it. To formalise the knowledge covered in many domains, both time and space are taken into account. A

Creative Design by Analogy

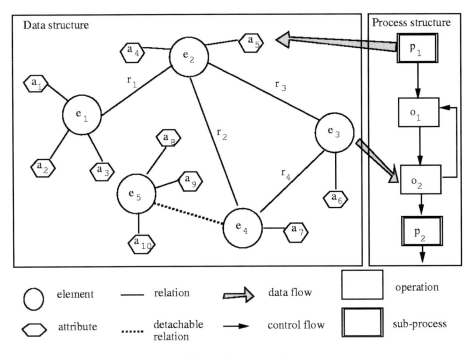

Fig. 15.1 Structure graph.

design structure is defined with data structure and process structure. Data structure is a formalism that represents the ordered and accessible relationships in the data (*i.e.*, elements and their attributes). The process represents a unique and finite set of events that is defined by its purpose or effect, which is achieved under certain conditions. The process performs operations on the data. Any of the data and process structure is a structure variable.

The design structure can be described with a graph G_s, which consists of five finite sets for elements, attributes, relationships, operations and processes; Fig. 15.1:

$$G_s = \{E, A, R, O, P\}, \tag{15.1}$$

where: A is a finite set of attribute variables (represented by hexagonal-shaped nodes linked to the elements, $\{a_i\}$ in Fig. 15.1); E is a finite set of element variables (represented by circular-shaped nodes in the graph, $\{e_i\}$ in Fig. 15.1); O is a finite set of operations (represented by single rectangles, $\{o_i\}$ in Fig. 15.1); P is a finite set of processes composed of operations (represented by double rectangles, $\{p_i\}$ in Fig. 15.1); and R is a finite set of relation variables between two elements (represented by undirected arcs; $R \subseteq E \times E$ $\{r_i\}$ in Fig. 15.1).

The structure is divided into two parts in Fig. 15.1: the data structure on the left-hand side, and the process structure on the right-hand side. An element is surrounded by a set of attributes, and elements are related if they are joined by an arc. The dotted line represents detachable relations. The operations are connected by a control flow (thin arrows) that shows the sequence of the process. An operation can have more than one branch according to the condition at the time of execution. The thick arrow shows the data flow that can come from an attribute or an element. This is not explicitly expressed in G_s; instead, it is specified in the operations.

15.2.1.1 Primitive element and structural element

Any design structure is composed of elements. An element in the system that cannot be divided is called a *primitive element*. A primitive element can be either a physical or logical entity. For example, mechanical engineering, civil engineering and architectural designs involve physical elements, whereas software-engineering designs and graphic designs deal with logical elements. A chair consists of six physical primitive elements: a seat, four legs, and a back. In a software design, an array is a logical data element.

Some elements group together and form a substructure or structural element that has well-defined characteristics. For example, a keyboard is a substructure of a computer. The keyboard has many keys, some light-emitting diodes and cable and is an input device. A structural element consists of a combination of primitive elements or structural elements.

Both primitive and structural elements are passive. They can be seen, accessed or manipulated by designers or reasoning processes. These passive elements are referred to as part of the *data structure* and similar to the concept found in software engineering.

The most important step in the design process is to choose elements for a design. These elements will affect the function of the design. For instance, a chair is not stable if it has three legs, unless these three legs are appropriately configured.

15.2.1.2 Attribute

An element has many properties, or attributes, *e.g.*, colour, shape, material, and so on. The value of an attribute may be continuous (*e.g.*, the width of the door is a continuous numerical value), or discrete (*e.g.*, the material from which the door is made can be wood or glass). The attribute itself or combined with others can play a key role in achieving certain behaviours. For example, the use of colour in a picture conveys information that achieves an effect and/or reflects a goal.

The attribute is characterised by an attribute variable and it is encompassed in the data structure. Not all attributes and their values are necessarily designed to achieve goals. Some are unintentional but can be transformed into an intentional goal if new goals are introduced.

15.2.1.3 Relationship

Besides the element attribute, the relation of the elements can also be used to achieve a function. The relationship between the elements is described with *relation variable* and is also included in the data structure.

If the elements are physical, the relationship between them is interconnected physically using topological or geometrical data. Examples include the configuration of rooms in a house and the interconnections between parts of a can opener in a mechanical-device design. A different topology or geometry of the rooms and the can opener might result in a different use for them.

Creative Design by Analogy

If the elements are logical entities, their relationship is abstract and is linked through internal or external processes. For example, in a simple string copy program in software design, two strings are defined as logical elements. One string is then copied to another by transferring the electronic data. The copy process relates two strings.

A relationship between elements can be fixed or changeable. One kind of changeable relation is that two elements might be tied up together but have loose ends: the relative position of the two elements can be changed, but they are always together. Another kind of changeable relation is detachable, *e.g.*, a key can be inserted into a lock or exists by itself. It is only required when one wants to open the lock.

15.2.1.4 Operation and process

Some designs, like software engineering and chemical control systems, have processes incorporated into the design structure. The action or operation is an element that receives information from passive elements and takes appropriate control action to adjust the structure behaviour. A process is formed by a sequence of actions (or operation) and subprocesses, each of which is referred to as a process unit. As this process is part of the design description, it is referred to as a process structure. Examples of the processes are controllers in chemical process-control systems and programs in software designs.

With the new technology introduced into our daily lives, many designs with mechanical operations have been automated. In other words, internal control processes or operative processes are incorporated into the design to substitute for the manual operations. The internal process belongs to the design description, whereas the mechanical operation is external.

15.2.1.5 Static and dynamic structure

A static structure means that elements, their attributes and their relationships are fixed, and no active element or process is included. The values of the structure variable are decided at the end of the design process and are used as the design description. They do not change with time. Although everything in reality changes gradually with time (*e.g.*, the colour of a door might fade over several years, or the shape of a car might be deformed in an accident), for practical purposes, only normal conditions are applied to the design and the quasi-static state approximation is assumed [8]. Quantitative or qualitative behaviour with approximations over some small time scale is called quasi-static.

In architecture and civil-engineering design, static structures are dominant. When a building is constructed, the number of floors and the positions of the windows and doors are fixed.

Dynamic structure means that elements, attributes and their relationship to one another can be changed. For example, glass colour (an attribute) in phototropic spectacles can darken in the sunlight; a door's structure is dynamic, as its position in relation to the doorframe (a relationship) changes when it opens or closes. Some substructures are detached from the main body of the design and are thus dynamic structures. For example, a lid can be either on or off on a cooking pot. Similarly, elements that enter or leave a design are defined as dynamic. For instance, data in a program can be created or deleted at run time. A process structure is itself dynamic as it has a sequence of operations, each executed at a specific time.

The range of values of a dynamic structure variable needs to be determined at time of design, but these values can be changed later at "run-time" or once the design is produced. For instance, the angle range of a door in relation to the door frame is between 0 to 90°, which is determined at design time, but the actual position is decided when it is used. Here, the angle is a structure relation variable.

The change of structure from one state to another must be caused directly or indirectly by internal or external effects. If the door is closed, it can be opened by an external force, *e.g.*, by either a person pushing it or a strong wind blowing it open. The force is the direct effect. In the process of opening a curtain, one pulls the string that goes through the rings. When the string moves, it pulls the rings together and, in turn, changes the area covered by the curtain. The change in the curtain is indirectly caused by the person pulling the string. Many pieces of mechanical and electronic equipment are designed using such dynamic structures.

15.2.2 Behaviour

Behind the design description are reasons for the design structure and the functionality of that structure. The behaviour of the design structure has an associated function and provides a causal explanation for the design. Both function and behaviour are considered to be deep, or high-level, knowledge. This section describes the behaviour representation.

Behaviour variables can be characterised in two ways.

- Structural (direct): a behaviour is derived from the structure itself without any external effect. For example, the floor of a room has an area as a behaviour variable, which is directly derived from the room's width and length.
- Exogenous (indirect): a specific kind of behaviour is shown when an external object is applied to a structure. For example, the water tap has an area that water can pass through, but water is not part of the water tap design, it is only related to the design. The water flow is an indirect behaviour variable controlled by the diameter of the water tap.

Behaviour is represented by behaviour type, behaviour variable, and qualitative causal relations. Generalised behaviour types are spatial, temporal, and aesthetic. The value of a behaviour variable is either deduced from a structure variable or an exogenous variable. Qualitative causal relations describe static and dynamic characteristics of behaviour. The relations among these entities are expressed by the following formalism:

$$B(T)\{q_b\} \leftarrow S\{q_s, Ex|O\}, \tag{15.2}$$

where B is a behaviour variable, Ex is an exogenous variable (external effect imposed onto design elements), O is an internal operation (operations may be required to make a design work), q_b is the qualitative state of B, q_s is the qualitative state of S, a structure variable, T is the behaviour type (spatial, temporal, or aesthetic), $\leftarrow$ indicates the causal direction, and {} denotes the set of descriptive symbols, including qualitative state, and exogenous variable or internal operation.

This expression not only reveals the dependency relation between the structure and the behaviour, but also describes the qualitative change from the structure to the behaviour, and relates the structure to the external effect. This behaviour type is

Creative Design by Analogy

encoded into this expression to give a more comprehensive explanation about the design.

If the behaviour is static, the corresponding behaviour variable B is dependent on a structure variable S with or without exogenous variable Ex. Therefore, the qualitative causal relation for static behaviour is turned into:

$$B(T) \leftarrow S\{Ex\}. \tag{15.3}$$

A static behaviour does not change with time. It is derived from a static structure. For example, a cup is static in the sense that neither its handle nor shape can be changed, once it has been designed. Therefore, the volume of the cup (as a behaviour variable) is fixed.

The external effect is optional, *i.e.*, a behaviour might or might not be affected by an external effect. In a cup design, the behaviour variable, heat insulation, depends on a structure variable, thickness of the cup, and an exogenous variable, the temperature of the hot water poured into the cup. Here, an external effect is applied.

If the behaviour is dynamic, the respective qualitative states (q_s, q_b) in the expression in Equation (15.2) are required to describe how the changes can occur. The qualitative state describes a stable status and change tendency in terms of qualitative measurement, such as increasing (+) and decreasing (−). For instance, the hinge angle of a door can be increased or decreased. A status can have discrete values or continuous values quantised into intervals. A discrete value with an interval for continuous values or a changing tendency makes a distinctive qualitative state.

For example, in a hand-dryer design, there are two main structure variables that keep a constant air temperature (*coil* and *thermostat switch*), an exogenous variable (*air*), and two behaviour variables (*temperature* and *thermostat status*). If the temperature of the hot air exceeds 80°F (qualitative state), the thermostat switch is turned off. When the temperature of the hot air decreases (qualitative state) below 80°F, the thermostat switch is turned on. The temperature with continuous values is then quantised into 0 and 1, where 0 corresponds to temperature below 80°F and 1 represents the status of temperature above 80°F. For the thermostat status, it has two discrete values: 0 for *Off* and 1 for *On*, see Table 15.1.

The qualitative causal relation also describes the link between two structure variables, and between two behaviour variables. If there are n related behaviour variables and m structure variables where the mth structure variable has a direct effect on the lth behaviour variable, the causal propagation is from structure to behaviour. This is referred to as **series propagation** and is expressed as:

$$B_n \leftarrow B_{n-1} \leftarrow \ldots B_1 \leftarrow S_m \leftarrow S_{m-1} \leftarrow \ldots \leftarrow S_1. \tag{15.4}$$

Note that B and S here are short forms for $B(T)\{q_b\}$ and $S\{q_s, Ex|O\}$ respectively. As the causation arrow is from right to left, the order of variables is numerated from

Table 15.1 Qualitative states for hand dryer.

Behaviour variable	Qualitative state
Heat coil temperature	+, −, 1 (>80 °F), 0 (<80 °F)
Thermostat status	1 (on), 0 (off)

right to left, *i.e.*, the variable with the smaller index is on the right. An example of the series propagation is that a cup's diameter S_1 can change the cup's bottom area B_1, which, in turn, changes the volume B_2 of the cup.

Multiple influences from structure to structure, from structure to behaviour, or from behaviour to behaviour represent another type of causal propagation called **parallel propagation**. If m structure variables affect one type of behaviour or if m behaviour variables cause another behaviour change, this can respectively be expressed by

$$B \leftarrow (S_m, S_{m-1}, \ldots, S_1) \qquad (15.5)$$

or by

$$B \leftarrow (B_n, B_{n-1}, \ldots, B_1). \qquad (15.6)$$

By combining series and parallel causal propagation, structures and behaviour can be represented as behaviour graphs that are similar to those of Sycara and Navinchandra [2]. An example of parallel propagation is that the insulation B_1 of a house can be affected by both thickness S_1 of the wall and the material S_2 of the wall.

The causal dependencies can be cyclic: *i.e.*, one variable A affects another variable B; meanwhile, B also affects A. An example of a behaviour graph, the hand-dryer design, is shown in Fig. 15.2. Initially, the thermostat switch that controls the heating coil is on (*i.e.*, 1). This causes the temperature of the blowing air to increase (*i.e.*, +). When the temperature reaches 80°F (*i.e.*, 1) the switch is turned off (*e.g.*, 0). Since the switch is off, the hot air becomes cooler and its temperature drops (*i.e.*, −). When the temperature decreases to below 80°F the switch is turned on again. The example for qualitative states is shown in Table 15.1.

In the behaviour graph, the arrow indicates the causal direction, and the small rectangle attached to the end of the arrow represents the qualitative state. The change from one behaviour state to another might be described by more than one qualitative causal relation. For instance, *temperatue* ← +1 ← *connection* and *temperature*

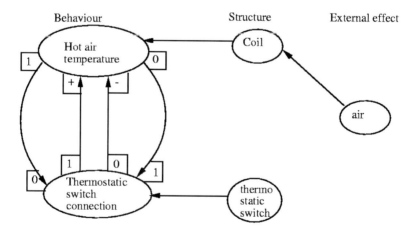

Fig. 15.2 Behaviour graph for hand dryer.

← –0 ← *connection* describe two distinctive qualitative changes. These changes are then combined into *temperature* ← +1 – 0 ← *connection*.

Unlike the influence graph, where a monotonic expression is used by marking the change tendency on the causation arrow (*e.g.*, $a \leftarrow + \leftarrow b$ means *a* increases while *b* increases) [2], the behaviour graph in Fig. 15.2 needs to describe the qualitative measurement at both ends of the causation arrow. This representation extends the capabilities of expressing the changes caused not only by other changes, but also by a certain stable status, *e.g.*, temperature above 80°F. Both behaviour variables and behaviour graphs can be used to retrieve analogous designs.

15.2.3 Function

Functions are usually shown as labels representing the purposes of an artefact. Existing design systems tend to limit functional descriptions to a few words due to the difficulties of interpreting a natural language. The function label then can be used to identify behaviour that achieves the corresponding function. However, a sophisticated parser with the capability of partial matching, and reordering of words, is required to process function labels. A dictionary of synonyms may also be required to handle more function labels of similar meanings. For example, an umbrella's function of blocking rain drops and a door's function of blocking access (*e.g.*, people, animals, *etc.*) can be expressed as the generalised function of blocking an object.

A list of functions can be specified for a design. Any function that is directly associated with a set of behaviours is called a primitive function. An overly abstracted function needs to be decomposed into several more specific primitive functions. For example, a house provides a living area; this, in turn, can be decomposed into sleeping area, eating area, *etc.* Eating area, for example, requires light provided by windows in the room, and heat insulation provided by the thickness of the walls and the material out of which they are made. Thus, provisions of light and heat insulation are two primitive functions.

Some functions of the design product are primary; others are secondary. Primary functions determine the name and the concept of the design product, whereas secondary functions are supportive, realised or user interpreted.

There is no uniform definition or representation of function. A widely accepted definition of function in design research is that a function is a relationship between the input and output of energy, material and information [9], or the manipulation of the fluxes of energy, material and information [10]. Taking a different approach, the function–behaviour–structure (FBS) model [7] treats function, behaviour and structure as individual concepts, and relates them to a causal relationship, or a dependency network [3,11,12].

Although design products differ from one domain to another, the way in which the structure or substructure achieves the design goal or function through behaviour can be classified across design domains. A function can be accomplished in the following generalised FBS path types or FBS types:

- FBS type I, achieved by a static behaviour (F ← B^s);
- FBS type II, achieved by a state of a dynamic behaviour (F ← B^d);
- FBS type III, achieved by a set of behaviours occurring at same time (F ← {*B*} ∪ {*B*});
- FBS type IV, achieved by a set of behaviours occurring in sequence (F ← ({*B*}, {*B*})$^+$).

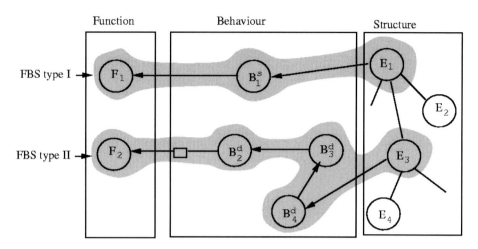

Fig. 15.3 FBS types I and II marked with hatching.

B^s and B^d represent static and dynamic behaviour states respectively. The B^d can be either a value or a range of values of the behaviour variable. ($\{B\} \cup \{B\}$) represents the coexistence of two sets of behaviour states referred to as parallel states, and ($\{B\}$, $\{B\}$) describes two sets of behaviour states occurring in sequence, referred to as series states. The superscript plus sign represents state repetition. ($\{B\}, \{B\})^+$ describes two sets of behaviour, one of which repeats after the other at least once.

FBS path types I and II are considered to be primitive function achievements, since each function associates with one behaviour variable; Fig. 15.3.

In FBS path type I (or FBS type I), one function is realised by static behaviour, which is caused by a static structure. The structure may also relate to external effects. If external effects are involved, they tend to be environmental, i.e., they relate to the design without the intention to change the structure. For example, in architectural design, a view can be provided by strategically placed windows. The static relationship between the windows (i.e., the design structure) and the surroundings (i.e., the environment effects) enables "*provide view*" functionality.

The FBS type II design has a function associated with a dynamic behaviour state caused by a dynamic structure. A dynamic structure variable has a valid range of values that can be changed by external effect or by a change in other structures. A specific value or a range of values determines a qualitative distinct state of the structure and the behaviour, and, in turn, provides a function. The function is achieved by a state or a qualitative value of a behaviour variable. The behaviour state (with a qualitative measurement) is represented by a small rectangle in Fig. 15.3.

For example, opening a door allows access, whereas closing the door impedes access and provides privacy. When the behaviour variable *access area* is zero, the door is closed; otherwise it is open. The physical position of the door is a dynamic structure variable. Depending on the operation applied to the door, different behaviours occur and different functions can be achieved.

However, in many designs, a function depends on more than one behaviour variable. The set of variables can be composed in parallel (FBS type III), in sequence (FBS type IV), or both, Fig. 15.4. A function achieved by repetition of several behaviour variables is seen as a special case of sequential behaviour.

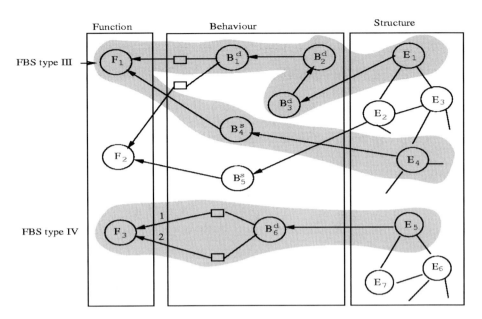

Fig. 15.4 FBS type III and IV marked with hatching.

The FBS type III design composes primitive FBS paths conjunctively. In many complex designs, a function is affected by more than one behaviour variable. A set of static behaviour and dynamic behaviour variables or a combination of these is present simultaneously. For instance, the function *resist-load* in a rigid frame design is achieved by two static behaviour variables – *mechanism bending* and *mechanism compression* – each of which is deduced from other behaviour variables [3].

The FBS type IV design has functions attained by a sequence of behaviour states represented by the indexes on the arrows in Fig. 15.4, *e.g.*, arrows with 1 and 2 towards function F_3. The function not only depends upon several behaviour states, but also upon their sequence, and some states may be required to repeat one after another. Repetitive behaviour variables are special sequences that allow the same behaviour states to reoccur several times after they have changed.

In the door example, although the *opened* and *closed* states provide the *access* and *private* functions, there is no incorporated information as to how the *opened* state is reached from the *closed* state. There should be two functions, the first of which allows the door to open (if it is closed) and the second that allows the door to close (if it is open). The opening function involves state change from closed to open, and the closing function does the reverse.

In general, the function representation can be described as

$$F \leftarrow B^s | B^d |(\{B\} \cup \{B\}) | (\{B\}, \{B\})^+ , \qquad (15.7)$$

where the modulus symbol | is interpreted as *or*.

Knowing the FBS type establishes a basis for ABD in between-domain designs. If two designs have the same FBS type, they are considered for analogical mapping because they have the same way of behaving to achieve a function. To some extent, the FBS type sets constraints upon retrieving analogous designs and elaborating analogical mapping.

Function consists of two parts: variable and FBS path. Whereas the function variable expresses intentions of a design using action and flux taxonomies, FBS path type generalises a design goal achievement path.

15.2.4 Qualitative causal knowledge

Although differences among different kinds (or domains) of designs vary from high-level abstraction (*e.g.*, function, behaviour) to low-level abstraction (*e.g.*, structure and words used in expressing knowledge), the causal links in the designs can be generalised into FBS models. Specifically, from a design perspective, an artefact is designed in such a way as to achieve functionality through an FBS path.

This inclusion of the function and behaviour in the design description and relating the three individual entities (*i.e.*, function, behaviour and structure) in one module reflects the causal knowledge in the design [7,11,13]. The path from structure to behaviour and then to function for an existing design shows the causal relationship of the design and is a one-way derivation:

$$F \leftarrow B \leftarrow S.$$

Relations between function, behaviour and structure are referred to as an **FBS path** or **causal knowledge** throughout this chapter.

If a structure is known, one can deduce its behaviour and abduce its anticipated function. Although behaviour can only be derived from the structure, the FBS path can trace individual entities in both directions. Well-understood behaviour derived from the structure can be specified by linking it with its function, thus eliminating the need for behaviour derivation every time the same structure is to be evaluated. This behaviour can be saved and used efficiently to relate its associated structure and function. Given a function, one can identify the behaviour that realises the function, and the structure that causes such behaviour. Given a structure, its expected behaviour and function can be found through the FBS path. When design experiences are stored with related causal knowledge or in an FBS path, a retriever can deduce both structure and the reason behind the design.

Compiling causal knowledge of a design into a knowledge base establishes a basis upon which we can reason about a design by analogy. A function does not necessarily associate with only one behaviour. Instead, it can be associated in many ways (*i.e.*, one function corresponds to many behaviours). Similarly, behaviour can be derived from more than one structure. The central idea of design by analogy is to capture this one-to-many mapping, in order to find different design descriptions. In other words, having specified a design function or behaviour, its design structure might be very different. The function and behaviour are considered to be analogical retrieval cues or searching indexes, which retrieve designs of a similar nature.

Qualitative causal knowledge summarises causal relations between the function and the behaviour qualitatively and between the behaviour and the structures. This is summarised by the expressions in Equations (15.7), (15.2) and (15.4)–(15.6).

Qualitative causal knowledge provides a basis for ABD as it includes the deep knowledge about what and how a behaviour achieves a function, and what and how a structure derives the relevant behaviour. This is especially useful during the analogy retrieval and analogy elaboration processes.

Creative Design by Analogy

15.2.5 Design prototype

Design prototypes are chunks of knowledge representing a class of generalised design experiences [11]. The content of a design prototype includes pertinent descriptions of the function, behaviour, structure and external effect, and various forms of knowledge that support the commencement and continuation of a design process.

Design prototypes have been used successfully in PROBER, a routine design system in the building envelope domain [14]. Through the dependency network that relates the function, behaviour and structure variables, PROBER identifies the structure element for design instantiation. The notion of design prototypes has also been developed for creative design [3] in a structural design domain. The dependency network is extended with generic causal relations, such as data dependency relations, existence dependency relations, and component-system relations, so that mutation and analogy techniques can be applied.

However, both design systems deal with static structures, like building and structural designs. To overcome the shortage of knowledge representation for designs with dynamic structures and to provide a uniform representation for between-domain analogies, this research enriches the notion of the design prototype by adding a new kind of knowledge called qualitative causal knowledge. This is a unified knowledge representation for structures with static and dynamic characteristics, and it describes generic behaviour–structure and function–behaviour relations.

Design prototypes include shallow operational knowledge, as well as deep causal knowledge. Whereas the shallow-level knowledge can be seen as knowledge that is sufficient for performing a task itself, the deep knowledge captures an underlying causal relationship between the function, behaviour and structure.

The design prototype is defined by a set of design variables (F, B, S) described in the previous sections and general knowledge about computational, context and, most importantly, qualitative causal knowledge.

A set of design variables and various kinds of knowledge described in the following formalism define the design prototype:

$$P = \{F, B, S, Ex, K_r, K_q, K_{cnt}, K_{cmp}, K_{qc}\}, \qquad (15.8)$$

where B is a finite set of behaviour variables, Ex is a finite set of exogenous variables, F is a finite set of function variables, K_{cmp} is the computational knowledge, K_{cnt} is the context knowledge, K_q is the qualitative knowledge, K_{qc} is the qualitative causal knowledge, K_r is the relational knowledge, and S is a finite set of structure variables.

Whereas function describes goals or intentions of the design artefact, structure specifies individual elements, their attributes, and relationships between them. The exogenous variables describe the environment affecting the structure of an artefact. The link between the structure, exogenous, and function is established by behaviour and represented by qualitative causal knowledge and relational knowledge.

The relational knowledge K_r represents the dependencies between function, behaviour and structure, whereas qualitative knowledge K_q describes the directional effect on other variables of changing a variable. The design prototype also contains computational knowledge K_{cmp}, which is the quantitative counterpart of qualitative knowledge, and context knowledge K_{cnt}, which identifies effects from external agents on the design in relation to exogenous variables. Distinguished from K_r, which focuses on the control of variables during the design process, K_{qc} concerns the dynamic characteristics of the design artefact once the design is made.

15.3 An ABD model

Though most popular analogical reasoning techniques in design involve transformational analogy and derivational analogy [1,3,15] as the design problem-solving process, this chapter introduces using analogy in design exploration to expand the searching space or modify processes state spaces [16]. When design requirements are given, design variables such as function, behaviour and structure variables are obtained from the existing design. From this known design state space an analogous design can be retrieved. By mapping the current existing design and the analogous design, useful information can be transferred by analogy to the current design in order to modify the existing state space. This design process bases itself on an existing design solution for a new design problem and explores alternative solutions to generate a creative design.

The computational model described in this chapter intends to cover between-domain analogies using the design prototype knowledge representation driven by function and behaviour. The model combines analogical problem solving and analogical exploration in the conceptual design process; see Fig. 15.5. The upper section of Fig. 15.5 shows the exploration process, and the lower section shows the problem-solving process. Both processes have retrieval and elaboration tasks, but their jobs differ.

To be more precise about the terms used throughout this chapter, three types of design are defined.

- *Target (or target concept) design* is an existing design retrieved from the given design name. For instance, designing a door can be a target concept design. A door must provide open and close functionality, but how to achieve the opening and closing behaviour can be modified for a new design of the door. The target design is an initial point to start the exploration process.
- *Source design* is an analogous design to the target concept design. A water tap is analogous to a door because the way of blocking access in a door is analogous to the way of blocking water flow in a water tap. The door and the water tap have behaviour matching.

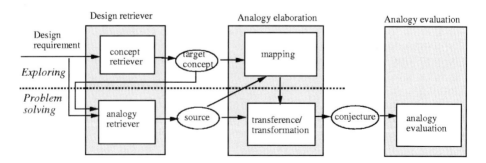

Fig. 15.5 Analogical problem solving and exploring processes.

- *New design* is the result of analogical mapping between the *target* and *source*, with knowledge transference from the source to the target. A new door can be designed with the transformed water tap structure.

The *design retriever* is divided into two tasks: a *concept retriever* and an *analogy retriever*. Whereas the *concept retriever* retrieves a *target* concept design with the given design name, the *analogy retriever* finds designs different from the new design, but with a similar function or behaviour. The primary functions, associated causal explanation or behaviour, and deduced features can then be used to retrieve analogous design. In the *analogy elaboration* process, mapping between the target and the source can be found. The correspondences are considered to be learned features that can be transferred and transformed from the source to the target as a guide in the problem-solving process. The result of the *analogy elaboration* is a design *conjecture* structure. This is the proposed new design and needs to be evaluated.

The design model proposed here aims to explore new design solutions from analogous designs. This model differs from the purely problem-solving approach in two respects: (1) useful features are derived from a target design concept and used to retrieve between-domain analogies; (2) adaptation and transformation of the source analogy solution is guided by mapping between the source and the target. This model compiles design prototypes into the knowledge base. Each design prototype contains the qualitative causal knowledge that provides a platform to explain how an old design achieves its functionality. It explores the solution space by using a target concept design that increases the possibility of retrieving between-domain analogies and which guides the adaptation process. In contrast to failure-triggered creative design models, this model uses a target concept design as a starting point.

This is an interactive system, where the computer and human designer assist each other to produce creative designs. The computer is good in searching, whereas the human is good in thinking and decision making. The feedback from extensive analogy retrieval and analogical mapping and transference by the computer can stimulate the human's creative thinking, and can guide the human designer to transform and evaluate from design conjecture to the new design.

15.3.1 Design retrieval process

There are two distinct tasks involved in *design retriever*. The first is to find a target design concept and the second is to retrieve an analogy source design. The computer plays a major role in searching desired designs from the knowledge base. The human designer may give the computer some cue where to start the searching.

Design requirements can be decomposed into function, behaviour, and/or structure, and usually they provide a design concept name (*i.e.*, what is to be designed: a door or a table). The *concept retriever* uses the concept name to find an old design or target concept design that is used to explain how an existing design works. The primitive functions and associated behaviours of this existing design are then used as analogy retrieval cues to search for analogous designs in different domains.

The *analogy retriever* receives input from design requirements and/or from the target concept design. These can be generalised as label indexes or graph indexes that retrieve analogous designs with the same function or behaviour. Although an analogy can be retrieved from the same design domain, this paper focuses on between-domain analogies and the issues related to them.

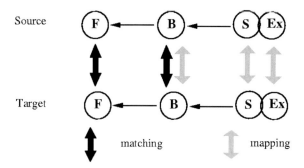

Fig. 15.6 Matching via function and behaviour; mapping via structure.

15.3.2 Analogy elaboration process

As part of an exploration process, mapping is used to identify correspondences between the source and target, whereas transferring analogy inferences indicates what can be used from the source design [17]. The aim of ABD is to find structures and external effect mappings (in terms of design variables) through function or behaviour matching; see Fig. 15.6. Matching means that two function variables, two behaviour variables, or two behaviour graphs are identical at an abstract level. Mapping means that two structure variables or two exogenous variables have the same associated functions or behaviours.

If two functions match, their associated mapping for behaviour and structure, plus external effect, can be found. If two behaviours (two behaviour variables or behaviour graphs) match, only structure plus external effect mapping can be obtained. In the case of matching behaviours, different structures might be found to attain different functions.

As part of the problem-solving process, the source design is transferred partially, and may be transformed, to the target, resulting in a new design. This adaptation process is guided by the structure mapping process. In other words, structure mapping identifies potential design variables that can be transferred from the source to the target.

These processes of mapping and transference are performed automatically by the computer. However, the transformation process may need a human designer's help if there is a lack of generalised knowledge covering different domains.

15.3.3 Mapping and transference

As part of an exploration process, mapping is used to identify correspondences between the source and target, whereas transferring analogy inferences indicates what can be used from the source design [17]. Mapping can be simple variable mapping. However, if a behaviour is complicated, the mapping can involve extensive computation to find mapping between two designs. The computer plays a very important role in the mapping process.

From a design point of view, the mapping process aims to find structure correspondences between the source and target designs. As mentioned in Sections

Creative Design by Analogy

15.2.1–15.2.4, a structure is composed of data structures and process structures. They have significant differences. A data structure is passive, but accessible. Its elements are ordered spatially, causally or logically. A process structure, on the other hand, is active and can act on the data, *i.e.*, it can change the data (if the data has a dynamic structure). The process structure involves temporal knowledge: the sequence of operations affects the sequence of the behaviour states. This difference between data and process structures suggests that mapping occurs from data to data structure or from process to process structure. The process structure needs to operate on the data structure and the process mapping is performed before the data mapping.

A mapping unit (MU) is the structure (including data and process) identified through the FBS path or qualitative causal knowledge, and contains one or more structure variables. A data structure MU appears in design FBS paths I, II or III, whereas a process structure MU is found in design FBS path IV. The mapping rules are domain independent and defined for these four types of FBS path. At the end of any mapping process, a list of correspondences for elements, attributes, relations and operations is recorded. Some will be selected for transference from the source to the target.

This structure mapping process that guides the analogy elaboration occurs at different levels, *e.g.*, the level of the individual variable (element, attribute, relation or process) and the level of the groups (behaviour or structure). Owing to the nature of between-domain analogy, mapping rules are defined without recourse to domain-dependent knowledge. A summary of the mapping and transferring rules is shown in Table 15.2.

Rule 1 applies when an attribute variable maps another attribute variable. A valid transfer involves adding the mapped attribute from the source to the target element, leading to a change in the mapped target attribute and its value. The corresponding external effect is also replaced. For instance, both a cup design and a clothes-hanger

Table 15.2 Summary of mapping and transferring rules.

Rule	Mapping condition	Mapping result	Transferring
Rule 1	$a_i \Leftrightarrow a_j$	$e_i \Leftrightarrow e_j$	a_s
Rule 2	$a_i \Leftrightarrow r_j$	$e_i \Leftrightarrow \{e_{j1}, e_{j2}, \ldots, e_{jn}\}$	a_s and e_s or r_s and $\{e_{s1}, e_{s2}, \ldots, e_{sn}\}$
Rule 3	$r_i \Leftrightarrow r_j$	$\{e_{i1}, e_{i2}, \ldots, e_{in}\} \Leftrightarrow \{e_{j1}, e_{j2}, \ldots, e_{jn}\}$	r_s and some $\{e_{s1}, e_{s2}, \ldots, e_{sn}\}$
Rule 4	$B_i^s \Leftrightarrow B_j^s$	apply **Static Structure Mapping** process	combined
Rule 5	$B_i^d \Leftrightarrow B_j^d$	apply **Dynamic Structure Mapping** process	combined
Rule 6	$B_i^s \Leftrightarrow B_j^s$ $B_i^d \Leftrightarrow B_j^d$	apply **Composite Behaviour Mapping** process	combined
Rule 7	Process structure for primary function	apply **Process Structure Mapping** process	Ex, O or S
Rule 8	Process structure for primary function	apply **Process Structure Mapping** process on initial and final state	Ex, O or S

design have a stability function. Both the attribute represented by the shape of the cup bottom and that represented by the hook shape of the hanger work to stabilise the product. Since these two designs operate in different contexts (the cup is placed on a flat surface, whereas the hanger is hung on a rod), the external elements and contexts should be considered for mapping. The hook in the hanger design is transferred to the cup, leaving the cup bottom to become any shape.

Rule 2 applies when an attribute variable and a relation variable are mapped. An example of this kind is the analogy between colour-changeable spectacles and a blind. The two designs have a similar functionality, *i.e.*, the control of light intensity passing through the objects. The special chemical material contained in the lens of the spectacles changes their opaqueness from full transparency to 50% transparency, for example. Similarly, the relation variable-angle between the laths and the vertical cord holding them in the blind design can also be changed. In this analogy, the lens and its material of the spectacles are transferred to the blind design. Although the transferred element and attributes cannot be used directly in the target design, they can influence the new design from a different point of view.

Rule 3 is for mapping two relation variables and the source elements, and their relations are transferred to the target. An example of this kind is the mapping that occurs between a wall-light fitting and a fountain pen. The light is stabilised by screwing it into a wall; a screw is secured on a vertical pad on the wall fitting. The fountain pen is stabilised by clipping it onto a pocket when carrying it. The relation between the screw and the vertical pad maps to the relation between the cap and the clip of the pen. Thus, if a new light fitting is to be designed, the clipping idea from the fountain pen can be borrowed.

The first three rules have been defined for one-to-one mapping. They are primitive rules that can be applied to the static and dynamic structure mapping processes invoked by **Rule 4** and **Rule 5**.

Rule 4 applies the **Static Structure Mapping** process if source and target have FBS path type I, in which their functions are deduced from static behaviours. This mapping process takes every pair of the structure variables from the source and target and compares them in a generalised form. The process produces a list of MUs. If one-to-one mapping is found for some pairs of single-variable mapping, the corresponding transference rule can be applied. Otherwise, the whole MU of the target is passed to the target at a higher level of abstraction. General knowledge is more successful if the pair has the same kind of structure, *e.g.*, both are attribute structures or both are relation structures. However, a pair in which one is an attribute and the other is a relation is more difficult to match using only structure information, because attributes and relations are different types of design feature.

Rule 5 applies the **Dynamic Structure Mapping** process (DSMP) if source and target have FBS path type II, in which their functions are caused by dynamic behaviours. It consists of a complicated process that finds the mapping at the element level. Owing to the dynamic behaviour of the design structure, behaviour semantics offer useful hints regarding the mapping. The mapping process uses qualitative causal knowledge to generalise behaviour–structure graphs constructed from FBS paths and related static structures. The mapping is executed using a subgraph isomorphism algorithm. At the transference, when an element in the target is substituted with another element from the source, the relations connected to this element must be adjusted. Depending on whether the adjacent element is also transferred from the source, the relation between the two transferred elements can also be copied. This rule is based on structure mapping theory [18], in which structural properties are mapped from a source domain to a target domain. This mapping theory suggests that a system of relations known to hold in the source also holds in the target, through causal links. The preservation of the behaviour that is generated from relations enables the detailed mapping between two designs to be traced.

Rule 6 combines **Rule 4** and **Rule 5**, for static and dynamic structures respectively, for FBS path type III in both source and target designs.

Rule 7 applies the **Process Structure Mapping** process (PSMP) when source and target are of FBS path IV and their corresponding functions are primary. Primary functions determine the name and the concept of the design, whereas secondary functions are supportive, realised, derived or user interpreted. If mapping of a primary function is required, the behaviour sequence must be mapped exactly. In process structure mapping, both internal processes and external operations imposed on the structure play key roles in composing the behaviour sequence. Therefore, not only design data structures, but also corresponding operations, or process structures, are required for mapping to produce new design. The transference of process structure requires checking consistency of the key variables, Ex, O and S. External effect Ex can be either an external operation (*e.g.*, insert a key in a lock) or an external element that relates to a structure variable or a behaviour variable (*e.g.*, light). The Ex for source can be transferred to the target if they involve the same kind of exogenous variable or the same kind of operation.

Rule 8 applies the PSMP when the source and target are of FBS path IV and their corresponding functions are secondary. The same process is used here as for **Rule 7**, except only initial and final states and their corresponding structures are required for mapping.

The algorithms to describe the rules can be found in Ref. [13].

15.3.4 Analogy evaluation process

The adapted design structure conjecture created by between-domain analogy needs to be evaluated using domain knowledge, according to the design requirements. New design variables introduced from the source to the target do not include associated domain knowledge; thus, it is difficult to justify the design conjecture. In such circumstances, automatic machine checking of the structure and behaviour is not realistic. Therefore, in a design exploration model, interaction between human and machine is recommended [19].

Owing to the absence of common-sense knowledge (or between-domain knowledge), transformation of the design conjecture is required using generalised knowledge in spatial, temporal and aesthetic. For instance, the range of a variable might require to be scaled. This scaling factor differs for spatial transformation rules and temporal transformation rules. The result of knowledge transference and transformation is a design structure conjecture that needs be evaluated. Interactive evaluation processes and visual design conjectures are proposed as important factors that overcome the lack of between-domain knowledge.

15.4 Design support system using analogy

The framework described in the previous sections for an ABD has been developed as DEsign Support System Using Analogy, DESSUA, of which the core system is implemented to demonstrate the feasibility, practicality and potentiality of the ABD model.

DESSUA is capable of representing design prototypes, carrying on design processes using analogies and exploring new design ideas.

DESSUA is implemented on Sun workstations. The analogy engine is written in IBUKI Common Lisp; the requirement processor, the prototype manager and the UI are programmed in C and XView. Part of the prototype manager is integrated with Protokit, implemented by M. Balachandran in the Key Centre of Design Computing, University of Sydney.

More than ten design domains have been implemented in DESSUA. They range from interior design, mechanical design, to electronic design and software design.

15.5 An example of designing a new door by behaviour analogy

An example of designing a new door using a dynamic behaviour analogy is described in this section. A door structure is dynamic in that the relation between a door and a doorframe is changeable. For different positions of the door, there are corresponding functions. These functions are achieved related to the status of a dynamic behaviour and are FBS type II. If a door has no internal process to control the opening and closing, this door only has data structure.

The following example demonstrates a door design using a dynamic behaviour analogy. The behaviour discovered through the FBS path of a target concept design's primary function provides the retrieval cue to find other analogous designs.

The input of the design requirements is entered as:

```
    NAME:        door
    OTHERS:      any behaviour analogy
```

As behaviour is selected for analogy retrieval, each of the behaviours associated with each of the primary functions is used as a retrieval cue to search for analogy candidates. Unlike function retrieval, which requires similarity of the function description, behaviour retrieval constraints specify that both design FBS paths have the same FBS type and matching behaviour graphs. Behaviour matching is not label matching *per se*, but rather a form of a semantic matching.

For demonstration purposes, the primary function *allow access* is chosen. This function is achieved by FBS type II (*i.e.*, achieved by a status of a dynamic behaviour). The following designs are retrieved for FBS type II: *curtain, flush tank, hot cold water faucet, locker, scales, water tap*.

These designs all have very different functions to the door's *allow access* function, but each of their functions is achieved by a dynamic behaviour status. For instance, when a physical object is put on the scales, the indicator shows a weight measure. This allows the design to achieve the measuring functionality.

Each of the FBS paths in the retrieved designs is used for behaviour matching. The dynamic structure mapping process (**Rule 5**) involves obtaining a behaviour graph for each potential FBS path, and the isomorphism between the target behaviour graph and each source behaviour graph [13].

The behaviour matching results for the door example are *curtain* and *water tap*. One common behaviour of these designs and the door design is that they all contain

Creative Design by Analogy

a behaviour variable capable of changing from 0 to a constant (>0). The two states (0 and >0) directly contribute to two functions.

Take *water tap* suggested from the retrieval process as an example. Although it is not relevant to door in terms of appearance, its behaviour is similar to the behaviour of the door, *i.e.*, they share the same qualitative causal relations. The script for matching the *water tap* design is as follows:

```
============ Target concept design =================
Obtaining DOOR's behaviour graphs......          ; Step 1, find BGSs
Initial BG elements: (AREA DOOR HINGE)           ; elements in behaviour graph
causal dir = ((2 1 ((+) (+)))                    ; indexes of elements are 0, 1, 2
             (1 0 ((+) (+))))                    ; 0:area, 1:door, 2:hinge
expand-list BG elements (AREA DOOR HINGE)        ; expanded with static element
              with (DOOR_FRAME)                  ; door_frame with index 3

causal entry = (3 2 ((S) (S)))                   ; causal rel from hinge to door_frame
causal entry = (2 3 ((S) (S)))                   ; causal rel from door_frame to hinge
causal entry = (2 1 ((+) (+)))                   ; causal rel from hinge to door
causal entry = (1 0 ((+) (+)))                   ; causal rel from door to hinge

Final BSG elements:
(AREA DOOR HINGE DOOR_FRAME)

BSG:                                             ; generalised qualitative causal
                                                 ; relation CC can be ++, —,
        0        0        0        0
       (CC)      0        0        0
        0       (CC)      0       (S)
        0        0       (S)       0

============ Source concept design =================
Obtaining WATER_TAP's behaviour graphs......
Initial BG elements: (WATER VALVE SCREW HANDLE)  ; 0:water, 1:valve, 2:screw, 3:handle
expand-list BG elements (WATER VALVE SCREW HANDLE) ; expanded with static
              with (PIPE)                        ; element pipe with index 4
causal entry = (4 3 ((S) (S)))
causal entry = (3 4 ((S) (S)))
causal entry = (3 2 ((+) (+)))
causal entry = (2 1 ((+) (+)))
causal entry = (1 0 ((+) (+)))

Final BSG elements:
(WATER VALVE SCREW HANDLE PIPE)
BSG:
        0        0        0        0        0
       (CC)      0        0        0        0
        0       (CC)      0        0        0
        0        0       (CC)      0       (S)
        0        0        0       (S)       0

=====================================================
Calculating the isomorphism of two BSGs:

b-list = (AREA DOOR HINGE DOOR_FRAME)            ; Step 3, find b-list
t-list = (WATER VALVE SCREW HANDLE PIPE)         ; Step 3, find t-list

tmp = (VALVE SCREW HANDLE PIPE), rm_elist = NIL
tmp = (SCREW HANDLE PIPE), rm_elist = (VALVE)
tmp = (HANDLE PIPE), rm_elist = (SCREW VALVE)
tmp = (PIPE), rm_elist = (SCREW VALVE)           ; Step 4, find possible delete element,
                                                 ; del-list
```

266   Engineering Design Synthesis

```
nth elem to delete = 1                          ; Step 5-9, try deleting each element
n = 1, t-list = (WATER SCREW HANDLE PIPE)       ; specified in del-list
                                                ; from t-list and compare t-matrix
                                                ; with b-matrix
nth elem to delete = 2                          ; when screw is deleted,
n = 2, t-list = (WATER VALVE HANDLE PIPE)       ; isomorphism is found.
                                                ; Step 10, no reverse dependencies,
                                                ; is found.
isom res = (((DOOR WATER_TAP)                   ; design name mapping
            ((AREA DOOR HINGE DOOR_FRAME)       ; element mapping
             (WATER VALVE HANDLE PIPE))

                    ((0      0       0       0)) ; graph matching
                     ((CC)   0       0       0)
                     (0      (CC)    0       (S))
                     (0      0       (S)     0))

                    ((0      0       0       0)
                     ((CC)   0       0       0)
                     (0      (CC)    0       (S))
                     (0      0       (S)     0))))

Mapping set is:
================================================================
                TARGET              SOURCE
================================================================
    NAME:
                DOOR                SIMPLE-WATER-TAP
    BEHAVIOUR:
                (ACCESS AREA)       (FLOW WATER)
    STRUCTURE:
        TARGET: ((IMPOSED HOLD_KNOB (RIGHT_MIDDLE DOOR))
                 (SCREWED_IN HINGE DOOR)
                 (OPEN_ANGLE DOOR DOOR_FRAME D (0-90 DEGREE))
                 (SCREWED_IN HINGE DOOR_FRAME))
        SOURCE: ((PERPENDICULAR-FIXED HANDLE SCREW)
                 (SCREWED VALVE SCREW)
                 (MOVE-ALONG VALVE SCREW D (0-2 CM))
                 (INSERTED-IN HANDLE PIPE))
    ELEMENT:
                R                       R
                (DOOR HINGE DOOR_FRAME) (VALVE HANDLE PIPE)
    EXTERNAL:
                WALL                WALL
----------------------------------------------------------------
```

The graphical representation of this mapping is shown in Fig. 15.7.
The following script shows the steps of the transference process:

```
====================
CANDIDATE INFERENCES:                           ; structure to be transferred
====================
STRUCTURE:
(MOVE-ALONG VALVE SCREW)
(SCREWED VALVE SCREW)
(PERPENDICULAR-FIXED HANDLE SCREW)
(INSERTED-IN HANDLE PIPE)

CONTEXT:                                        ; context to be transferred
(FIXED_ON PIPE WALL)

=================================================
FOR CONJECTURE FROM ANALOGICAL INFERENCES, ANSWER:  ; Start evaluation
'Y' for ACCEPT;
'M' for MODIFICATION;
'N' for REJECTION.
=================================================
```

Creative Design by Analogy

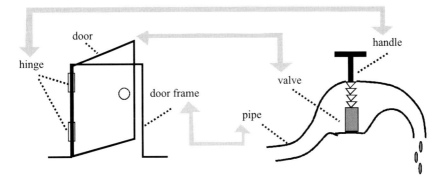

Fig. 15.7 Structure mapping between door and water tap.

```
(INSERTED-IN HANDLE DOOR_FRAME) ? n        ; rejected of rel. and handle
(PERPENDICULAR-FIXED HANDLE SCREW) ? y     ; take as it is,
                                           ; transferred all
((MOVE-ALONG DOOR SCREW)) ? y              ; accepted although how we're
                                           ; going to is unknown
((SCREWED DOOR SCREW)) ? y                 ; accepted
((FIXED_ON DOOR-FRAME WALL)) ? y           ; accepted context

============
CONJECTURES:                               ; a new structure is created
============
STRUCTURE:
(PERPENDICULAR-FIXED HANDLE SCREW)
(SCREWED DOOR SCREW)
(MOVE-ALONG DOOR SCREW)

CONTEXT:
(FIXED DOOR-FRAME WALL)
```

Relations described in the structure of the water tap prototype, which contain the mapping elements, are extracted and are transferred to the target domain. From the qualitative causal knowledge we know that the valve position changes by turning the handle to different angles, which in turn causes the screw to rotate, thus pushing down the valve or pulling it up. The relations linking these elements and some elements are copied directly from the source to the target, but need to be adjusted by the designer. As the conjectures are suggestions for the creative design, a lot of space exists to extend or refine the design. For example, the conjecture from the water tap design gives many potential designs, as shown in Fig. 15.8.

A designer rarely thinks about water taps when designing a door, but this system can assist this by recourse to similar behaviour in both domains.

15.6 Conclusion

At the core of a computational model for analogical reasoning in different design domains is knowledge representation. From the analysis of designs in different domains, it can be seen that, although the focus of each design may be different, design purposes or functions can be achieved in various, but limited, ways. Thus,

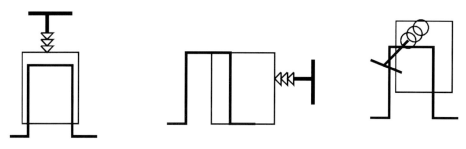

Fig. 15.8 Hints from water tap.

associations between function, behaviour and structure (F, B, S) or goal achievement paths can be generalised. In addition, design behaviour is represented by qualitative causal knowledge in an abstract form. As analogous designs do not share superficial commonalities, the causal relation provides the deep understanding of the design and provides a platform for analogical reasoning. The generalisation of function, behaviour and structure is included in a formalism proposed in this chapter, but only function and behaviour are used for retrieving analogous design from different domains. The formalism represents both designs with static and dynamic characteristics and distinguishes designs with components, or elements, related in space or in time. For example, a mechanical design has its elements configured in space, whereas a software design (*e.g.*, a program) is structured in time. The design prototype is used to represent this formalism.

A design process typically starts with a set of requirements, which give a conceptual idea of the design. This design, referred to as the target concept design, along with its analogous design, referred to as the source design, establish a platform for the exploration of a new solution. Both target concept design and source design must be represented in the formalism described in this chapter during the design process. Mappings between the target concept and source designs are elaborated through general rules (**Rule 1**–**Rule 8**). Of these mappings, only useful ones are transferred from the source to the target, substituting the corresponding elements and/or relations of the target concept design, thus resulting in a design conjecture. This conjecture is then evaluated according to the original requirements. A problem identified in the evaluation process of the design conjecture generated from two different domains is that the knowledge about an individual design exists, but it cannot be used for evaluation purposes.

Analogy is one technique in lateral thinking, which is a process of *being deliberately and self-consciously unreasonable in order to provoke a new pattern* [20]. As transference of analogical inference from one domain to another is involved, the design conjecture could be unreasonable due to the difference in two design domains in terms of space, time and aesthetics. This conjecture might be wrong, but it exists only temporarily in order to achieve a correct solution.

The proposed computational model is developed into a design-support design system. DESSUA can produce creative ideas during the synthesis design process and, in particular, can potentially inspire the designer to think creatively. This interactive system is not an automatic design tool; it provides an exploration medium. It is capable of reasoning about between-domain analogies.

References

[1] Bhansali S, Harandi M. Synthesizing UNIX programming using derivational analogy. Knowledge Systems Laboratory, Report, no. KSL 92-02, Stanford University, Stanford, 1992.
[2] Sycara K, Navinchandra D. Index transformation techniques for facilitating creative use of multiple cases. In: Gero JS, editor. IJCAI-91 Workshop on AI in Design, University of Sydney, Sydney, 1991; 15–20.
[3] Zhao F, Maher M. Using network-based prototypes to support creative design by analogy and mutation. In: Gero JS, editor. Artificial Intelligence in Design'92. Dordrecht: Kluwer, 1992; 773–93.
[4] Dershowitz N. Programming by analogy. In: Michalski RS, Carbonell JG, Mitchell TM, editors. Machine learning II: an artificial intelligence approach. Los Altos (CA): Morgan Kaufmann, 1986; 393–421.
[5] Huhns MN, Acosta RD. Argo: an analogical reasoning system for solving design problems. MCC Technical Report, Number AI/CAD-092-87, Microelectronics and Computer Technology Corporation, Houston, 1987.
[6] Maiden NA, Sutdiff AG. Exploiting reusable specifications through analogy. Commun ACM 1992;35(4):55–64.
[7] Sembugamoorthy V, Chandrasekaran B. Functional representation of devices and compilation of diagnostic problem-solving systems. In: Kolodner J, Riesbeck C, editors. Experience, memory and reasoning. Hillsdale (NJ): Erlbaum, 1986.
[8] DeKleer J, Brown JS. A qualitative physics based on confluences. Artif Intell 1984;24:7–83.
[9] Rodenacker W. Methodisches Konstruieren. Berlin: Springer, 1971.
[10] Wirth R, O'Rorke P. Representing and reasoning about function for failure modes and effects analysis. Working Notes, AAAI-93 Workshop on Reasoning about Function, Washington, DC, 1993; 188–94.
[11] Gero JS. Design prototypes: a knowledge representation schema for design. AI Mag 1990;11(4):26–36.
[12] Kumar AN, Franke D, Hodges J, McDonell JK, Stiklen J, Upadhyaya SJ, editors. Working Notes, AAAI-93 Workshop on Reasoning about Function, Washington, DC, 1993.
[13] Qian L. Design by analogy. Ph.D. thesis, Department of Architectural and Design Science, University of Sydney, Sydney, 1995.
[14] Tham KW. A model of routine design using design prototypes. Ph.D. thesis, Department of Architectural and Design Science, Faculty of Architecture, University of Sydney, 1991.
[15] Carbonell JG. Derivational analogy: a theory of reconstructive problem solving and expertise acquisition. In: Michalski RS, Carbonell JG, Mitchell TM, editors. Machine learning II: an artificial intelligence approach. Los Altos (CA): Morgan Kaufmann, 1986; 371–92.
[16] Gero SJ. Towards a model of exploration in computer-aided design. In: Gero JS, Tyvgy E, editors. Formal design methods for CAD. Amsterdam: North-Holland, 1994; 315–36.
[17] Kedar-Cabelli ST. Analogy-from a unified perspective. In: Helman DH, editor. Analogical reasoning-perspectives of artificial intelligence, cognitive science, and philosophy. Dordrecht (The Netherlands): Kluwer, 1988; 65–104.
[18] Falkenhainer B, Forbus KD, Gentner D. The structure-mapping engine: algorithm and examples. Artif Intell 1989–90;41:1–63.
[19] Kolodner J. Improving human decision making through case-based decision making. AI Mag 1991;(summer):52–68.
[20] DeBono E. Lateral thinking: a textbook of creativity. London: Ward Lock Educational, 1970.

Design patterns and creative design 16

Sambasiva R. Bhatta and Ashok K. Goel

Abstract Design patterns specify generic relations among abstract design elements. In the domain of physical devices, design patterns, called generic teleological mechanisms (or GTMs), specify generic functional relations and abstract causal structure of a class of devices. We hypothesise that GTMs are productive units of analogical transfer in device design. We describe a normative theory of analogical design called model-based analogy (or MBA) based on this notion. Whereas design patterns provide a content account of analogical transfer, MBA provides a process account of acquisition, access, and use of GTMs. In particular, MBA shows how structure–behaviour–function models of specific designs enable the acquisition of design patterns, and how goals of adapting familiar designs to meet new design requirements result in the access, transfer, and use of previously learned GTMs. We describe how the IDeAL system evaluates the MBA theory.

16.1 Background, motivations and goals

Design patterns are oft-studied entities in the design literature. Christopher Alexander [1], for example, analysed evolutionary design of village centres in rural India in terms of design patterns. We characterise design patterns as specifying generic relations among abstract design elements. The relations are generic in that they are independent of any specific design situation, and the elements are abstract in that they do not refer to any specific physical structure. Let us consider the domain of physical devices, *i.e.*, the domain of teleological artefacts instantiatable in physical form. In this domain, we characterise design patterns as specifying generic functional relations and abstract causal structure of a class of devices. The abstract concept of feedback in control systems provides one example of such a functional and causal pattern; a generic mechanism for transforming translational motion into rotational motion is another example. The design pattern of feedback specifies both the generic function it achieves (*e.g.*, regulation of a device output, given possible fluctuations in the device input) independent of any specific design situation, and the abstract causal structure that achieves it (*e.g.*, transmission of information about fluctuations in the device output to a device control input) without reference to the physical structure of any particular device. We call these functional and causal design patterns *generic teleological mechanisms* (or GTMs).

We posit that design patterns are productive units of analogical transfer in design. That is, design patterns are what gets learned from one design situation and transferred to another situation. In particular, we hypothesise that GTMs are productive units of analogical transfer in device design. We have developed a normative theory of conceptual device design called *model-based analogy* (or MBA) based on this notion. Whereas GTMs provide a content account of analogical transfer, MBA provides a process account that specifies how the transfer occurs, *i.e.*, how GTMs are acquired, accessed, and used in analogical design. In particular, it shows how structure–behaviour–function (SBF) models of specific designs enable the acquisition of GTMs, and how goals of adapting familiar designs to meet new design requirements result in the access and use of previously learned GTMs. We have instantiated and evaluated the MBA theory in an operational computer program called IDeAL. In this chapter, we briefly describe the MBA theory, focusing on the hypothesis about GTMs forming a productive unit of analogical transfer in device design.

16.2 MBA

The computational process of MBA takes as input a specification of a target design problem in the form of the functional requirements and structural constraints on a desired design, and gives as output a solution in the form of a structure that realises the specified function(s) and also satisfies the structural constraints. In addition, MBA gives an SBF model that explains how the structure realises the desired function. Figure 16.1 illustrates a part of the MBA process that pertains to GTMs. In this figure, the boxes represent the subtasks of analogical design, such as the retrieval of source analogue and analogical transfer; the arrows between the boxes represent the process flow between the subtasks; the ovals represent the knowledge in memory, such as design analogues; and the dotted ovals represent methods to achieve the

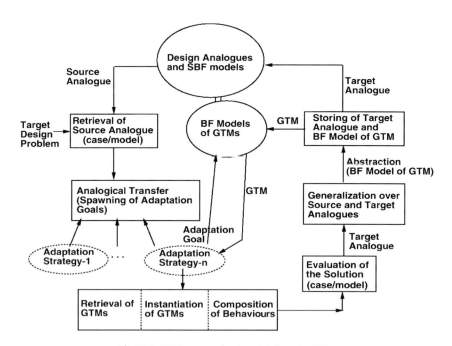

Fig. 16.1 IDeAL's process of analogical design using GTMs.

subtasks. A stored design analogue in this process specifies (i) the functions delivered by the known design, (ii) the structure of the design, and (iii) a pointer to the causal behaviours of the design (the SBF model). The design analogues are indexed both by the functions that the stored designs deliver and by the structural constraints they satisfy.

MBA considers a source analogue to be an exact match for a target problem if both the input states and output states of the functions in the analogue and the problem are identical along with the property values and their constraints. If and when IDeAL is unable to find a source analogue that exactly matches the target problem, it spawns reasoning goals for adapting the source design. Different types of functional difference between the target and the source lead to different types of adaptation goal, some requiring only simple modifications (such as parameter tweaks) and some others requiring more complex modifications (such as topological changes). In order to control the reasoning involved in making complex modifications, MBA requires knowledge that can encapsulate the relationships between candidate modifications and their causal effects. In device design, design patterns, and, in particular, GTMs, provide such knowledge. Therefore, MBA uses the knowledge of GTMs in modifying device topology in the source design. IDeAL evaluates a candidate design by qualitative simulation of the model it generates by instantiating an applicable GTM.

16.2.1 SBF models of devices

IDeAL represents its comprehension of specific design cases (*i.e.*, device models) in an SBF language [2]. This language provides conceptual primitives for representing and organising knowledge of the structures, behaviours, and functions of a device. In this representation, the **structure** of a device is viewed as constituted of *components* and *substances*. Substances have *locations* in reference to the components in the device. They also have *behavioural properties*, such as *voltage* of *electricity*, and corresponding *parameters*, such as 1.5 Vs, 3 V, *etc*. Figure 16.2a, b illustrates the SBF model of a simple amplifier and Fig. 16.2c, d that of an inverting amplifier. For each device, the structure, its function, and the behaviour that achieves the function are shown. For exposition, the structure here is illustrated diagrammatically. IDeAL, however, internally represents the structure using symbols, not diagrams. The structure of a device in SBF models is represented hierarchically in terms of its constituent structural elements and relations among them such as *part-of, includes*, and *parallelly connected*. For instance, the structure of the simple amplifier (shown in Fig. 16.2a) is represented symbolically as having components *resistor* (R_{in}) and operational amplifier (*Op-Amp*), and the structural relation *serially connected* between them at the location V_-.

A **function** in the SBF models is a behavioural abstraction and is represented as a schema that specifies the behavioural state the function takes as input, the behavioural state it gives as output, and a pointer to the internal causal behaviour of the design that achieves the function. The pair of states indicated by GIVEN and MAKES in Fig. 16.2b shows the function "Amplify Electricity" of the simple amplifier. Both the input state and the output state are represented as *substance schemas*. Informally, the function specifies that the amplifier takes as input electricity with a voltage of V_{in} volts (*i.e.*, 1 V) at i/p and gives as output electricity with a voltage of V_{out} volts (*i.e.*, 100 ± 20 V, where 100 is the average value and 20 is the fluctuation around the average

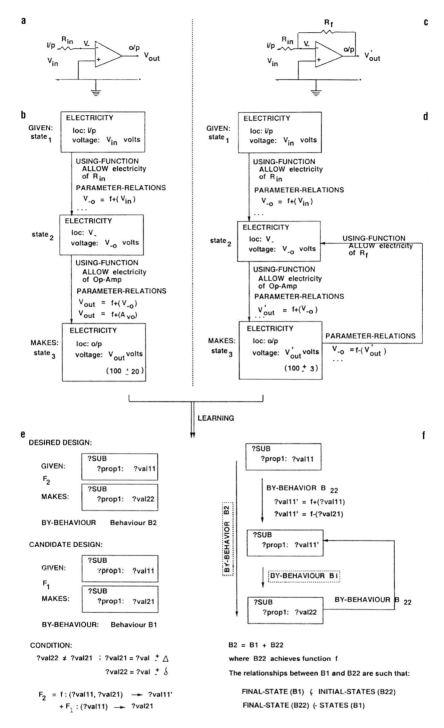

Fig. 16.2 Learning of feedback mechanism. **a** A simple amplifier. **b** Behaviour "Amplify Electricity" of the simple amplifier. **c** An inverting amplifier with op-amp. **d** Behaviour "Controlled Amplify Signal" of the inverting amplifier with op-amp. **e** Functional difference that the feedback mechanism reduces. **f** Behaviour modification that the feedback mechanism suggests.

value) at o/p. Note that while the representation of a specific design may specify fluctuations in terms of quantitative tolerance limits (*e.g.*, the voltage is 100 ± 20 V), the representation of a design pattern would specify fluctuations in terms of qualitative abstractions, such as *small*, *medium* or *large*, independent of specific quantitative values.

The internal causal **behaviours** in the SBF model of a device explicitly specify and explain how the functions of structural elements in the device get composed into device functions. The annotations on the state transitions express the *causal*, *structural*, and *functional contexts* in which the transformation of state variables, such as substance, location, properties, and parameters, can occur. Figure 16.2b shows the causal behaviour that explains how electricity applied at the input location i/p of the simple amplifier is transformed at the output location o/p. The annotation USING-FUNCTION in the transition $state_2 \rightarrow state_3$ indicates that the transition occurs due to the primitive function "ALLOW electricity" of operational amplifier (op-amp). The PARAMETER-RELATIONS associated with this transition $state_2 \rightarrow state_3$ indicate how the output voltage V_{out} is dependent on the voltage at location V_- and the open loop gain A_{V_o} of the op-amp. The two transitions $state_1 \rightarrow state_2$ and $state_2 \rightarrow state_3$ together explain how the input voltage V_{in} is transformed into the output voltage V_{out} of the simple amplifier and how it achieves the amplification function. The difference between the behaviours of the simple amplifier and the inverting amplifier is that in the latter the output voltage alters the intermediate input and thus determines the overall output voltage. In specific terms (Fig. 16.2d), the voltage at location V_- is not only dependent on V_{in}, but also on the output voltage V_{out} that is fed back to the location V_-, and the voltage at V_- in turn determines the overall output voltage V'_{out} of the inverting amplifier.

16.2.2 Design patterns

IDeAL represents GTMs as behaviour–function (BF) models using a subset of the SBF language as above. The SBF representation of a GTM encapsulates two types of knowledge: knowledge about the patterns of differences between the functions of known designs and desired designs that the GTM can help reduce; and knowledge about patterns of modifications to the internal causal behaviours of the known designs that are necessary to reduce the differences. That is, it specifies relationships between patterns of functional differences and patterns of behavioural modifications to reduce those functional differences. For example, Fig. 16.2e, f respectively show these two types of knowledge for a partial model of the feedback mechanism.[1] Figure 16.2e shows the patterns of functions F1 and F2 respectively of a candidate design available and the desired design, and the conditions under which the mechanism is applicable. Because of the tasks for which they are used in MBA, the GTMs are indexed by the patterned functional differences such as shown in Fig. 16.2e (*i.e.*, the fluctuations in the output substance property values in the candidate design function and the desired design function respectively are large and small as denoted by Δ and δ around an average value ?*val*). The model of the feedback indicates that the desired behaviour B_2 can be achieved by modifying the candidate behaviour B_1 through setting up the indicated causal relationships between the latter (*i.e.*, B_1) and

---

[1] Feedback can be open loop or closed loop. The feedback mechanism described here is one type of closed-loop feedback in which the output substance, feedback substance, and the input substance are all the same.

the additional behaviour B_{22} (which achieves the subfunction of F_2 in addition to F_1 characterised as f in the conditions of the mechanism). In particular, the feedback mechanism suggests the addition of a causal link (as indicated by B_{22}) from a change in the output substance state to a change in an earlier state (input state or intermediate state) in the candidate behaviour so that the effective input to the device is modified. Figure 16.2f shows the relationships in the model of the feedback that IDeAL learns from the two designs of amplifiers.

In more specific terms, the feedback mechanism as represented in Fig. 16.2e, f is an abstraction of similarities and differences between the SBF models of the simple amplifier (Fig. 16.2b) and the inverting amplifier (Fig. 16.2d). It suggests that one can achieve the function F_2 (?val11 → ?val22, where ?val22 fluctuates small) by composing the behaviour B_1 of a similar function F_1 (?val11 − ?val21) and the behaviour B_{22} that transforms ?val11 into the intermediate input val11′ by taking a sample of ?val21 (the output of F_1, which fluctuates large) so that the modified intermediate input val11′ in turn produces the overall desired output of ?val22. Note that val11′ is dependent both on the input val11 and the output val21 (which fluctuates large and which is fed back). The resulting, overall output of B_2, however, will be val22 (which fluctuates small).

16.3 Acquisition of GTMs

Let us consider the situation in which IDeAL's analogue memory contains the design of a simple amplifier (Fig. 16.2a, b). Note that the output of this device is dependent on the open loop gain (A_{V_o}, a device parameter) of the op-amp and is typically very high (ideally infinity) and unstable. Consider the scenario where IDeAL is given the problem of designing an amplifier whose function is to deliver a specific, controllable output electricity – an output that does not fluctuate much. That is, an output electricity with a voltage value, V'_{out} volts (*i.e.*, 100 ± 3 V, where 100 is the average value and 3 is the fluctuation allowed around the average value) given an input electricity of $V_{in} = 1$ V. See MAKES and GIVEN states in Fig. 16.2d. IDeAL uses the specified function as a probe into its memory of analogues and retrieves the design of the simple amplifier (Fig. 16.2a, b) because the two functions are similar (*i.e.*, the input states are identical and the output states differ only in a parameter value and a constraint on that value). Suppose now that IDeAL only has a simple strategy, such as replacing a component in a past design, to deliver new functions. Given the model of the simpler amplifier as shown in Fig. 16.2b, IDeAL cannot localise the modification needed to reduce the difference between the source and the target, and hence it fails. Under this failure, if an oracle presents the correct design (the diagram of the structure of the new device is shown in Fig. 16.2c) and the SBF model of the new device (shown in Fig. 16.2d), IDeAL learns a model of the feedback mechanism (shown in Fig. 16.2e, f) that encapsulates the abstracted functional differences and the abstracted behavioural relationships. IDeAL learns the feedback mechanism from the two designs of amplifier (one without feedback and the other with feedback) by differential diagnosis and model-based learning: given the SBF models of the two designs of amplifier in a problem-solving context, IDeAL focuses on the parts relevant to the context and compares them state-by-state to determine what differences in the device behaviours might be responsible for the functional differences; once it identifies them, it abstracts the behavioural and functional relationships from the device-specific models to form the generic BF model of the feedback

Design Patterns and Creative Design

GTM shown. Bhatta and Goel [3,4] provide a detailed account of the learning task and method.

16.4 Analogical transfer based on GTMs

Let us now consider a design problem in the domain of mechanical controllers presented to IDeAL. The new problem has a functional specification that, given the substance angular momentum with a magnitude of L_i and clockwise direction at an input location (gyroscope), the device needs to produce the angular momentum with a magnitude L'_o proportional to the input and the same direction at a specified output location. It also specifies the constraint that the output cannot fluctuate much around an average value (*i.e.*, $L'_o = L_{avg} \pm \delta$, where δ is small). This is the problem of designing a gyroscope follow-up [5].

Let us consider the knowledge condition in which the design of a device (Fig. 16.3) that transfers angular momentum from a gyroscope to an output shaft location is available in IDeAL's analogue memory (or is given explicitly as part of the adaptive design problem). Given an input angular momentum of magnitude L_i and clockwise direction at the input (gyroscope) location, this device produces a proportional angular momentum of magnitude L_o and of clockwise direction at the output shaft location; however, L_o fluctuates over a large range, *i.e.*, $L'_o = L_{avg} \pm \Delta$, where Δ is large. Fig. 16.3 shows the design of the available device: its function, structure and internal causal behaviour. IDeAL retrieves (if not given explicitly) the design of gyroscope control system available in its memory because the desired function matches with the function of this design. That is, the function of this available device is similar to the function of the desired design in that the input states are identical and the output states differ in a parameter value (*i.e.*, magnitude of angular momentum) and a constraint on that value.

Now, the task for IDeAL is to modify the available design of gyroscope control system to deliver the desired function. Simple modifications, such as replacing a component in the given design analogue, will not result in a device that can solve the new design problem because there is no single component in the device that seems responsible for the large fluctuations and that which may be selected for modification. The issue becomes if and how IDeAL can automatically modify the device topology using the knowledge of GTMs.

IDeAL first retrieves the relevant GTM from its memory: it uses the difference in the functions of the candidate and desired designs as a probe into its memory because it indexes the mechanisms by the functional differences and the decomposability conditions on the desired functions. It retrieves the feedback mechanism because the current functional difference, namely the fluctuation in the output property being large vs. small (*i.e.*, Δ versus δ), is the same as the difference that the feedback mechanism reduces, which is specified in a device-independent manner. Then, it tries to match the decomposability condition on the desired function in the feedback mechanism (see Fig. 16.2e for the condition $F_2 = f : (?val11, ?val21) \rightarrow ?val11' + F_1 : (?val11) - ?val21$) with the desired function in order to find the subfunctions f (or g) that need to be designed for and composed with the candidate function. By performing this match, as guided by the language of SBF models, IDeAL finds the subfunction $f : (L_i, L'_o \rightarrow L'_{ww}$, *i.e.*, it needs to design for a structure that takes two inputs, angular momentum with a magnitude of L_i and angular momentum with a magnitude of L'_o and gives as output an angular momentum of L'_{ww} in the opposite

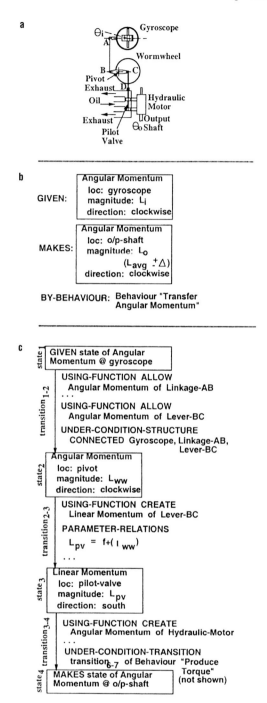

Fig. 16.3 Simple gyroscope follow-up (without feedback control). **a** Gyroscope Follow-up w/o Feedback. **b** Functional specification of "Gyroscope Follow-up" w/o Feedback. **c** Behaviour "Transfer Angular Momentum" of the "Gyroscope Follow-up w/o Feedback".

Design Patterns and Creative Design

> Input:
> - M_1, the SBF Model of the Design Analogue, and its Function F_1.
> - F_2, the desired function.
> - G, a GTM (retrieved by matching $F_2 \sim F_1$).
>
> Output:
> - M_2, the SBF Model of the new device that achieves F_2.
>
> Procedure:
>
> begin
>
> (1) Select the behaviour B_1 in M_1 relevant to F_1.
>
> (2) Bind the initial & final states of B_1 to the appropriate GIVEN and MAKES states of the subfunctions f and g in G.
>
> (3) if $\exists$ an unbound state variable in f or g
>
>   then backtrace B_1 to find states in B_1 that may be modified, considering the bindings from step **2**.
>
>   (3.1) if $\exists$ multiple candidate states for modification
>
>   then Select the state that is nearest to the final state in B_1.
>
>   (3.2) Compute values of unbound state variables in f and g based on the selected state, $(F_2 \sim F_1)$, and PARAMETER-RELATIONS in B_1.
>
> (4) if $\exists$ multiple GIVEN or MAKES states in f or g
>
>   then Check if $\exists\, b \in B_1$ that achieves the transformation from any of the GIVEN states to any of the MAKES states in f or g.
>
>   (4.1) if *yes*
>
>   then $f' = $ rest of the transformation in f.
>
>   (i.e., $<$ (GIVEN-states(f) - initial-state(b)), (MAKES-states(f) - final-state(b)) $>$.)
>
> $g' = $ rest of the transformation in g.
>
>   (i.e., $<$ (GIVEN-states(g) - initial-state(b)), (MAKES-states(g) - final-state(b)) $>$.)
>
> (5) Retrieve subdesigns for f' and g'.
>
>   (5.1) if $\exists$ no subdesigns for f' or g' then FAIL.
>
>   (5.2) else
>
>   (5.2.1) Adapt the retrieved subdesigns for f' and g' (if necessary).
>
>   (5.2.2) Compose $B_{f'}$, the behaviour for f', and $B_{g'}$, the behaviour for g', with B_1 as per the relationships in G.
>
>   (5.2.3) Propagate the resulting changes in state variables forward in B_1 and in the dependent behaviours in M_1.
>
> end

Fig. 16.4 IDeAL's method for instantiating a GTM.

direction at the location of pivot in the candidate design. Next IDeAL instantiates the retrieved GTM in the context of the target problem. The algorithm for IDeAL's process of instantiating the GTMs is shown in Fig. 16.4. When the abstractions are GTMs, this process involves designing for the subfunction(s) determined by matching the applicability conditions of the mechanism (in steps 2 and 3 of the algorithm) and composing the new sub-behaviour(s) with the behaviour of the candidate design as per the relationships specified in the retrieved mechanism (in step 5). Let us walk through the algorithm as it applies to the current example. Step 1 is to select the behaviour relevant for the function of the available design. Since the function in an SBF model of a device directly points to the behaviour relevant for that function, this step is trivial. In the current example, B_1 is the behaviour shown in Fig. 16.3c. Step 2 is to identify bindings for variables in the retrieved GTM, in particular, in the subfunctions to be designed for. Some of the bindings for the state variables are obtained while doing the matching for the retrieval of the GTM itself. As described above, in the current example, IDeAL finds the subfunction f to be $(L_i, L'_o) \to$?$val11'$ because ?$val11$ is the value of the property (whose output values in the desired and retrieved functions are different) in the initial state of B_1 and ?$val21$ is the value in the final state of B_1. As in this example, even after step 2, some other variables, such as ?$val11'$, still need to be bound with specific values from the behaviour of the available design. Step 3 is exactly for doing that: the idea is to trace the relevant behaviour of the available design, B_1, backwards from the final state to the initial state, and identify the intermediate states that are possible candidates for the states of the subfunctions. In the current example, IDeAL needs to find a candidate state from B_1 that could be the output state of the subfunction f. As it traces back the behaviour shown in Fig. 16.3, the first state to be considered is $state_3$. But since it describes a substance (linear momentum) different from what the substance (angular momentum) is from the bindings in step 2, this state cannot be a candidate. Next, it considers $state_2$ which is the only state left and which is a candidate for the output state of f. If $state_3$ were to describe angular momentum, it would also have been a candidate. In such a case, IDeAL would have chosen $state_3$, the state nearest to the final state of B_1. The rationale in this is that the modification selected should cause as minimal disturbance in the candidate behaviour as possible, which means modifying the state as near to the final state as possible in order to solve the problem. Since $state_2$ is selected in the current example, we get the binding for ?$val11'$ from this state, and we have all parts of the subfunction specified.

Since the subfunction has multiple states, step 4 is relevant. In the current design scenario, the subfunction that IDeAL needs to design really has two parts (as it takes two inputs and produces one output): one that specifies the need for transferring angular momentum from the input location to the pivot location, and the other for transferring angular momentum from the output shaft location to the pivot location. Applying step 4, we can find that $transition_{1-2}$ really covers the transformation ?$val11$ $\to$?$val11'$ and the remaining transformation (f') in f is ?$val21 \to$?$val11'$. That is, the first part is already designed for in the candidate design as the behaviour segment $state_1 \to state_2$ (Fig. 16.3c) achieves it.

Therefore, in successfully instantiating the mechanism in the candidate design of gyroscope follow-up, IDeAL only needs to find a behaviour (and a structure) that accomplishes the second part of the subfunction (f') given the context of the first transformation. Let us consider the knowledge condition in which IDeAL has the knowledge of a component (called worm) whose function is to transfer an input angular momentum to an output location with the magnitude proportional to the output component and the direction dependent on the direction of threading on the worm. This component reverses the direction of the input angular momentum. In step 5, given the subfunction f', IDeAL retrieves that component because the desired part of the subfunction matches with the component's function. It substitutes the appropriate parameters in the behaviour of the retrieved design (i.e., worm) to generate a behaviour for the desired subfunction. Then it composes that behaviour (i.e., B_{22}) with the behaviour of the candidate design (i.e., B_1) as per the specification of the causal relationships in the feedback mechanism (as in Fig. 16.2f) to propose a behaviour (shown in Fig. 16.5c) for achieving the desired function. Note that the resulting modification is non-local, in that it modifies the device topology. It finally propagates the changes in states resulting from composing the subdesign's behaviour with B_1 forward to the final state or until a state is revisited.

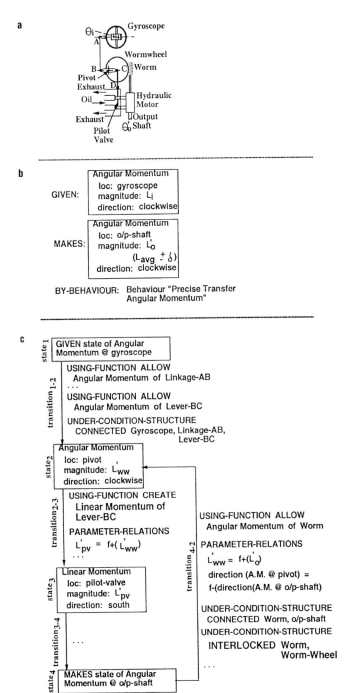

Fig. 16.5 Gyroscope follow-up instantiating the feedback mechanism. **a** Gyroscope Follow-up with Feedback. **b** Functional specification of the desired "Gyroscope Follow-up" w/o Feedback. **c** Behaviour "Precise Transfer Angular Momentum" of the desired "Gyroscope Follow-up with Feedback".

16.5 Evaluation

IDeAL provides a test-bed for experimenting with the MBA theory. We conducted several kinds of experiment with IDeAL that evaluate the MBA theory for its acquisition, access, and use of GTMs. Each experiment typically contained two steps. The first step involved giving IDeAL a pair of designs, one without any instance of GTM and the other with an instance of a GTM, and testing IDeAL's ability to learn a BF representation of the GTM instantiated in one of the two input designs. In the second step, IDeAL is given a design problem, from a different domain in some cases, such that it would need to access and use a previously learned GTM in order to solve the given problem. We verified if it could autonomously recognise the applicability of a GTM and successfully access and use it to solve the given problem. We conducted 12 such experiments with different combinations of design sources and target problems from four different design domains involving 28 distinct designs. We tested IDeAL's learning of six different GTMs and its use of three of them. In all these cases, IDeAL was successful in learning GTMs, and in accessing and using them in solving design problems. The behaviour of IDeAL in these experiments led us to conclude the results in the following four dimensions.

1. *Computational feasibility and efficacy*: IDeAL successfully addresses the multiple tasks in the MBA theory, e.g., the tasks of learning, accessing, transferring and using GTMs. It thus demonstrates the computational feasibility and efficacy of the MBA theory.
2. *Uniformity of representations*: the different tasks in the MBA theory impose constraints on one another. They also impose different constraints on the design knowledge representations. IDeAL uses the same SBF language for addressing the different tasks. The GTMs, for example, are represented in a BF language that is a subset of the SBF language. The design analogues too are indexed in the SBF vocabulary.
3. *Generality of domains*: as mentioned above, IDeAL presently contains about 30 design analogues from four different device domains, namely the domains of simple electric circuits, heat exchangers, electronic circuits, and complex mechanical devices (such as momentum controllers and velocity controllers). This includes the design problem used as an illustrative example in this paper, which was taken from a classical textbook on mechanical design [5].
4. *Generality in terms of different GTMs*: IDeAL presently covers six different GTMs – cascading, four different types of feedback, and one type of feedforward.

16.6 Related research

Design patterns are often discussed in the design literature as a basic unit of design knowledge. Gamma *et al.* [6], for example, have advocated the construction of libraries of reusable design patterns for supporting interactive object-oriented programming. Both their study and ours appear to be based on the same notion: design patterns form productive units of transfer and reuse of design knowledge.

SYN [7] and DSSUA [8] are two recent analogical design systems (in contrast to case-based design). The former is a module within the FABEL system, a case-based computer-aided design environment intended to assist human architects in designing spatial layouts of air-circulation systems in buildings. It represents its design cases as graph-based topological models, and the models are organised in a tree of design concepts, where a design concept corresponds to a maximal common subgraph between the designs under the concept. SYN's design concepts may be viewed as topological design patterns. However, it does not abstract any pattern from specific designs; rather, it assumes the patterns as given. Also, its use of the patterns is limited to simple within-domain analogies.

DESSUA is based on the notion of design prototypes. Like design patterns, design prototypes too specify functional relations and causal structures in a class of devices, but, unlike a design pattern, a design prototype also specifies the generic physical structure of the device class. Whereas a design prototype is a generalisation over design cases such that a case is an instance of a prototype, a design pattern is an abstraction over design prototypes such that a design prototype is a subclass of a design pattern. DESSUA uses an analogical process similar to that of the structure-mapping engine (SME) [9] to abstract causal behaviours at transfer time. In contrast, in MBA, design patterns are abstracted at storage time and acquired for potential reuse. The design patterns are indexed by the problem-solving goals stated in terms of functional differences between two design situations. This aspect of MBA is similar to purpose-directed analogy [10].

PHINEAS [11], which evolves from SME, also uses high-level abstractions for establishing correspondence between the source and the target situations. But it provides neither any content account of the high-level abstractions nor a process account of their acquisition. MBA provides both a content account of generic abstractions in relation to device design, and also a process account of the acquisition, access, and use of the abstractions.

Case-based theories of design mostly involve direct transfer of the structure of familiar designs to new design situations. That is, the transfer is not mediated by high-level abstractions. In some case-based theories, *e.g.*, see Shinn [12], however, high-level abstractions do enable case reminding, but still play little role in analogical transfer. Also, the high-level abstractions in these case-based theories are generalisations over features of a problem, and do not specify relations that characterise a problem and its solution. Finally, design adaptations in case-based design are, in general, limited to local (typically parametric) design modifications.

The MBA theory evolves from an earlier theory of case-based design called adaptive modelling [13,14]. The adaptive-modelling theory described case-specific SBF models. It showed how case-specific SBF models enable local (*i.e.*, parametric or componential) modifications to source designs for solving target design problems. It also showed how case-specific SBF models of new designs can be acquired by adapting the models of known designs. Stroulia and Goel [15] described how case-independent generic models enable topological modifications to source designs in the same domain as that of target problems. Bhatta and Goel [3,4] showed how generic models can be acquired by abstraction over case-specific SBF models. The MBA theory completes the circle by showing how generic models mediate analogical transfer of design knowledge from the source domain to a target domain (*e.g.*, from amplifiers to gyroscopes).

16.7 Conclusions

That analogy plays a central role in creative design is a standard cliché in design research. MBA provides a normative computational theory of analogical design based on the notion of design patterns. In the domain of physical devices, it shows that functional and causal design patterns, called GTMs, constitute productive units of analogical transfer of design knowledge. It also shows that SBF models of specific devices enable the acquisition of GTMs, that problem-solving goals pertaining to the adaptation of a familiar design to meet new functional requirements access relevant GTMs, and that the instantiation of the GTMs in the context of a partial design enables its completion.

Acknowledgements

This paper has benefited from numerous discussions with members of the Intelligence and Design research group at Georgia Tech. This work has been supported in part by research grants from NSF (IRI-92-10925 and DMI-94-20405) and ONR (research contract N00014-92-J-1234).

References

[1] Alexander C. Notes on the synthesis of form. Harvard University Press: 1964.
[2] Goel A, Bhatta S, Stroulia E. Kritik: an early case-based design system. In: Maher ML, Pu P, editors. Issues in case-based design. Hillsdale (NJ): Erlbaum, 1997.
[3] Bhatta S, Goel A. From design experiences to generic mechanisms: model-based learning in analogical design. Artif Intell Eng Des Anal Manuf 1996;10:131–6.
[4] Bhatta S, Goel A. Learning generic mechanisms for innovative strategies in adaptive design. J Learn Sci 1997;6(4):367–96.
[5] Hammond PH. Feedback theory and its applications. London (UK): The English University Press, 1958.
[6] Gamma E, Helm R, Johnson R, Vlissides J. Design patterns: elements of reusable object-oriented software. Addison-Wesley: 1995.
[7] Borner K, Pippig E, Tammer E, Coulon C. Structural similarity and adaptation. In: Proceedings of European Workshop on Case-Based Reasoning, Lausanne, Switzerland, 1996; 58–75.
[8] Qian L, Gero J. A design support system using analogy. In: Gero J, editor. Proceedings of the Second International Conference on AI in Design. Kluwer Academic: 1992; 795–813.
[9] Falkenhainer B, Forbus K, Gentner D. The structure-mapping engine: algorithm and examples. Artif Intell 1989;41:1–63.
[10] Kedar-Cabelli S. Toward a computational model of purpose-directed analogy. In: Michalski R, Carbonell J, Mitchell T, editors. Machine learning II: an artificial intelligence approach. Los Altos (CA): Morgan Kaufmann, 1988; 284–90.
[11] Falkenhainer B. Learning from physical analogies: a study in analogy and the explanation process. Ph.D. thesis, University of Illinois, Department of Computer Science, Urbana, IL, 1989.
[12] Shinn HS. Abstractional analogy: a model of analogical reasoning. In: Kolodner J, editor. Proceedings of the DARPA Workshop on Case-Based Reasoning, Clearwater Beach, FL, 1988; 370–87.
[13] Goel A. A model-based approach to case adaptation. In: Proceedings of the Thirteenth Annual Conference of the Cognitive Science Society, Chicago, 1991; 143–8.
[14] Goel A. Model revision: a theory of incremental model learning. In: Proceedings of the Eighth International Conference on Machine Learning, Chicago, 1991; 605–9.
[15] Stroulia E, Goel A. Generic teleological mechanisms and their use in case adaptation. In: Proceedings of the Fourteenth Annual Conference of the Cognitive Science Society, Bloomington, IN, 1992; 319–24.

FAMING: supporting innovative design using adaptation – a description of the approach, implementation, illustrative example and evaluation

17

Boi Faltings

Abstract A popular saying claims that "innovation is 1% inspiration and 99% perspiration". The goal of using computers in design should be to automate the perspiration that makes up the bulk of the work in design. We consider supporting innovative design processes that can be structured into three steps: *discovery* of a new technique, *understanding* it, and *generalising* it to fit the problem at hand. The method we describe leaves the discovery phase to the human designer, but automates the understanding and generalisation phases that involve most of the perspiration. As the main example to demonstrate the technique, we present the FAMING system for designing part shapes in two-dimensional elementary mechanisms, also called *kinematic pairs*. We believe that the results can be generalised to other domains with similar characteristics, particularly for any problem where geometry plays an important role.

17.1 Introduction

Most computer programs concern *deductive* tasks, such as *analysis*, where a single answer follows from the input data. In contrast, design is an *abductive* problem, which often has infinitely many correct answers. As an example, consider the function of an escapement, the central element of mechanical clocks. Its function is to regulate the motion of a *scape wheel* to advance one tooth per oscillation of a pendulum. Since the pendulum has a fixed period of oscillation, this means the wheel moves at constant speed and can drive the hands of a clock. There are many mechanisms that can be used to implement an escapement function. Figure 17.1 shows three different mechanisms that satisfy this function, and many more variations could be designed.

Because of the abductive nature of design, an apparently closed problem, like designing part shapes for an escapement mechanism, still leaves much room for innovation and creativity; even though the problem has been studied for hundreds of years, novel devices are still being designed in the watch industry to this day. In this paper, we present the Functional Analysis of Mechanisms for Inventing New Geometries (FAMING)[1] program for supporting creative design of part shapes in higher kinematic pairs. This domain has proven to be particularly interesting for

---

[1] FAMING also means "invent" in Mandarin Chinese.

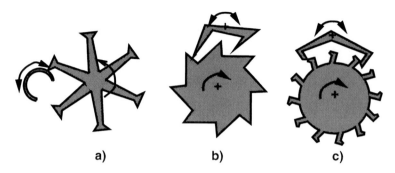

Fig. 17.1 Three different kinematic pairs implementing an escapement function.

research because it demands creative design solutions but can still be formalised with a small knowledge base.

The only known method for solving abductive problems in general is to search the space of potential solutions. For example, there is a very successful program that invents novel automatic transmissions for automobiles by searching the space of all physically possible topologies allowed by a certain technology [1]. This technique is successful in producing innovations because (i) a computer can search through large numbers of candidates, and (ii) a computer is much better than people at providing the correct analysis of a device's behaviour. Whereas transmissions can be modelled by a fixed set of parameters, kinematic pairs require more sophisticated models and design methodologies. Before describing the approach we have developed to support shape design, we are going to review several design paradigms and their applicability to design of kinematic pairs.

17.1.1 Model-based design

Work in *model-based design* [2–4] has attempted to develop systems that systematically search the space of combinations of a set of *models* to produce novel designs. A model is a structural element with a fixed behaviour. For achieving the desired function, the models can be *inverted* and thus propose a potential structure. For example, the system of Williams [2] proposes a design for refilling a punch-bowl by composing qualitative models of pipes and the behaviour of liquids. Subramanian [5] describes a program that can propose kinematic chains. These methods work well when the interactions between the individual models can be identified precisely (the *no-function-in-structure* principle of qualitative modelling [6]).

Kinematic pairs achieve their function through the possible contacts between elements of the part shapes. Because all contacts have to be integrated into a common geometry, they interact in very unstructured ways: any contact can *subsume* any other one and inhibit it. This makes it impossible to model kinematic function in a context-free formalism that would allow a model-based design strategy. Joskowicz and Addanki [7] have attempted to apply methods of model-based design to kinematic pairs. However, their method is capable of obtaining a solution only if by accident the specifications define a set of non-interacting contact relations. It is incapable of producing devices, such as escapements or ratchets, that require interaction of contacts between very different parts of the devices.

17.1.2 Prototype-based design

The highly context-dependent nature of mechanical shape design means that the space of mechanisms with interesting functions is sparse. If designs that are good for particular functions are to be described by combinations of predicates, this often means that predicates are invented specifically for one particular device. Rather than identifying combinations of properties that are responsible for a particular function, it is more appropriate to model *prototypes* [8] of complete designs.

Prototypes are usually parameterised to allow their integration in varied design contexts. For many kinematic pairs, one can identify several key parameters that can be varied without changing the function. There exist commercial tools for supporting such design, for example ProEngineer [9] or ICAD [10]. However, these tools only allow instantiation of preprogrammed prototypes and thus do not permit innovation without involving a programmer.

17.1.3 Case-based design

Case-based reasoning [11] is a technique where past solutions are reused or adapted to solve new problems. In *case-based* design [12], specific design precedents are reused for new problems. This approach maintains the advantages of the prototype-based approach, but simplifies knowledge acquisition, as a library of cases can be built up by simply recording earlier designs. FAMING, the system we describe in this chapter, uses geometric models of mechanism part shapes as its cases.

Because designs can never be reused exactly, a key issue in case-based design is how to *adapt* a case to a new problem. For adaptation, it is crucial to know what changes can be made to a design case without perturbing its function. This knowledge, called *adaptation knowledge*, must be provided in addition to the model of the structure stored in the design case [13].

17.1.4 Annotating cases with functional models

In FAMING, adaptation knowledge is provided by associating with each case an *interpretation* in terms of structure–behaviour–function (SBF) models [14]. An SBF model represents an *understanding* of the case: it defines how each element of the structure is responsible for aspects of the behaviour, and how aspects of the behaviour are, in turn, responsible for the function of the device. SBF models for case adaptation were first used in the KRITIK program of Goel [15], which adapts a nitric acid cooler design to cooling sulphuric acid. In FAMING, the user must only provide a formal expression of the kinematic functions that are considered important in the case. The program then automatically constructs the intermediate behavioural model and the links between it and structure on one side and function on the other.

In order to express function correctly, the SBF model must be a *qualitative* model. A qualitative model [16] differs from a quantitative one in that it specifies properties that hold over ranges of parameter values. This makes it possible to express functions that hold over a *range* of situations, such as "block any counterclockwise motion". Qualitative models also make it possible to determine *all* possible behaviours of a device, not just a particular snapshot valid for certain input parameters. This is necessary for expressing negative specifications, such as "parts A should never move counterclockwise". Finally, a third reason for using qualitative

SBF models is that their discrete nature allows problem solving by search among all possibilities.

17.1.5 Case adaptation using SBF models

FAMING translates the understanding embodied in the SBF model to the structural level for use in adapting the case to new specifications. This *structure–behaviour* (S–B) inversion is possible owing to the use of qualitative behaviour models. It consists of mapping each property in the behaviour model to *constraints* on the shapes that ensure that the property is satisfied. The resulting set of constraints on the structural model allows *generalisation* of the structure in the case. Additional constraints on the behaviour, such as numerical bounds, can also be formulated as structural constraints.

Finally, the case is *adapted* to fit a new problem. A first solution is obtained either by *combination* of the structures in several cases, or by *modification* of a single case's structure. Both operations are carried out while maintaining the validity of all constraints defined in the generalisation stage. Since the composition may require taking into account further compositional constraints, these are discovered through renewed qualitative simulation and corrected by modification operators.

17.1.6 Innovation in case-based design

Innovation and creativity can arise in case-based design through adaptation and combination of cases that might yield very different results from what was previously known. As observed for example by Wills and Kolodner [17], another powerful source of innovation is the *reinterpretation* of an existing case using a different SBF model. Figure 17.2 shows an example where a ratchet device is reinterpreted using a different SBF model than originally intended. Combination of two such devices results in an innovative design of a forward–reverse-mechanism, a device that transforms an oscillating input motion into a rotation and which advances two steps forward and one step backward with each oscillation (this example is described in detail in Ref. [18]). The design obtained using FAMING is much simpler than those found in the

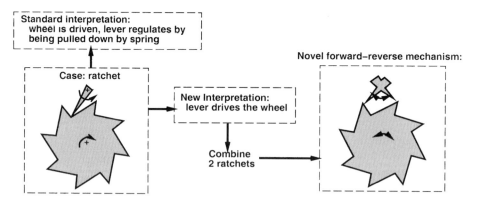

Fig. 17.2 When viewed by a designer, artefacts are interpreted in terms of SBF models. Innovation often results from reinterpreting cases with different models than originally intended.

literature and thus a truly creative solution. When the set of known shapes is large, such reinterpretation is the source of much of the innovation in design.

17.1.7 FAMING: an interactive design tool

Truly creative design involves a large amount of knowledge and is well beyond the capacity of computers today. FAMING is an interactive tool that leaves the 1% inspiration – choice of cases, identification of interesting functions – to the human and focuses on automating the 99% perspiration required to produce a complete and correct result. Using FAMING relieves the human designer of tedious tasks and thus allows improved productivity.

17.2 Qualitative SBF models used in FAMING

Figure 17.3 shows the structures that implement the SBF models in FAMING, and how they are related through reasoning processes. The structure of a device is rep-

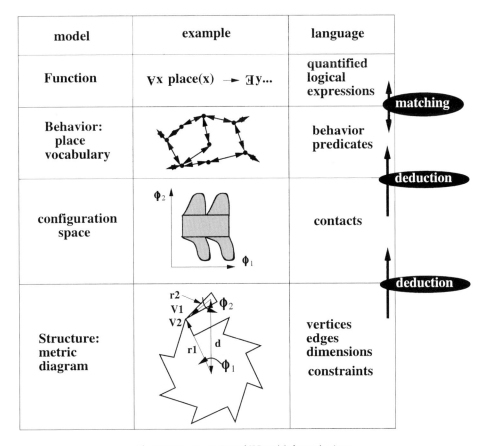

Fig. 17.3 Representations of SBF models for mechanisms.

resented by a *metric diagram*, a geometric model of vertices and connecting edges. From the geometric model, deductive algorithms are used to compute a *place vocabulary*, a complete model of all possible qualitative behaviours of the device. The place vocabulary can be formalised as a qualitative representation of the device's *configuration space*. Configuration space is a compact representation of the constraints on the part motions. Its usefulness for modelling mechanism kinematics has been argued for example by Sacks and Joskowicz [19]. In order to be fast and reliable, FAMING directly computes a qualitative configuration space, called a *place vocabulary* and modelled using *behaviour predicates*. The *function* of a device is formulated using logical expressions on these behaviour predicates. Functional expressions can thus be *matched* automatically to the qualitative behaviour.

17.2.1 Structure: metric diagram

Shapes are represented using a metric diagram. The metric diagram consists of a symbolic structure that defines vertices, edges and metric parameters for the positions of the vertices. In our current implementation, the metric diagram is restricted to polygons, but see Ref. [20] for ways to extend it to include circular arcs. A metric diagram represents two objects, each of which has one well-defined degree of freedom.

Interesting aspects define *shape features*, which may involve both objects. For example, the fact that the top of the ratchet's lever is able to touch the wheel (v_1 touching v_2 in Fig. 17.3) depends on a shape feature. It is defined by:

- a set of vertices and edges, in this case v_1 and v_2;
- the metric parameters associated with them, in this case ϕ_1, r_1, ϕ_2 and r_2;
- a set of constraints that must hold simultaneously for the shape feature to be present, in this case $\{|d - r_1| < r_2\}$.

17.2.2 Qualitative behaviour

Textbooks explain kinematic behaviour qualitatively by sequences of *kinematic states*. A kinematic state is defined by a contact relation and directions of part motion. Examples of kinematic states of a ratchet device are shown in Fig. 17.4.

In qualitative physics terminology, a graph of kinematic states and transitions is called an *envisionment*. It can be computed based on a place vocabulary, a graph where each node represents a different combination of contact relationships, and each arc represents a potential transition between them. The envisionment is obtained by combining each node of the place vocabulary with assumed motions and keeping only the states and transitions consistent with the external forces and motions. We have developed and implemented complete algorithms to compute place vocabularies for arbitrary two-dimensional higher kinematic pairs in fixed-axis mechanisms. These have been used to compute envisionments for a large variety of practical mechanisms, such as a mechanical clock [21,22].

17.2.2.1 Qualitative motions

Each part in a kinematic pair has exactly one degree of freedom, so that the motion of a kinematic pair can be modelled by a vector of two parameters. Qualitatively, each value is modelled by its *sign*, +, 0 or −, so that a qualitative motion is a *qualitative*

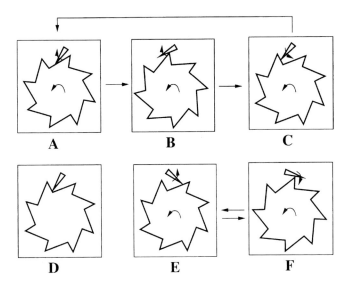

Fig. 17.4 Examples of kinematic states and transitions in a ratchet. States A–C represent a cycle of behaviour in which the wheel can turn counterclockwise. State D can be reached from state A by reversing the wheel's motion; it blocks any further clockwise motion. States E and F represent a cycle that does not allow the ratchet function and is normally avoided by applying a counterclockwise force on the lever.

vector consisting of two such signs. As a shorthand, we shall use "*" to denote the set of values {+,0,–}.

Owing to the fact that the information in a qualitative model is incomplete, the qualitative motion will often be *ambiguous*, i.e., admit several different vectors. For each state x of the device, we define a set $\mathcal{M}(x)$ that specifies all qualitative motion vectors possible in state x.

17.2.2.2 External influences

A kinematic pair can be actuated either by applying a force or momentum, in which case there could be an opposing force that prevents motion, or by forcing a particular motion, which cannot be prevented by any counteracting force. We assume that externally imposed motions and forces are independent of the device state, and represent them by sets of qualitative vectors $\mathcal{M}_{ext}$ and $\mathcal{F}_{ext}$. The set $\mathcal{M}_{ext}$ specifies all motions that are *consistent* with the external influence. The set $\mathcal{F}_{ext}$ denotes the set of actual force vectors that might be applied, and is often ambiguous because a function might be required under a range of circumstances.

17.2.2.3 Place vocabulary

The place vocabulary represents the set of possible contact relations. We represent place vocabularies using a set of behaviour predicates, which characterise places, their features and their connectivity. For a kinematic pair, the place vocabulary defines a graph containing three types of kinematic state, corresponding to two, one and no contacts, and identified by the behaviour predicates **point-place**(x), **edge-**

place(*x*) and **face-place**(*x*), which specify that *x* is a place with two, one or zero contacts respectively.

For each place, the place vocabulary defines the allowed qualitative directions of motion. The predicate **allowed-motion**(*x,d*) specifies that motion *d*, where *d* is a qualitative vector, is possible everywhere in place *x*. It is often more useful to make use of the set of *admissible motions* $\mathcal{M}_{adm}(x) = \{d|\textbf{allowed-motion}(x,d)\}$. For each link between states, the directions of motion that can cause a transition are defined by the predicate **transition**(*x,y,d*), which specifies that motion *d* can cause a transition from place *x* to *y*.

17.2.2.4 Behaviour = envisionments of kinematic states

A qualitative *behaviour* of a kinematic pair is an ordered sequence of kinematic states. A kinematic state is characterised by a particular contact relationship (place) and a qualitative motion. The set of all possible behaviours of a kinematic pair can be modelled by connecting all possible kinematic states in a graph whose arcs represent all possible transitions. Such a graph is called an envisionment [6,23] and is the result of the qualitative simulation procedure. The fact that the envisionment represents *all* possible behaviours, not just the behaviour for a certain input, is an important advantage of qualitative simulation over numerical techniques.

The envisionment of a kinematic pair is obtained from the place vocabulary in two steps. First, for each place we compute the set of *consistent* motions. Next, for each pair of kinematic states such that their underlying places are connected, we compute the set of possible transitions.

We define the *envisionment functions* $\mathcal{M}$ and $\mathcal{T}$ as the combination of all these considerations. $\mathcal{M}(x, \mathcal{F}_{ext}, \mathcal{M}_{ext})$, applied to a place *x*, returns the minimal set of qualitative motions in the place *x*, and $\mathcal{T}(x, \mathcal{F}_{ext}, \mathcal{M}_{ext})$ returns the set of possible transitions from place *x* to other places.

17.2.3 A language for specifying function

There has been much recent work within the model-based reasoning community on formalisms for representing function in design [24]. Sembugamoorthy and Chandrasekaran [25] have investigated the notion of a *functional representation*. Iwasaki et al. [26] have developed a language called CFRL based on qualitative physics. Another major work is that of Tomiyama and coworkers [27], who define function using notions of qualitative process theory. All these proposals consider function to be a *causal* relation between an *environment* (or context) and a particular *behaviour*, and are concerned mainly with vocabularies for specifying these causal connections. Behaviour is defined, for example, as a set of relations between parameters (functional representation [25]), a set of active processes and views [27], or a precise sequence of states, each specifying a particular set of relations [26].

In kinematic pairs, there is only one form of causality, that of pushing on a part contact. In any particular state, the *function* of the device is given by the inference rules defined by the part contacts, expressed by the **allowed-motion** behaviour predicates. Many important functions, however, are properties of *sequences* of states. For example, a function might be that from any initial state, motion in a particular direction will eventually lead to a particular state, or that a certain behaviour cannot occur anywhere.

Such specifications can only be specified as logical *conditions* on the set of *all possible behaviours*. These are formulated using the behaviour predicates and allow quantification over states and qualitative motions and forces.

We therefore formulate functions in two levels.

- A *functional feature* defines a property of a particular state or set of states, and thus always takes at least one state as an argument. Furthermore, it takes the external influences $\mathcal{M}_{ext}$ and $\mathcal{F}_{ext}$ as implicit additional arguments. Functional features are similar to causal process descriptions (CPDs) used in CFRL [26].
- A *device function* defines a property of the entire behaviour. It consists of a logical expression in functional features where all states are bound by quantifiers, and a specification of the $\mathcal{M}_{ext}$ and $\mathcal{F}_{ext}$ assumed for this device function.[2]

For example, some functional features that our current prototype system uses are:

- **blocking-place(x):**
  $(\forall d \in \mathcal{M}_{ext}) \neg$**allowed-motion**$(x,d)$
  (a place blocks motions if all external motions are disallowed);
- **possible-path(S):** $S = (x_0, x_1, x_2, \ldots, x_n)$
  $(x_0 = x_n) \vee (\forall i < n) x_{i+1} \in \mathcal{T}(x_i, \mathcal{F}_{ext}, \mathcal{M}_{ext})$
  (S is a path from place x_0 to place x_n whenever there is a sequence of places with transitions between them under at least one assumed motion).

A place vocabulary can only fulfil the required functions if the number of states and their connectedness is sufficient. Reasoning about such *topological* features is difficult in the place vocabulary itself, since it is based only on individual states. We use an explicit representation of the *kinematic topology* [28] of the mechanism to formulate properties relating to the topology of behaviour. An example of a functional feature defined on the basis of kinematic topology is:

- **cycle-topology(C,d_1,d_2)** – if the first or second object has rotational freedom, the cycle C involves d_1 rotations of the first or d_2 rotations of the second object,

which can be defined formally using similar behaviour predicates as those that define place vocabularies.

As an example, the functions of a ratchet can be defined as follows:

- for $\mathcal{M}_{ext} = \{(+,*)\} \wedge \mathcal{F}_{ext} = \{(*,+)\}$:
  $(\exists C)$**cycle**$(a,C) \wedge$ **cycle-topopogy** $(C,1,0) \wedge$
  $\neg(\exists x)$**blocking-space**$(x) \wedge$ **possible-path**(a,x)
  (assuming that the wheel turns counterclockwise and the lever is forced onto it, there is a cycle of states where the wheel can rotate, and no reachable blocking state from any starting state a)
- for $\mathcal{M}(x) = \{(-,*)\} \wedge \mathcal{F}_{ext} = \{(*,+)\}$:
  $(\forall y)[(\exists S = (a, \ldots, y))$**possible-path**$(S)] \Rightarrow$
  $\{\neg(\exists C)$**cycle**$(y,C) \wedge (\exists z)$(**blocking-state**$(z) \wedge$ **possible-path**$(y,z))\}$
  (assuming that the wheel turns clockwise, no reachable state leads to a cycle and all states can eventually lead to a blocking state).

---

[2] Recall that we assume the external influences to be independent of the mechanism state.

Note that owing to the ambiguities inherent in qualitative envisionments, they overgenerate behaviours. Therefore, it is only possible to define *necessary*, but never *sufficient* specifications of behaviour and, consequently, function. For example, we can express the specification that clockwise motion leads to a blocking state only in an indirect manner: if there is no possibility to cycle, and there is at least one reachable blocking state, the device must eventually reach this state.

17.2.3.1 Quantitative constraints on behaviour

In many cases, purely qualitative specifications are insufficiently precise to specify a device function. For example, the specification of the forward–reverse mechanism (shown in Fig. 17.2) must mention the fact that the forward motion is to be two times the reverse motion. Such constraints refer to particular configurations of the device, represented in the place vocabulary as zero-dimensional **point-places**. In order to allow their specification, the functional language contains the function **component**(x,i) which returns the ith coordinate of the configuration represented by **point-place** x. Constraints involving precise positions can be formulated on these coordinates.

17.3 Inverting the FBS model

Adapting a case C to a novel problem requires

- understanding what aspects of the device are relevant to the interesting function, *i.e.*, constructing an *interpretation* of its behaviour B in terms of the functional specification F, and
- using this understanding to construct a generalisation that either does not change these essential aspects, or changes them in the way that is intended. This is accomplished by an abductive *inversion* of behaviour to structure (S–B inversion), which defines a set of *constraints* on the structural model.

17.3.1 Matching behaviour to functional specification

F is a quantified logical expression of behaviour predicates. C implements F by its behaviour B; therefore, there exists at least one instantiation of the quantified variables in F with individuals of B such that the behavioural predicates of B satisfy F. Replacing all the quantified variables in F using this substitution, we obtain a *conjunction* of instances of behaviour predicates that define the way that F is implemented in C:

functional feature $F \Rightarrow$ behaviour-pred$_1 \wedge$ behaviour-pred$_2 \wedge \ldots$.

As an example, suppose that the functional specification of a device contains the conditions that there exists a blocking-place:

for $\mathcal{M}_{ext} = \{(-,*)\}, \mathcal{F}_{ext} = \{(*,*)\}$:
$(\exists x)(\textbf{point-place}(x) \wedge \textbf{blocking-place}(x))$.

Suppose, furthermore, that the designer has selected a device that has a place P that qualifies as a blocking place. Unification of the functional specification with the place vocabulary substitutes P for x, thus transforming it into

blocking-place(P).

Replacing the **blocking-place** predicate by the full expression in its definition and expanding the quantification over all motions, we obtain the following conjunction of behaviour predicates:

$\neg$**allowed-motion**$(P,(-,-)) \wedge$
$\neg$**allowed-motion**$(P,(-,0)) \wedge$
$\neg$**allowed-motion**$(P,(-,+))$.

The presence of quantification in F is essential for allowing innovation and creativity: if F did not contain any quantified individuals, it could at most admit a finite number of equivalent[3] behaviours that could be enumerated in a straightforward way. In general, finding all conjunctive propositions that satisfy a quantified logical expression is a non-computable problem, thus putting creativity beyond the scope of algorithms.

17.3.2 S–B inversion

Each behaviour predicate in the conjunction is implemented by particular aspects of the object shapes. Using a trace of their computation, it is possible to determine the limits up to which the behaviour predicates remain valid. These limits define *constraints* on the shapes. The constraints, taken together, define a *qualitative shape feature* that is associated to the functional feature. That is:

behaviour-pred$_1 \wedge$ behaviour-pred$_2 \ldots \Rightarrow$
constraints on shapes $\Rightarrow$ shape feature.

Thus, the behaviour predicate $\neg$**allowed-motion**$(P,(-,-))$ can be translated into constraints on position of vertex v_1 when the device is in configuration P. The **point-place**(P) predicate translates into conditions for the particular touch being physically possible, *i.e.*, not ruled out by other contacts.

Reversing the causal chain of the analysis thus establishes a mapping from functional features to shape features, and we call such a process causal inversion. More details on the mapping between shape and qualitative behaviour can be found in Ref. [29].

17.4 Case adaptation

The final stage of the design process is to adapt the case(s) to the new problem. Cases can be adapted either by combining several cases, or by incrementally modifying one

---

[3] In the sense that the aspects satisfying the function are identical.

single case. Either way, the interpretation of the cases defines a set of *structural constraints*. Some of these structural constraints fix the existence and connectivity of structural variables, and others restrict their relative values. Taken together for all desired functions, the structural constraints define a *constraint satisfaction problem* whose solution is the adapted case.

The constraint network for combining shape features is dynamic and involves many non-linear constraints. No reliable and efficient method exists for solving such constraint networks. However, the cases themselves already provide partial solutions. Recent studies have shown that that iterative repair methods seem to be very efficient for solving large constraint satisfaction problems [30,31]. Cases can be used as initial solutions for such repair algorithms, as proposed by Pu and Purvis [32]. In the same spirit, our prototype solves the CSP by a combination of case *combination*, where value combinations that solve partial problems are combined into a new solution, and case *modification*, where an iteration of local repairs involving single variables is used to refine an initial assignment of values incrementally. The *topology* of the constraint network is decided by querying the user: in case combination, the user is asked to identify which parts of the shapes can be reused in the two devices. If the problem turns out to be unsatisfiable, the user is asked to provide additional degrees of freedom by adding further variables.

In this section, we illustrate both case combination and modification for the example of designing an *escapement* mechanism. An escapement is a device where the constant-period oscillation of a pendulum regulates the rotation of a wheel such that it advances by one tooth for each oscillation. This function can be formally specified by stating that both extreme positions of the pendulum fall within **blocking-places** for the wheel's motion, and that there are paths between successive instances of these blocking places:

1. $\exists \mathcal{X} = [x_0, x_1, x_2, \ldots, x_{n-1}]$
   $(\forall x_i)$**partial-blocking-place**$(x_i,(+,*)) \wedge$ **cycle-topology**$(\mathcal{X},1,0)$.
2. $\exists \mathcal{Y} = [y_0, y_1, y_2, \ldots, y_{n-1}]$
   $(\forall y_i)$**partial-blocking-place**$(y_i,(+,*)) \wedge$ **cycle-topology**$(\mathcal{Y},1,0)$.
3. For $\mathcal{M}(x) = \{(*,0),(*,-)\} \wedge \mathcal{F}_{ass} = \{(+,*)\}$:
   $(\forall x_i \in \mathcal{X})$**possible-path**$(x_i, y_i), y_i \in \mathcal{Y}$
   (when the pendulum swings from right to left or is stationary, there exists a path from place x_i to y_i).
4. for $\mathcal{M}(x) = \{(*,0),(*,+)\} \wedge \mathcal{F}_{a33} = \{(+,*)\}$:
   $(\forall y_i \in \mathcal{Y})$**possible-path**$(y_i, x_{\mathrm{mod}(i+1,n)}), x_i \in \mathcal{X}$
   (when the pendulum swings from left to right or is stationary, there exists a path from place y_i to the place following x_i in the cycle $\mathcal{X}$).
5. for $\mathcal{M}(x) = \{(*,0)\}$: $(\forall x)\neg(\exists C)$**cycle**$(x,C)$
   (the wheel is prevented from rotating a full cycle whenever the pendulum does not move).

17.4.1 Case combination

Assume that the designer has noticed that a ratchet device, when used in the environment of an escapement, can implement the desired **partial-blocking-places**. The designer decides to compose two ratchet devices and identifies their blocking states as the interesting functions, thus creating constraints on their composition. Figure 17.5 (left) shows a trace of the design process.

FAMING: Supporting Innovative Design Using Adaptation

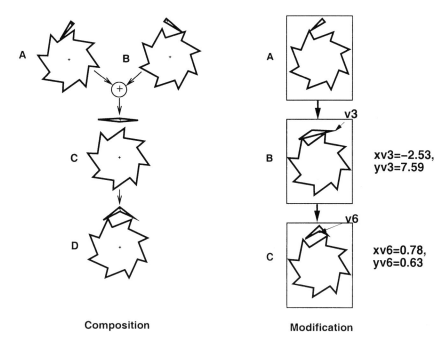

Fig. 17.5 Design of an escapement by (left) composing two ratchet devices and (right) incrementally modifying a single ratchet.

The functional features have been mapped into two shape features, each defined as a set of constraints on the metric diagrams of single ratchet devices A and B. A first composition (C) results in a non-functional device, as there is a cycle of states where the lever does not move, but the wheel is free to turn; this is a contradiction to specification (5). The transitions in this cycle of states have conditions associated with them. By S–B inversion, these are translated into constraints on the positions of the vertices on the lever. Solving the modified constraint satisfaction problem results in design (D), which satisfies all the specifications.

A general and complete, but still somewhat inefficient, method for solving the complex constraint satisfaction problems resulting from kinematic constraints has been proposed by Haroud and Faltings [33]. FAMING currently uses a more pragmatic solution using incremental refinement operators, called *modification operators*, that refine an initial candidate solution by modifying one parameter at a time.

17.4.2 Modification operators

A constraint satisfaction with continuous variables defines a feasible region within which all combinations of variable assignments satisfy the constraints. Assume for example that we have a problem with three variables x_1, x_2 and x_3 and the constraints:

$C1: x_1 + x_2 > x_3$ and $C2: |x_1 - x_2| > x_3 - 10$.

Assume that there is an initial solution candidate where $x_1 = 2$, $x_2 = 26$ and $x_3 = 30$. This candidate does not satisfy $C1$. We can now generate three modification operators for this problem, one for each variable, by *projecting* the constraints into *feasible intervals* of each variable, assuming that all other variables retain the same values. For example, if we would like to change x_1, we project both constraints onto x_1:

$C1': x_1 > 4$ and $C2': x_1 > 56 \vee x_1 < 6$

and thus obtain the bounds of its feasible intervals as $[4 \ldots 6]$, $[56 \ldots \infty]$. A modification operator would now change x_1 into one of the feasible intervals.

Though some important successes with applying iterative refinement methods to constraint satisfaction problems have been reported in the literature [30,31], these methods are incomplete: the correct solution might require simultaneous modification of several parameters. In FAMING, we avoid the most serious incompleteness problems by always considering simultaneous modification of *pairs* of parameters that define the position of a vertex.

Topological modifications are proposed when the constraints formulated by metric predicates are contradictory: adding an additional vertex gives two additional degrees of freedom to resolve the contradiction. A vertex can be removed if placing it on the straight line between its two neighbours is consistent with all metric constraints.

Modification operators are indexed by their effect on the place vocabulary: *changing* the appearance of a state in the place vocabulary, or making a state *appear* or *disappear*. Based on matches between possibilities and active goals, the system computes a finite set of potentially applicable modification operators for proceeding with the design.

Generation and application of modification operators must be controlled by domain knowledge to avoid excessive search. Since it is very difficult to formulate such domain knowledge, our current system asks the user to choose the discrepancy to modify among a list of suggestions, the modification operator to apply among a list of suggestions, and any topological changes that might be required to create additional degrees of freedom.

One possible trace of an incremental modification where an escapement is obtained from a ratchet mechanism is illustrated in Fig. 17.5 (right).

The ratchet (device A) already provides a cycle of blocking states that can be used to satisfy either the functional specification (1) or (2) of the escapement. However, it does not satisfy specification (5), and specifications (3) and (4) cannot be evaluated.

The system matches the cycle of blocking states of the ratchet to specification (1), and we assume that the user chooses first to satisfy specification (2). The first subgoal is then to create the cycle of places it requires, in a way that they satisfy the **partial-blocking-place** property. The user chooses to change the position of vertex v_3 among the variables proposed by the system.

Solving the constraints added to resolve the discrepancies with specifications (2), (3) and (4) results in the values $x_{v_3} = -2.53$, $y_{v_3} = 7.59$, as shown by B in Fig. 17.5.

17.4.3 Discovering and satisfying compositional constraints

Combination of shape features often implies novel interactions that result in additional compositional constraints. In kinematics, the only interactions we have to consider are subsumptions, where one shape features makes the contact of another impossible or alters the way it occurs. Although it is possible to formulate all possible compositional constraints for guaranteeing that a particular device is subsumption-free, their number grows as $O(n^d)$, where n is the number of possible contacts and d is the number of degrees of freedom of the device. In order to limit this complexity, our prototype generates compositional constraints only when they have been observed to be violated in a proposed solution.

For example, simulating version B of the escapement shown in Fig. 17.5 shows that, owing to a subsumption, the required transitions between blocking states are still impossible. In this case, the additional constraint makes the system of constraints contradictory[4] and FAMING requires the user to introduce a new degree of freedom. We assume that the user chooses to introduce a new vertex v_6 between v_2 and v_3 to create this new degree of freedom. The system proposes to place v_6 at $x_{v_6} = 0.78$ and $y_{v_6} = 0.63$ to satisfy all constraints, shown in C in Fig. 17.5. Simulation shows that there are no subsumptions that produce unexpected behaviour, and, furthermore, specification (5) also turns out to be satisfied. User interaction is indispensable for controlling the modification process; had the system started by attempting to satisfy specification (5), for example, a long and not very fruitful search would have resulted. The intuitions behind such choices appear very complex, and we doubt that they could be formulated in a sufficiently concise way.

17.5 Conclusions

Human intelligence has evolved to help people survive in natural environments: it is very good at handling natural, rounded shapes, reasoning with approximate information, and similar tasks. Modern engineering operates in a highly constrained world of artificial shapes with sharp edges, precise and well-controlled interactions, and many constraints to be respected. Human intuition is not well suited to this environment, and its creativity cannot be exploited to its fullest.

By automating the low-level details, FAMING provides the designer with a "mechanical intuition" not normally possessed and which makes it much easier to create novel devices. Using its numerical precision, the program provides the designer with the capability to envision and thus design devices whose behaviour depends on precise shapes and interactions.

The same ideas apply in other domains. We have implemented the CADRE system [13,34,35], which uses model-based case adaptation to relieve architects from tedious and time-consuming work on details of building and installation geometry, making it possible to explore a much larger range of creative possibilities than previously possible.

As stated above, FAMING can provide a designer with an intuition about mechanism behaviour that would not normally be possessed. Figure 17.6 shows three arrangement of gears that are almost indistinguishable to people, but implement very different functions. FAMING makes it possible not only to detect these possibilities, but to search for them in a systematic manner. As an example of the practical impact of the system, we have reproduced the development of a new escapement design for a major Swiss watch manufacturer within one afternoon. In current practice, this development has taken 6 months for a team of several people!

---

[4] Meaning that no solution can be reached by modifying a single vertex; there may still be a solution by modifying several vertices simultaneously, which is not found by the system.

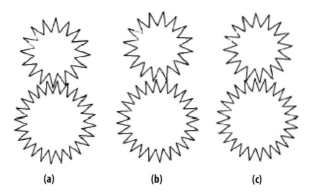

Fig. 17.6 Three similar-looking gear devices with very different behaviours: **a** functions as a normal gear transmission; **b** the transmission ratio from the top wheel to the bottom is half that of the opposite direction; **c** motion can only be transmitted from the top wheel to the bottom one, the other direction will cause a jam.

The work we have presented presents several novel contributions for artificial intelligence (AI) in design research. In spite of the apparent simplicity of kinematic pairs, the scope of possible functions that can be constructed with them turned out to be much richer than had previously been addressed in research on knowledge-based computer-aided design systems. Consequently, certain aspects of the formalism we developed to represent qualitative function formally is significantly more powerful than previous proposals, such as CFRL [26]. The infinite matching possibilities between function and behaviour allowed in our functional language also provides an explanation for why mechanism design is not a closed space of possible devices, but allows endless possibilities for creativity and innovation.

Earlier work on designing mechanism part shapes [7] was not sufficiently general to design devices of practical interest, and FAMING is the first program to allow goal-directed design of mechanism part shapes. It could be integrated with systems for conceptual design of kinematic chains, *e.g.*, see Ref. [5]. The techniques can be generalised to other domains where qualitative models have been investigated: thermodynamics [36], non-linear control systems [37], electronic circuits [6], as well as to many other engineering problems [38].

The example of FAMING shows that the case-based design paradigm is not limited to routine design problems, as is often assumed. In fact, in the domain addressed by FAMING, case-based design proves to be a better tool for supporting innovation than methodologies such as model-based design, which are intended for innovative design but are impractical because of computational intractability. This is for two reasons. First, cases provide coherent starting points for design solutions and thus avoid much of the computational complexity imposed by geometric considerations. Second, and maybe more importantly, cases provide a communication framework where the designer can bring his extensive knowledge and intuitions to bear on the design process. We believe that the combination of case- and model-based reasoning as shown in FAMING is a promising way to make AI techniques work for real-world design problems.

References

[1] Nadel BA, Lin J. Automobile transmission design as a constraint satisfaction problem: first results. In: Tong C, Sriram D, editors. Artificial intelligence in engineering design. Academic Press, 1992.
[2] Williams B. Interaction-based invention: designing novel devices from first principles. In: Proceedings of the National Conference on Artificial Intelligence, Boston, 1990.
[3] Neville D, Weld DS. Innovative design as systematic search. In: Proceedings of the AAAI, Washington, 1993.
[4] Sycara K, Navinchandra D. Index transformation techniques for facilitating creative use of multiple cases. In: Proceedings of the 12th IJCAI, Sydney, 1991.
[5] Subramanian D. Conceptual design and artificial intelligence. Proceedings of the 13th International Joint Conference on Artificial Intelligence, Chambery, France, 1993.
[6] DeKleer J, Brown JS. A qualitative physics based on confluences. Artif Intell 1984;24:7–83.
[7] Joskowicz L, Addanki S. From kinematics to shape: an approach to innovative design. In: Proceedings of the AAAI, St. Paul, 1988.
[8] Gero J. Design prototypes: a knowledge representation schema for design. AI Mag 1990;11(4).
[9] James M. Prototyping CAD. In: CAD/CAM, ProEngineer manual. PTC Corporation, Needham (MA, USA), 1991 [http://www.ptc.com/products/proe/].
[10] Wagner MR. Understanding the ICAD system. ICAD Inc., 1990.
[11] Kolodner JL. Case-based reasoning. Morgan-Kaufmann, 1993.
[12] Pu P, editor. Case-based design [special issue]. Artif Intell Eng Des Anal Manuf 1993;7(2).
[13] Hua K, Faltings B. Exploring case-based building design – CADRE. Artif Intell Eng Des Anal Manuf 1993;7(2):135–43.
[14] Gero JS, Lee HS, Tham KW. Behavior: a link between function and structure in design. In: Brown D, Waldron M, Yoshikawa H, editors. Intelligent computer-aided design. North Holland, 1992.
[15] Goel AK. A model-based approach to case adaptation. In: Proceedings of the 13th Annual Conference of the Cognitive Science Society, 1991; 143–8.
[16] DeKleer J, Weld D, editors. Readings in qualitative reasoning about physical systems. Morgan-Kaufmann, 1990.
[17] Wills LM, Kolodner JL. Towards more creative case-based design systems. In: Proceedings of the 12th National Conference of the AAAI. Morgan-Kaufmann, 1994.
[18] Sun K, Faltings B. Supporting creative mechanical design. In: Gero J, Sudweeks F, editors. Artificial Intelligence in Design '94. Kluwer, 1994.
[19] Sacks E, Joskowicz L. Automated modeling and kinematic simulation of mechanisms. Comput Aided Des 1993;25(2).
[20] Faltings B. Qualitative kinematics in mechanisms. Artif Intell 1990;44(1):89–120.
[21] Forbus K, Nielsen P, Faltings B. Qualitative spatial reasoning: the CLOCK project. Artif Intell 1991; 51(3):417–72.
[22] Nielsen P. A qualitative approach to rigid body mechanics. Ph.D. thesis, University of Illinois, 1988.
[23] Forbus K. Qualitative process theory. Artif Intell 1984;24.
[24] Chakrabarti A, Holden T, Blessing L. Representing function in design. In: Workshop at the 3rd International Conference on AI in Design, Lausanne, 1994.
[25] Sembugamoorthy V, Chandrasekaran B. Functional representation of devices and compilation of diagnostic problem-solving systems. In: Kolodner J, Riesbeck C, editors. Experience, memory and reasoning. Lawrence-Earlbaum Associates, 1986.
[26] Iwasaki Y, Fikes R, Vescovi M, Chandrasekaran B. How things are intended to work: capturing functional knowledge in device design. In: Proceedings of the 13th International Joint Conference on Artificial Intelligence, Chambery, 1993.
[27] Umeda Y, Takeda H, Tomiyama T, Yoshikawa Y. Function, behavior and structure. In: Gero JS, editor. Applications of artificial intelligence in engineering V, vol. 1. Springer, 1990.
[28] Faltings B, Baechler E, Primus J. Reasoning about kinematic topology. In: Proceedings of the 11th International Joint Conference on Artificial Intelligence, Detroit, 1989.
[29] Faltings B. A symbolic approach to qualitative kinematics. Artif Intell 1992;56(2):139–70.

[30] Selman B, Levesque H, Mitchell D. A new method for solving hard satisfiability problems. In: Proceedings of the 10th National Conference of the AAAI. Morgan-Kaufman, 1992.
[31] Minton S, Johnston M, Phillips A, Laird P. Minimizing conflicts: a heuristic repair method for constraint satisfaction and scheduling problems. Artif Intell 1992;58.
[32] Pu P, Purvis L. Formalizing case adaptation in a case-based design system. In: Gero J, Sudweeks F, editors. Artificial Intelligence in Design '94. Kluwer, 1994.
[33] Haroud D, Faltings B. Global consistency for continuous constraints. Proceedings of the European Conference on Artificial Intelligence. Wiley, 1994.
[34] Hua K, Smith I, Faltings B, Shih S, Schmitt G. Adaptation of spatial design cases. In: Gero J, editor. Artificial Intelligence in Design '92. Kluwer, 1992.
[35] Hua K, Faltings B, Smith I. CADRE: case-based geometric design. Artif Intell Eng 1996;10(2):171–84.
[36] Forbus KD, Whalley PB. Using qualitative physics to build articulate software for thermodynamics education. Proceedings of the National Conference of the AAAI. Morgan-Kaufman, 1994.
[37] Zhao F. Extracting and representing qualitative behaviors of complex systems in phase space. Artif Intell 1994;69(1):51–92.
[38] Milne R, Travfie-Massuyes L. Qualitative reasoning for design & diagnosis applications, tutorial. In: National Conference of the AAAI, Washington, 1993.

Transforming behavioural and physical representations of mechanical designs[1]

Susan Finger and James R. Rinderle

Abstract We are exploring the use of formal grammars to represent two distinct but interconnected attributes of mechanical designs: geometry and behaviour. By creating a formal description of a limited set of behaviours and geometry for mechanical design specifications and corresponding descriptions of physical components, we can generate the description of a physical system that takes advantage of the multiple behaviours of its components. The goal of our research is to create a transformational strategy by which the design specifications for a mechanical system can be transformed into a description of a collection of mechanical components. For any given design specification, many different physical systems could potentially meet that specification. Hence, many transformations potentially could be applied to a particular design specification. We are interested in those transformations of the design specifications that preserve the original behaviour of the design and that guide the design toward a solution that meets all the requirements. In this paper, we explore a set of transformations that enable us to move from specifications to topological configurations and from topological configurations to geometric configurations.

18.1 Introduction

During the design process, a designer transforms an abstract functional description for a device into a physical description that satisfies the functional requirements. In this sense, design is a transformation from the functional domain to the physical domain [1,2]; however, the basis for selecting appropriate transformations and methods for accomplishing transformations are not well understood. The implicit basis for design transformations in circuits [3], software [4], and some architectural applications [5] results in a type of modularity that is not well suited to mechanical devices [6].

Our approach is based on the following assertions:

- the behavioural *requirements* of mechanical systems can be represented using the language of a graph grammar;
- the behavioural *characteristics* of components can be represented using the language of a graph grammar;
- the physical characteristics of designs and components can be represented using formal topological and geometric models;

---

[1] A version of this paper originally appeared in the Proceedings of the First International Workshop on Formal Methods in Design, Manufacturing, and Assembly, Colorado Springs, January 1990, pp 133–51.

- the behavioural and physical graphs of components can be linked parametrically or algorithmically;
- the behavioural specifications graph can be formally transformed into a description of a physical system with associated behavioural and geometric representations.

The goal of our research is to create a transformational strategy by which the design specifications for a mechanical system can be transformed into a description of a configuration of mechanical components. Both behavioural and physical *requirements*, as well as behavioural and physical *characteristics*, of the available mechanical components must be represented to execute such a transformational approach to design. Because the interactions of components are important in our synthesis strategy, the representation of the behaviours of mechanical components must be linked to the representation of their physical characteristics; *i.e.*, we are concerned with modelling the relationship between form and function of components. We are also interested in the emergence of additional behaviours as components are combined during the design process. Finally, we need a strategy that enables us to transform an abstract description of the desired behaviour of a device into a description that corresponds to a collection of physical components.

To realise the goal of formalising the transformation from the behavioural to the physical domain, we have explored a small domain within mechanical design: the domain of gearbox design. Clearly, one reason for selecting this domain is that gearbox design is a well-understood highly parameterised area of mechanical design. Nevertheless, we believe that our representation and transformation formalism will be applicable in other mechanical design domains, particularly to the class of design problems that we call configuration design. By configuration design, we mean designs composed from standard component families but for which allowable *configurations* are not specified *a priori*.

The representation of a design includes a specified configuration of a set of components and their associated geometrical and behavioural representations.[2] Partially complete designs also consist of components and junction structures and some number of behavioural primitives not yet associated with a component. Since the components have associated with them both behaviours and physical characteristics, the distinction between the representation of designs and the representation of components is in the representation of the physical *configuration* and its associated behaviour.

18.2 Related work

Our research builds upon research from several different areas, including bond graphs, representation of mechanical behaviour, grammars for shape representation, and configuration design. In this section, we briefly discuss related research in these areas.

---

[2] The representation of a design can also include manufacturing, assembly, or maintenance information, but we are not concerned with these in this chapter.

18.2.1 A brief introduction to bond graphs

Our underlying representation for behaviour is based on bond graphs. Bond graphs, which were created by Paynter [7], provide a convenient and uniform representation for the dynamic behaviour of a broad class of physical systems, including those within the mechanical, electrical, hydraulic, thermal, and biological domains. Bond graphs have been used in a broad range of application areas, including vacuum cleaners [8], robotic manipulators [9], and torque converters [10]. A very brief introduction to bond graphs is given here. For a complete discussion of bond graphs and their uses, see Karnopp and Rosenberg [11].

Bond graphs enable mechanical and hydraulic systems to be represented in a manner equivalent to electric circuit diagrams. For example, a spring in a mechanical device acts like an electrical capacitor by storing and releasing energy. One of the most powerful attributes of bond graphs is that they can be used to model integrated electrical, mechanical, and hydraulic systems. Using bond graphs, physical systems are represented as a graph of lumped-parameter, idealised elements. Power is the currency of bond graphs; power flows through the bonds (edges) in the graph, and power is dissipated, stored, supplied, and transformed at the ports (vertices) in the graph.

The ports, or vertices, of bond graphs are divided into three categories.

- 1-port elements dissipate power, store energy, and supply power. Dampers, springs, and masses are the mechanical elements represented by the passive 1-port elements. Force (effort) and velocity (flow) sources are represented as active 1-ports.
- 2-port elements transform power. Transformers are 2-port elements that represent an imposed proportionate relationship between analogous quantities, *e.g.*, a gear pair constrains rotational speeds. Gyrators are 2-port elements that impose a proportional relationship between dual quantities, *e.g.*, a torque converter constrains the relationship between torque and angular velocity. Power is conserved across a 2-port.
- *N*-port elements represent the structure of the system corresponding to the connections among the elements. There are two types of *N*-port element: 0-junctions and 1-junctions, which correspond respectively to "same force" and "same velocity" connections. Equivalents of Kirchoff's laws apply to *N*-port elements: the sum of the efforts incident on a 1-junction is zero (the bonds share a common flow); the sum of the flows incident on a 0-junction is zero (the bonds share a common effort). *N*-port elements are power conserving.

For a more detailed discussion of our use of bond graphs for synthesis, as opposed to modelling, see Ref. [12].

18.2.2 Representation of function and behaviour

Mechanical engineers tend to use the words function and behaviour interchangeably. Qualitative physicists make a distinction between these words; *i.e.*, the design's function is what it is used for, whereas its behaviour is what it does. For example, the function of a clock is to display the time, but its behaviour might be the rotation of hands. Similarly, a motor may be designed to function as a prime mover, but it can also function as a doorstop because it has additional behaviours due to its mass. In this paper, *function* is used to indicate the subset of *behaviours* that are required for the device to perform satisfactorily.

The representation of the function and behaviour of mechanical designs has been explored by, among others, Lai and Wilson [13], Crossley [14], and Pahl and Beitz [15]. The function structures of Pahl and Beitz provide a graphical system for laying out the functions of a design. In this system, functions, such as "mix" or "deliver", are arranged in a graph to represent the overall function of the design. This is similar to our specifications graph, discussed below; however, there are several important differences. Pahl and Beitz's work does not discuss how to integrate form specifications or functional constraints with this functional representation. Therefore, though the function structures are used to study functional configurations in Pahl and Beitz's synthesis strategy, there is little guidance for transforming the function structure to a physical description of the device. Lai and Wilson created a formal, English-language-based system, called FDL, for representing the function and structure of mechanical designs. In FDL, nouns and verbs are used to create sentences that represent the function of a design, and design rules operate directly on the nouns and verbs in the sentence. Allowable verbs, for example "fasten", do not have physical or mathematical representations and so their meaning is determined by the rules that use them. Though the FDL language can represent the function and form of a design, it provides no assistance in transforming a functional description into a physical description.

More closely related to our approach for representing behaviour is the work of Fenves and Baker [5], and Ulrich and Seering [16]. Fenves and Baker created a spatial and functional representation language for structural designs. They use operators that execute a known grammar to generate architectural layouts, as well as structural and functional configurations; however, they assume that the layout and structure are independent if they are generated sequentially. Ulrich and Seering also use a formal representation of function based on bond graphs; however, their goal is to create a system in which the bond graphs for single-input, single-output dynamic systems can be automatically synthesised and transformed into physical components. Using a strategy they call *design and debug*, they transform a graph of design requirements directly to functionally independent physical components. Reconfiguration for function sharing is performed after the components have been selected. Prabhu and Taylor [17] have also used bond graphs to generate conceptual parametric designs.

18.2.3 Grammars for representation of geometry

Formal grammars that can generate and parse valid strings in a language have proven useful in a number of fields, most notably linguistics and computer science. Recently, interest has been growing on the use of formal grammars in engineering design. Our work draws primarily from these engineering applications of grammars. We are specifically concerned with graph grammars, the class of grammars that operates on graphs. Tutorials on graph grammars and their applications are given in Ehrig [18] and Nagl [19].

Stiny [20], who was among the first to apply grammars in design, created shape grammars based on the formalisms of computational linguistics [21]. Architects, in particular, have been interested in shape grammars, using them to generate families of floor plans, ornamentation, and building layouts [22–25]. Woodbury and coworkers [26,27] have created grammars to generate solid models for designs. Pinilla *et al.* [28] have created a grammar that can be used to describe and represent the geometric features of a design. They use a non-manifold topological representation of a design to create a general, but formal, representation of form features.

18.2.4 Configuration design

Our research combines configuration design and parametric design because we generate both the structure of the design and the individual components. Many design systems, such as HI-RISE [29], AIR-CYL [30], and VT [31], utilise either a set of predetermined decompositions of the structure of a design, or utilise design methods that generate, as part of the design process, a decomposition belonging to such a set. Therefore, all of the designs generated by these systems will share, at a relatively low level, a structural similarity. For design domains in which the most desirable structures can be enumerated or explicitly decomposed in advance, this approach has proven useful. However, for many design problems the structural decomposition of the solution cannot be predetermined. For a more complete discussion of configuration design, see Finger and Dixon [32,33].

18.3 Representation of behaviour of specifications and components

Our underlying representation for behaviour is based on bond graphs [7]. Using bond graphs, we can construct a formal grammar that gives us a general representation of classes of mechanical behaviour. In common practice, bond graphs are constructed to model the behaviour of physical systems. We use bond graphs not only to model the behaviour of physical systems, but also to represent behaviour in the abstract, as with a design specification. Thus, a device configuration can be generated by transforming a specification bond graph into a functionally equivalent graph that corresponds to a configuration of available components. The types of graph transformation used are those that decompose, aggregate, and redistribute graph primitives. A bond graph grammar for representing the behavioural requirements of a mechanical system is presented in Ref. [12].

A major advantage of using bond graphs to represent design requirements is that we can define transformation rules that alter the structure of the bond graph but that do not alter the dynamic behaviour of the system represented by the graph. The implications of this statement are important. Because we can transform the specifications graph to represent many different physical systems, we do not impose an initial structure or configuration on the physical design; *i.e.*, we do not require an *a priori* decomposition of the design specifications.

18.3.1 Representation of design specifications

We observe that system specifications for engineering designs are of two types: *behavioural* and *physical*. That is, some specifications describe at an abstract level the desired behaviour of the overall system, whereas others describe physical restrictions or requirements on the final design. For example, the requirements for a vibration absorber might include the behaviour of the device in terms of frequency and rejection ratio and might also specify physical properties, such as allowable size and weight. The physical and behavioural specifications express the design objective. Since the specifications are given without regard to design configuration, they are independent of each other. (That is, they are independent unless the requirements are contradictory, for example by virtue of physical law.) Physical specifications for a design may or may not be given, whereas at least some behavioural specifications must be given, since the behavioural specifications express the central aspect of the design objective.

Although the behavioural and physical specifications are independent in the functional domain, they are coupled in the *physical* domain, because any physical arrangement results in a specific set of behaviours. In the physical domain, the physical and behavioural characteristics of individual components depend on one another, and the behaviour of the whole design depends strongly on the configuration and interaction of components. The representation of interactions is essential for our purposes, since we are investigating the effects of, and the means for, achieving functional integration in design.

Physical and behavioural specifications and characteristics can be represented as combinations of abstract primitives. They are abstract because each primitive corresponds to only one behavioural or physical characteristic. Individually they do not correspond to any particular component or configuration of components, but collectively they may represent the design specifications or the form and behaviour of components. The set of primitives must satisfy two important criteria. One requirement is that the set of primitives chosen be complete; *i.e.*, all relevant behavioural and physical characteristics must be representable by some combination of primitives. The other requirement is that the number of primitives must be small, although not necessarily minimal. The latter requirement minimises combinatorial problems associated with representing a single behaviour in different ways. In addition, the primitive behavioural and physical elements should enable commonly used components and typical specifications to be represented easily.

18.3.2 Representation of behavioural requirements of mechanical systems

When the behavioural primitives correspond to bond graph elements, the behavioural specifications can be represented easily in a bond graph. This specification bond graph is not unique; indeed, any behaviourally equivalent graph is an acceptable representation of the specifications. The specifications graph represents only the desired behaviour of the design.

Deriving a correct specifications graph for a design problem is, in itself, a major research problem. Initially, for transmission design, we assume that deriving a specifications graph from the problem statement is straightforward. Later, as less constrained design domains are explored, we will look at the problem of constructing the specification graph from the design requirements.

For example, consider the design of a simple gearbox that is to have a single input shaft and two output shafts. The first output shaft is to be offset a prescribed amount from the input shaft and is to operate at a 20:1 speed reduction relative to the input shaft. The second output shaft operates at a 240:1 reduction and must be in-line with the input shaft. The direction of shaft rotation is not relevant. This kinematic specification can be expressed by the bond graph shown in Fig. 18.1.

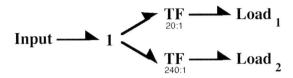

Fig. 18.1 Behavioural specification for a gearbox.

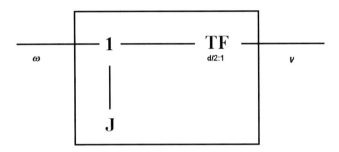

Fig. 18.2 Behaviour graph for a single spur gear.

18.3.3 Representation of behavioural characteristics of components

One goal in transforming the specification graph is to create a correspondence between the required behaviours and the available components; therefore, it is convenient to represent the behavioural characteristics of components with the same basic structure that is used for specifications, *i.e.*, bond graphs. Each component has a bond graph associated with it that represents its behaviour in *isolation*. As an example, consider a single spur gear as a component. The behaviour graph for the component is shown in Fig. 18.2. In this example, a relationship is imposed between flow quantities rotational speed ω on the left and surface speed v on the right. The graph also includes the gear inertia.

One crucial difference between specifications and components is that the behaviours and the physical characteristics of a component are inherently linked; no single characteristic, either behavioural or physical, can be obtained in isolation. So, for example, the spur gear in Fig. 18.2 may be selected for its power transformation characteristics, but its inertial characteristics are implicitly selected as well. Section 18.3.4 addresses some of the issues in linking the representation of the behavioural and physical characteristics of components.

18.3.4 Representation of designs

A complete design is the specified configuration of a set of components. Since the components have associated with them both behaviours and physical characteristics, the distinction between the representation of designs and the representation of components is in the representation of the behavioural and physical configuration. The behavioural configuration fundamentally consists of the kinematic connections, *e.g.*, mounting to a frame, rigid connections, or rolling. Most of the common kinematic arrangements can be categorised [34] and reduced to a bond graph junction structure. A 1-junction, for example, represents a common translational velocity. In this way, the behaviour of complete devices may be represented in terms of the behavioural bond graphs of components and a number of bond graph junction structures representing kinematic connections. Partially complete designs also consist of components and junction structures and some number of behavioural primitives not yet associated with a component.

Physical components, like specifications, have both behaviours and physical characteristics. Again, the behaviours and physical characteristics of a component are inherently linked; no single characteristic can be obtained in isolation. Each physical component is represented as an object that has a behavioural representation, a physical representation, and an explicitly represented interaction between the two. The relationships can take the form of design equations, analytical models that relate geometric characteristics to behavioural characteristics, or database entries that prescribe a relation between the physical and the behaviour characteristics. For example, the weight of a helical coil spring, which is a physical characteristic, is proportional to the product of stiffness and the square of allowable deflection, which are behavioural characteristics. These relationships are often critical during the design.

It is important to note that the completeness of the behavioural and physical representations determines the extent to which additional behaviours can be exploited. For example, with a worm gear, if only the transformation ratio and shaft orientations are represented, then only the effects of these behaviours can be exploited. If the self-locking characteristic of the worm gear is important, it must be represented explicitly or it must emerge from the representation of more fundamental characteristics, which in this case would be Coulombic friction. Bond graph and similar representations are useful in this context, since these approaches provide the mechanisms to explicate useful characteristics that emerge from more fundamental physical interactions.

18.4 Transformation of specifications into physical descriptions

The specification graph is composed of abstract behavioural primitives that do not yet correspond to any physical components. To arrive at a design, we transform the specification graph, without changing function, to obtain a graph that more nearly corresponds to a collection of components. In this section, we discuss the principles that we use in transforming the specifications into a physical system. The transformation process, which is described in detail by Hoover and Rinderle [35], is guided by the function integration and incidental behaviour principles and by knowledge of the available components. It provides an approach to finding configurations that satisfy the physical requirements, thus eliminating blind search. This approach has been successful for the design of mechanical power transmissions and serves to illuminate several important issues for extending this technique to other domains.

18.4.1 Behaviour-preserving transformations

The specification graph is not unique. Many equivalent graphs can represent the same behaviour. By transforming the specification graph, without altering behaviour, we

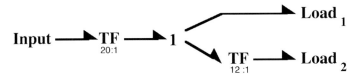

Fig. 18.3 Functionally equivalent graph using a common power path.

can explore design alternatives that have the same behaviour. Some of the resulting configurations result in physically desirable designs and some do not. In addition to knowing the general rules for behaviour-preserving transformations, we need to know which transformations to apply and the sequence of application. Guidance in selecting which behaviour-preserving transformation to apply comes from the physical requirements of the system, from the physical characteristics of the components, and from the relationship between geometry and behaviour of the components.

For the gearbox design, more economical designs often have maximum commonality of the power paths. Therefore, one type of behaviour-preserving transformation of interest is one that preserves the dynamic behaviour but which has a different kinematic configuration. For example, Fig. 18.3 is a behaviour-preserving transformation of the specification graph from Fig. 18.1.

18.4.2 Component-directed transformations

Because mechanical designs are characterised by a high degree of integration, transformations should be directed toward this goal. Integration in a design requires the appropriate utilisation of the behavioural and physical characteristics of its components. Therefore, intelligent application of transformations requires utilising both the behavioural and physical representations of the components to guide the selection process. Transformations selected in this way are *component-directed transformations* because their selection and use are directed by knowledge of the available components. It is the class of component-directed transforms that ultimately enables the selection of a single component to fulfil multiple functions, *e.g.*, selecting a worm gear to execute speed reduction, shaft-offset and right-angle functions, as described more fully by Hoover and Rinderle [35].

As the library of components consists only of spur gears, and because the speed ratio that can be obtained with a *pair* of spur gears is limited to $8:1$, the specification graph shown in Fig. 18.3 cannot be transformed directly into a collection of components. A component-directed transform can be applied to split the transformer elements into elements that can be mapped into individual pairs of spur gears. The result of one such transformation is shown in Fig. 18.4. Alternative transformations might result in more stages of reduction or different reduction ratios. Methods to direct the grammar, *i.e.*, to select the most advantageous transforms, are critical and remain a topic of current interest. Rinderle [36] extended the formalism of attribute grammars for this purpose. By augmenting the language that describes the behaviour of components with engineering parameters, the design transformations can be directed to generate alternatives that meet both the behavioural and physical requirements and which are preferable to other alternatives. Zawicky [37] used this approach to generate more nearly optimal gearbox designs using a grammar.

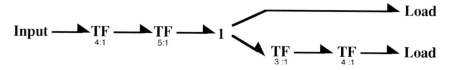

Fig. 18.4 Graph with kinematics equivalent to Fig. 18.3.

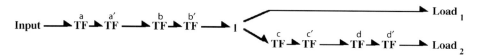

Fig. 18.5 Transformation into spur gear equivalents.

The transformation shown in Fig. 18.4 imposes a relationship between two rotational speeds that is achieved using a pair of spur gears. Applying a lower-level, component-directed transform results in a behaviourally equivalent graph in which the transformer elements correspond to individual spur gears, as shown in Fig. 18.5. This graph preserves the required kinematic behaviour of the gearbox. If individual transformer elements are interpreted as gears, then the graph implies topological relationships among gears; *i.e.*, it implies which gears mesh. The topological information, however, is not complete. The graph shown in Fig. 18.5 does not indicate how gears are arranged on shafts. Both the geometrical and topological information associated with the shaft matrix are absent.

18.5 The shaft matrix

We are now interested in investigating alternatives that have the same kinematic functionality but different physical configurations. The power path topology of Fig. 18.5 can be represented in a matrix format. The individual TF elements that correspond to individual gears are labelled a, a', b, b', etc. Using a matrix-like format, each row of the matrix corresponds to a unique rotational speed. Each column of the matrix corresponds to a unique pitch line velocity at a pair of mating gears. Since the gears a and a' mesh with each other, they have a common velocity at their pitch diameters and so are represented in the same column of the matrix. Similarly, a' and b share a rotational speed and, therefore, fall within the same row of the matrix. The shaft matrix is readily completed by inspection, as shown in Table 18.1. Note that column exchanges have no impact on the implicit commonalities among gear surface speeds or shaft rotational speeds. Similarly, exchanging rows has no impact on power path topology or the functional kinematics.

In a sense, this matrix is a canonical form for the gearbox and could be implemented with a shaft and associated bearings for each of the five rows in the matrix. Although implementation may be direct, it is not economical owing to the absence of any shafting commonality. Elements with different rotational speeds may indeed correspond to the same shaft position. For example, a gear stack may rotate relative to its supporting shaft, relying on the shaft only to maintain its posi-

Table 18.1 Shaft matrix corresponding to Fig. 18.5.

a			
a′	b		
	b′	c	
		c′	d
			d′

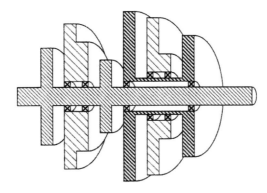

Fig. 18.6 Nested shafts.

tion in space. Similarly, shafts may be nested. One such configuration is shown in Fig. 18.6.

Although shafts may be shared, the allowable mechanical configurations impose certain nesting requirements on shafts that manifest themselves as topological constraints. These constraints can be represented and enforced by delimiting the elements in each row of the shaft matrix, as shown in Table 18.2.

The column position within the matrix can be interpreted as corresponding to a physical position. For example, column one can be interpreted as the left end

Table 18.2 Shaft matrix with nesting constraints.

⟨a⟩			
[a′	b]		
	{b′	c}	
		(c′	d)
			⟨d′⟩

Table 18.3 Shaft matrix transformed so that the input and output shafts are in-line.

⟨a⟩			⟨d′⟩
[a′	b]		
	{b′	c}	
		(c′	d)

Table 18.4 Shaft matrix transformed to reduce the number of shafts.

⟨a⟩			⟨d′⟩
[a′	b]	(c′	d)
	{b′	c}	

of the gearbox, and the last column is at the right end. Columns of the matrix can be interchanged to obtain input and output gears at the proper physical locations. The shaft matrix shown in Table 18.2 is already an acceptable configuration because the input a is nearest the left of the gear box, and the two outputs d' and c are accessible to the right edge. We now seek commonality in the shafting configuration to maintain not only the power path topology, but also the physical access of input and output shafts to the appropriate position on the gearbox. Rotational elements with different speeds can be combined on a single shaft by combining two rows whenever the appropriate shaft matrix elements are null. Gear d' corresponds to the 240:1 speed shaft and must be in-line with the input shaft. Combining the first and last rows of the shaft matrix shown in Table 18.2 results in the shaft matrix shown in Table 18.3.

Because one requirement of the original specification was that the 20:1 speed shaft be offset from the input shaft, the b'–c row cannot be combined with the first row; however, the second and fourth rows can be combined on a common shaft, as shown in Table 18.4. Note that this shaft matrix represents shafts at three physical locations on which all eight gears reside. Note also that the number of rows in this shaft matrix cannot be reduced further; therefore, this configuration has the minimum number of shaft locations for this specific power path topology. Direct translation of the shaft matrix in Table 18.4 is shown in Fig. 18.7.

Alternative power path topologies may result in fewer required shafts. Using similar methods, the original kinematic specification from Fig. 18.1 may be transformed into a functionally equivalent specification, as shown in Fig. 18.8. If each of the transformers is interpreted to be a pair of spur gears, then this configuration will consist of ten individual spur gears compared with the previous configuration involving only eight. However, when this power path topology is mapped onto the shaft matrix and the number of shafts is minimised, a configuration using only two shafts can be found. The final shaft matrix and the physical configuration corresponding to this alternative are shown in Table 18.5 and Fig. 18.9.

Transforming Behavioural and Physical Representations of Mechanical Designs

Table 18.5 Final shaft configuration for Fig. 18.8.

⟨a⟩	{b′	c	d′}	⟨e′⟩
[a′	b]	(c′)	[d	e]

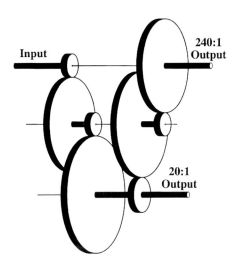

Fig. 18.7 Shaft and gear configuration corresponding to Table 18.4.

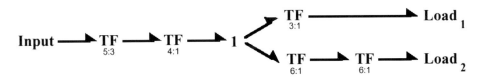

Fig. 18.8 Functionally equivalent specification.

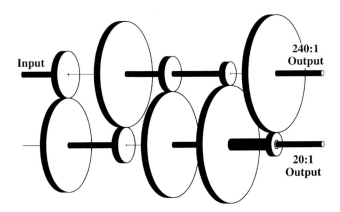

Fig. 18.9 Gear and shaft configuration for Table 18.5.

18.6 Conclusions

In this paper, we have presented a transformational paradigm for design in which design specifications are transformed into physical descriptions. We focus on the class of configuration design in which designs are composed from standard component families, but for which no *a priori* configuration is assumed. In the domain of simple gear design, we first apply behaviour-preserving transformations to maximise common power paths. We then apply component-directed transformations that preserve behaviour and power path topology and that make the best use of component characteristics. Finally, we apply transformations that preserve the kinematic topology and that result in compact layouts and shaft access. At each stage, many alternative transformations are valid, each corresponding to a design alternative. We select and apply transformations to obtain synergistic interactions among components.

Acknowledgements

The authors are pleased to acknowledge the support of the Design Theory and Methodology Program of the National Science Foundation (NSF Grants DMC-84-51619 and DMC-88-14760) and the Engineering Design Research Center at Carnegie Mellon University (NSF Grant CDR-85-22616).

References

[1] Mostow J. Toward better models of the design process. AI Mag 1985;6(1):44–57.
[2] Rinderle JR. Measures of functional coupling in design. Ph.D. dissertation, Massachusetts Institute of Technology, 1982.
[3] Steinberg L, Langrana N, Mitchell T, Mostow J, Tong C. A domain independent model of knowledge-based design. Technical report AI/VLSI Project Working Paper No. 33, Rutgers University, 1986.
[4] Wirth N. Program development by stepwise refinement. Commun ACM 1971;14(4):221–7.
[5] Fenves SJ, Baker NC. Spatial and functional representation language for structural design. In: Expert systems in computer-aided design. Elsevier North-Holland, 1987 [IFIP 5.2].
[6] Rinderle JR. Implications of function–form–fabrication relations on design decomposition strategies. In: Proceedings of the Computers in Engineering Conference, Chicago. New York: ASME, 1986,193–8.
[7] Paynter HM. Analysis and design of engineering systems. Cambridge (MA): MIT Press, 1961.
[8] Remmerswaal JAM, Pacejka HB. A bond graph computer model to simulate vacuum cleaner dynamics for design purposes. J Franklin Inst 1985;83–92.
[9] Margolis DL, Karnopp DC. Bond graphs for flexible multibody systems. Trans ASME 1979;101:50–7.
[10] Hrovat D, Tobler WE. Bond graph modeling and computer simulation of automotive torque converters. J Franklin Inst 1985;93–114.
[11] Karnopp D, Rosenberg R. System dynamics: a united approach. New York: Wiley, 1975.
[12] Finger S, Rinderle J. A transformational approach to mechanical design using a bond graph grammar. In: Proceedings of the First ASME Design Theory and Methodology Conference, American Society of Mechanical Engineers, Montreal, Quebec, 1989.
[13] Lai K, Wilson WRD. FDL: a language for function description and rationalization in mechanical design. In: Proceedings of the Computers in Engineering Conference, New York. New York: ASME, 1987;87–94.
[14] Crossley E. Make science a partner. Mach Des 1980;52(April 24):50–5.
[15] Pahl G, Beitz W. Engineering design. London: The Design Council/Springer, 1984.

[16] Ulrich KT, Seering WP. Synthesis of schematic descriptions in mechanical design. Res Eng Des 1989; 1(1):3–18.
[17] Prabhu DR, Taylor DL. Some issues in the generation of the topology of systems with constant power-flow input-output requirements. In: Rao SS, editor. Advances in design automation. New York: American Society of Mechanical Engineers, 1988;41–8.
[18] Ehrig H. Tutorial introduction to the algebraic approach of graph grammars. In: graph-grammars and their applications to computer science. Lecture note series. New York: Springer, 1987;3–14.
[19] Nagl M. Set theoretic approach to graph-grammars. In: Graph-grammars and their applications to computer science. Lecture note series. New York: Springer, 1987;41–54.
[20] Stiny G. Pictorial and formal aspects of shape and shape grammars. Basel: Birkhauser, 1975.
[21] Chomsky N. Syntactic structures. Atlantic Highlands (NJ): Humanities Press, 1957.
[22] Stiny G, Mitchell WJ. The palladian grammar. Environ Planning B 1978;5:5–18.
[23] Downing F, Flemming U. The bungalows of Buffalo. Environ Planning B 1981;8:269–93.
[24] Koning H, Eizenberg J. The language of the prairie: Frank Lloyd Wright's prairie houses. Environ Planning B 1981;8:295–323.
[25] Flemming U. More than the sum of parts: the grammar of Queen Anne houses. Environ Planning B 1987;14:323–50.
[26] Heisserman J, Woodbury R. Solid grammars for generative geometric design. In: NSF Engineering Design Research Conference, University of Massachusetts, Amherst, MA, 1989.
[27] Woodbury R, Carlson K, Heisserman J. Geometric design spaces. In: Second IFIP WG 5.2 Workshop on Intelligent CAD, IFIP, Cambridge, UK, 1988.
[28] Pinilla JM, Finger S, Prinz FB. Shape feature description and recognition using an augmented topology graph grammar. In: NSF Engineering Design Research Conference, University of Massachusetts, Amherst, MA, 1989;285–300.
[29] Maher ML. HI-RISE and beyond: directions for expert systems in design. Comput Aided Des 1985;17:420–7.
[30] Brown D, Chandrasekaran B. Expert systems for a class of mechanical design activity. In: Gero J, editor. Knowledge engineering in computer-aided design. North Holland, 1985;259–90.
[31] Marcus S, Stout J, McDermott J. VT: an expert elevator designer that uses knowledge-based back-tracking. Technical report CMU-CS-86-169, Department of Computer Science, Carnegie Mellon University, 1986.
[32] Finger S, Dixon JR. A review of research in mechanical engineering design, part I. Res Eng Des 1989; 1(1):51–67.
[33] Finger S, Dixon, J. R. A review of research in mechanical engineering design, part II. Res Eng Des 1989;1(2):121–37.
[34] Reuleaux F. The kinematics of machinery. New York: Dover Publications, 1876 [reprinted in 1963].
[35] Hoover SP, Rinderle JR. A synthesis strategy for mechanical devices. Res Eng Des 1989;1(2):87–103.
[36] Rinderle JR. Attribute grammars as a formal approach to melding configuration and parametric design. Res Eng Des 1991;2(3)137–46.
[37] Zawicky K. On the use of grammars for configuration design. MS thesis, Mechanical Engineering Department, Carnegie Mellon University, 1994.

Automatic synthesis of both the topology and numerical parameters for complex structures using genetic programming   19

John R. Koza

Abstract This chapter demonstrates that genetic programming can automatically create complex structures from a high-level statement of the structure's purpose. The chapter presents results produced by genetic programming that are from problem areas where there is no known general mathematical technique for automatically creating a satisfactory structure. The results include automatically synthesising (designing) both the topology (graphical arrangement of components) and sizing (component values) for two illustrative analog electrical circuits and automatically synthesising both the topology and tuning (component values) for a controller. Genetic programming not only succeeds in producing the required structure, but the structure is competitive with that produced by creative human designers. The claim that genetic programming has produced human-competitive results is supported by the fact that the automatically created results infringe on previously issued patents, improve on previously patented inventions, or duplicate the functionality of previously patented inventions.

19.1 Introduction

The design of a complex structure typically entails two steps. First, the topology of the structure must be identified. After the topology is identified, optimal (or near-optimal) numerical values can be sought for the elements comprising the structure.

In this chapter, we demonstrate this two-step process with examples involving the design of several types of complex structure.

For example, if one is seeking an analog electrical circuit whose behaviour satisfies certain prespecified high-level design goals, one must first ascertain the circuit's topology. Specifically, the *topology* of an electrical circuit comprises:

- the total number of electrical components in the circuit;
- the type of each component (*e.g.*, resistor, capacitor, transistor) at each location in the circuit;
- a list of all the connections between the leads of the components.

After the topology is established, one must then discover the sizing of each electrical component in the circuit. Specifically, the *sizing* of a circuit consists of the component value(s) for each component of the circuit that requires a component value.

A similar two-step process is required if one is seeking the design of a controller whose behaviour satisfies certain prespecified high-level design goals. Controllers are typically composed of signal-processing blocks, such as integrators, differentiators, leads, lags, delays, gains, adders, inverters, subtractors, and multipliers [1]. The purpose of a controller is to force, in a meritorious way, the actual response of a system (conventionally called the *plant*) to match a desired response (called the *reference signal*).

The topology of a controller comprises:

- the total number of signal processing blocks in the controller;
- the type of each block (*e.g.*, integrator, differentiator, lead, lag, delay, gain, adder, inverter, subtractor, and multiplier);
- the connections between the inputs and output of each block in the controller and the external input and external output points of the controller.

The *tuning* (sizing) of a controller consists of the parameter values associated with each signal-processing block.

A parallel situation arises in connection with networks of chemical reactions (metabolic pathways). The concentrations of substrates, products, and intermediate substances participating in a network of chemical reactions are modelled by non-linear continuous-time differential equations, including various first-order rate laws, second-order rate laws, power laws, and the Michaelis–Menten equations [2]. The concentrations of catalysts (*e.g.*, enzymes) control the rates of many chemical reactions in living things. The *topology* of a network of chemical reactions comprises:

- the total number of reactions;
- the number of substrates consumed by each reaction;
- the number of products produced by each reaction;
- the pathways supplying the substrates (either from external sources or other reactions in the network) to each reaction;
- the pathways dispersing each reaction's products (either to other reactions or external outputs);
- an identification of whether a particular enzyme acts as a catalyst.

The *sizing* for a network of chemical reactions consists of all the numerical values associated with the network (*e.g.*, the rates of each reaction).

A similarly vexatious situation arises if one is seeking the design of an antenna whose behaviour satisfies certain prespecified high-level design goals.

19.2 Genetic programming

Genetic programming is an automatic method for solving problems. Specifically, genetic programming progressively breeds a population of computer programs over a series of generations by starting with a primordial ooze of thousands of randomly created computer programs and using the Darwinian principle of natural selection, recombination (crossover), mutation, gene duplication, gene deletion, and certain mechanisms of developmental biology.

Genetic programming breeds computer programs to solve problems by executing the following three steps:

1. Generate an initial population of compositions (typically random) of the functions and terminals of the problem.
2. Iteratively perform the following substeps (referred to herein as a generation) on the population of programs until the termination criterion has been satisfied:
   (A) execute each program in the population and assign it a fitness value using the fitness measure;
   (B) create a new population of programs by applying the following operations. The operations are applied to program(s) selected from the population with a probability based on fitness (with reselection allowed).
      (i) Reproduction: copy the selected program to the new population.
      (ii) Crossover: create a new offspring program for the new population by recombining randomly chosen parts of two selected programs.
      (iii) Mutation: create one new offspring program for the new population by randomly mutating a randomly chosen part of the selected program.
      (iv) Architecture-altering operations: select an architecture-altering operation from the available repertoire of such operations and create one new offspring program for the new population by applying the selected architecture-altering operation to the selected program.
3. Designate the individual program that is identified by result designation (e.g., the best-so-far individual) as the result of the run of genetic programming. This result may be a solution (or an approximate solution) to the problem.

Genetic programming is described in a number of books and videotapes by Koza and coworkers [3–8]. Genetic programming is an extension of the genetic algorithm [9] in which the population being bred consists of computer programs.

Genetic programming starts with an initial population of randomly generated computer programs composed of the given primitive functions and terminals. The programs in the population are, in general, of different sizes and shapes. The creation of the initial random population is a blind random search of the space of computer programs composed of the problem's available functions and terminals.

On each generation of a run of genetic programming, each individual in the population of programs is evaluated as to its fitness in solving the problem at hand. The programs in generation zero of a run almost always have exceedingly poor fitness for non-trivial problems of interest. Nonetheless, some individuals in a population will turn out to be somewhat more fit than others. These differences in performance are then exploited so as to direct the search into promising areas of the search space. The Darwinian principle of reproduction and survival of the fittest is used to select probabilistically, on the basis of fitness, individuals from the population to participate in various operations. A small percentage (e.g., 9%) of the selected individuals are reproduced (copied) from one generation to the next. A very small percentage (e.g., 1%) of the selected individuals are mutated in a random way. Mutation can be viewed as an undirected local search mechanism. The vast majority of the selected individuals (e.g., 90%) participate in the genetic operation of crossover (sexual recombination) in which two offspring programs are created by recombining genetic material from two parents.

The creation of the initial random population and the creation of offspring by the genetic operations are all performed so as to create syntactically valid, executable programs. After the genetic operations are performed on the current generation of the population, the population of offspring (i.e., the new generation) replaces the old generation. The tasks of measuring fitness, Darwinian selection, and genetic operations are then iteratively repeated over many generations. The computer program resulting from this simulated evolutionary process can be the solution to a given problem or a sequence of instructions for constructing the solution.

The dynamic variability of the size and shape of the computer programs that are created during the run is an important feature of genetic programming. It is often difficult and unnatural to try to specify or restrict the size and shape of the eventual solution in advance.

The individual programs that are evolved by genetic programming are typically multi-branch programs consisting of one or more result-producing branches and zero, one, or more automatically defined functions (subroutines).

The *architecture* of such a multi-branch program involves:

1. the total number of automatically defined functions;
2. the number of arguments (if any) possessed by each automatically defined function;
3. if there is more than one automatically defined function in a program, the nature of the hierarchical references (including recursive references), if any, allowed among the automatically defined functions.

Architecture-altering operations enable genetic programming automatically to determine the number of automatically defined functions, the number of arguments that each possesses, and the nature of the hierarchical references, if any, among such automatically defined functions.

Additional information on genetic programming can be found in books such as Banzhaf et al. [10]; books in the series on genetic programming from Kluwer such as by Langdon [11], Ryan [12], and Wong and Leung [13]; in edited collections of papers, such as the *Advances in Genetic Programming* series of books from MIT Press [14–16]; in the proceedings of the Genetic Programming Conference held between 1996 and 1998 [17–19]; in the proceedings of the annual Genetic and Evolutionary Computation Conference (combining the annual Genetic Programming Conference and the International Conference on Genetic Algorithms) held starting in 1999 [20,21]; in the proceedings of the annual Euro-GP conferences held starting in 1998 [22–24]; at web sites such as www.genetic-programming.org; and in the *Genetic Programming and Evolvable Machines* journal (from Kluwer).

19.3 Automatic synthesis of analog electrical circuits

The design process entails creation of a complex structure to satisfy user-defined requirements. The field of design is a good source of problems that can be used for determining whether an automated technique can produce results that are competitive with human-produced results. Design is a major activity of practising engineers. Since the design process typically entails tradeoffs between competing considerations, the end product of the process is usually a satisfactory and compliant design, as opposed to a perfect design. Design is usually viewed as requiring creativity and human intelligence.

The field of design of analog and mixed analog–digital electrical circuits is especially challenging because (prior to genetic programming) there has been no previously known general technique for automatically creating the topology and sizing of an analog circuit from a high-level statement of the circuit's design goals. As Aaserud and Nielsen [25] noted

> [M]ost ... analog circuits are still handcrafted by the experts or so-called 'zahs' of analog design. The design process is characterized by a combination of experience and intuition and requires a thorough knowledge of the process characteristics and the detailed specifications of the actual product.
>
> Analog circuit design is known to be a knowledge-intensive, multiphase, iterative task, which usually stretches over a significant period of time and is performed by designers with a large portfolio of skills. It is therefore considered by many to be a form of art rather than a science.

Although it might seem difficult or impossible automatically to create both the topology and numerical parameters for a complex structure merely from a high-level statement of the structure's design goals, genetic programming is capable of automatically creating both the topology and sizing (component values) for analog electrical circuits composed of transistors, capacitors, resistors, and other components merely by specifying the circuit's behaviour. For purposes of illustrating this point, we discuss:

- a lowpass filter circuit using a fitness measure based on the frequency-domain behaviour of circuits;
- a computational circuit employing transistors and using a fitness measure based on the time-domain behaviour of circuits.

Genetic programming can be applied to the problem of synthesising circuits if a mapping is established between the program trees (rooted, point-labelled trees with ordered branches) used in genetic programming and the labelled cyclic graphs germane to electrical circuits. The principles of developmental biology provide the motivation for mapping trees into circuits by means of a developmental process that begins with a simple embryo. For circuits, the initial circuit typically includes a test fixture consisting of certain fixed components (such as a source resistor, a load resistor, an input port, and an output port), as well as an embryo consisting of one or more modifiable wires. Until the modifiable wires are modified, the circuit does not produce interesting output. An electrical circuit is developed by progressively applying the functions in a circuit-constructing program tree to the modifiable wires of the embryo (and, during the developmental process, to succeeding modifiable wires and components). A single electrical circuit is created by executing the functions in an individual circuit-constructing program tree from the population. The functions are progressively applied in a developmental process to the embryo and its successors until all of the functions in the program tree are executed. That is, the functions in the circuit-constructing program tree progressively side-effect the embryo and its successors until a fully developed circuit eventually emerges. The functions are applied in a breadth-first order.

The functions in the circuit-constructing program trees are divided into five categories:

1. topology-modifying functions that alter the topology of a developing circuit;
2. component-creating functions that insert components into a developing circuit;
3. development-controlling functions that control the development process by which the embryo and its successors become a fully developed circuit;

4. arithmetic-performing functions that appear in subtrees as argument(s) to the component-creating functions and specify the numerical value of the component;
5. automatically defined functions that appear in the automatically defined functions and potentially enable certain substructures of the circuit to be reused (with parameterisation).

19.3.1 Lowpass filter circuit

A simple *filter* is a one-input, one-output electronic circuit that receives a signal as its input and passes the frequency components of the incoming signal that lie in a specified range (called the *passband*) while suppressing the frequency components that lie in all other frequency ranges (the *stopband*).

A *lowpass filter* passes all frequencies below a certain specified frequency, but stops all higher frequencies.

A filter may be constructed from inductors and capacitors. The desired lowpass LC filter has a passband below 1,000 Hz and a stopband above 2,000 Hz. The circuit is driven by an incoming AC voltage source with a 2 V amplitude. The *attenuation* of the filter is defined in terms of the output signal relative to the reference voltage (half of 2 V here). In this problem, a voltage in the passband of exactly 1 V and a voltage in the stopband of exactly 0 V is regarded as ideal. The (preferably small) variation within the passband is called the *passband ripple*. Similarly, the incoming signal is never fully reduced to zero in the stopband of an actual filter. The (preferably small) variation within the stopband is called the *stopband ripple*. A voltage in the passband of between 970 mV and 1 V (*i.e.*, a passband ripple of 30 mV or less) and a voltage in the stopband of between 0 and 1 mV (*i.e.*, a stopband ripple of 1 mV or less) is regarded as acceptable. Any voltage lower than 970 mV in the passband and any voltage above 1 mV in the stopband is regarded as unacceptable.

19.3.1.1 Preparatory steps for lowpass filter circuit

Before applying genetic programming to a problem of circuit design, seven major preparatory steps are required: (1) identify the embryonic circuit; (2) determine the architecture of the circuit-constructing program trees; (3) identify the primitive functions of the program trees; (4) identify the terminals of the program trees; (5) create the fitness measure; (6) choose control parameters for the run; and (7) determine the termination criterion and method of result designation. A detailed discussion concerning how to apply these seven preparatory steps to a particular problem of circuit synthesis (such as a lowpass filter) is found in Koza *et al.* [7, chapter 25].

The initial circuit used on a particular problem depends on the circuit's number of inputs and outputs. All development originates from the modifiable wires of the embryo within the initial circuit. An embryo with two modifiable wires (Z0 and Z1) was used for the lowpass filter circuit.

The architecture of each circuit-constructing program tree depends on the embryo. There is one result-producing branch in the program tree for each modifiable wire in the embryo.

The function set for each design problem depends on the type of electrical components that are to be used for constructing the circuit.

For the problem of synthesising a lowpass filter, the function set included two component-creating functions (for inserting inductors and capacitors into a developing circuit), topology-modifying functions (for performing series and parallel divisions and for flipping components), one development-controlling function ("no operation"), functions for creating a via (direct connection) to ground, and functions for connecting pairs of points. That is, the function set $\mathcal{F}_{\text{ccs-initial}}$ for each construction-continuing subtree was

$$\mathcal{F}_{\text{ccs-initial}} = \{\text{L}, \text{C}, \text{SERIES}, \text{PARALLEL0}, \text{PARALLEL1}, \text{FLIP}, \text{NOOP}, \text{T\_GND\_0},\\ \text{T\_GND\_1}, \text{PAIR\_CONNECT\_0}, \text{PAIR\_CONNECT\_1}\}.$$

For additional details of these functions (and other aspects of this problem), see Koza et al. [7].

The initial terminal set $\mathcal{T}_{\text{ccs-initial}}$ for each construction-continuing subtree was

$$\mathcal{T}_{\text{ccs-initial}} = \{\text{END}, \text{SAFE\_CUT}\}.$$

The initial terminal set $\mathcal{T}_{\text{aps-initial}}$ for each arithmetic-performing subtree consisted of

$$\mathcal{T}_{\text{aps-initial}} = \{\Re\}.$$

where $\Re$ represents floating-point random constants from -1.0 to $+1.0$.

The function set $\mathcal{F}_{\text{aps}}$ for each arithmetic-performing subtree was

$$\mathcal{F}_{\text{aps}} = \{+, -\}.$$

The terminal and function sets were identical for all result-producing branches for a particular problem.

The evolutionary process is driven by the *fitness measure*. Each individual computer program in the population is executed and then evaluated, using the fitness measure. The nature of the fitness measure varies with the problem. The high-level statement of desired circuit behaviour is translated into a well-defined measurable quantity that can be used by genetic programming to guide the evolutionary process.

The evaluation of each individual circuit-constructing program tree in the population begins with its execution. This execution progressively applies the functions in each program tree to the initial circuit, thereby creating a fully developed circuit. A netlist is created that identifies each component of the developed circuit, the nodes to which each component is connected, and the value of each component. The netlist becomes the input to our modified version of the 217,000-line SPICE (Simulation Program with Integrated Circuit Emphasis) simulation program [26]. SPICE then determines the behaviour of the circuit.

Since the high-level statement of behaviour for the desired circuit is expressed in terms of frequencies, the voltage VOUT is measured in the frequency domain. SPICE performs an AC small signal analysis and reports the circuit's behaviour over five decades (between 1 and 100,000 Hz) with each decade being divided into 20 parts (using a logarithmic scale), so that there are a total of 101 fitness cases.

Fitness is measured in terms of the sum over these cases of the absolute weighted deviation between the actual value of the voltage that is produced by the circuit at the probe point VOUT and the target value for voltage. The smaller the value of fitness, the better. A fitness of zero represents an (unattainable) ideal filter. Specifically, the standardised fitness is

$$F(t) = \sum_{i=0}^{100} (W(d(f_i), f_i) d(f_i)),$$

where f_i is the frequency of fitness case i, $d(x)$ is the absolute value of the difference between the target and observed values at frequency x, and $W(y,x)$ is the weighting for difference y at frequency x.

The fitness measure is designed so as not to penalise ideal values, to slightly penalise every acceptable deviation, and to heavily penalise every unacceptable deviation. Specifically, the procedure for each of the 61 points in the three-decade interval between 1 and 1,000 Hz for the intended passband is as follows:

- if the voltage equals the ideal value of 1.0 V in this interval, the deviation is 0.0;
- if the voltage is between 970 mV and 1 V, the absolute value of the deviation from 1 V is weighted by a factor of 1.0;
- if the voltage is less than 970 mV, the absolute value of the deviation from 1 V is weighted by a factor of 10.0.

The acceptable and unacceptable deviations for each of the 35 points from 2,000 to 100,000 Hz in the intended stopband are similarly weighed (by 1.0 or 10.0) based on the amount of deviation from the ideal voltage of 0 V and the acceptable deviation of 1 mV.

For each of the five "don't care" points between 1,000 and 2,000 Hz, the deviation is deemed to be zero.

Many of the random initial circuits, and many that are created by the crossover and mutation operations in subsequent generations, cannot be simulated by SPICE. These circuits receive a high penalty value of fitness (10^8) and become the worst-of-generation programs for each generation.

The population size was 640,000.

The problem was run on a medium-grained parallel Parsytec computer system consisting of 64 80 MHz PowerPC 601 processors arranged in an 8×8 toroidal mesh with a host PC Pentium-type computer. The distributed genetic algorithm was used in which 64 demes (semi-isolated subpopulations) of 10,000 resided at each node of the parallel computer. On each generation, four boatloads of emigrants, each consisting of $B = 2\%$ (the migration rate) of the node's subpopulation (selected on the basis of fitness) were dispatched to each of the four adjacent processing nodes.

19.3.1.2 Results for lowpass filter circuit

Figure 19.1 shows the frequency-domain behaviour of an illustrative lowpass filter in which the boundary of the passband is at 1,000 Hz and the boundary of the stopband is 2,000 Hz. The horizontal axis represents the frequency of the incoming signal and ranges over five decades of frequencies between 1 and 100,000 Hz on a logarithmic scale. The vertical axis represents the peak voltage of the output and ranges between 0 to 1 V on a linear scale. Figure 19.1 shows that when the input to the circuit consists of a sinusoidal signal with any frequency from 1 to 1,000 Hz, the output is a sinusoidal signal with an amplitude of a full 1 V. Figure 19.1 also shows that, when

Automatic Synthesis of Both the Topology and Numerical Parameters

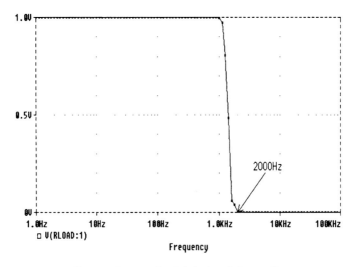

Fig. 19.1 Frequency-domain behaviour of a lowpass filter.

the input to the circuit consists of a sinusoidal signal with any frequency from 2,000 to 100,000 Hz, the amplitude of the output is essentially 0 V. The region between 1,000 and 2,000 Hz is a transition region where the voltage varies between 1 V (at 1,000 Hz) and essentially 0 V (at 2,000 Hz).

Genetic programming is capable of automatically creating both the topology and sizing (component values) for lowpass filters (and other filter circuits, such as highpass filters, bandpass filters, bandstop filters, and filters with multiple passbands and stopbands).

A filter circuit may be evolved using a fitness measure based on frequency-domain behaviour. In particular, the fitness of an individual circuit is the sum, over 101 values of frequency between 1 and 100,000 Hz (equally spaced on a logarithmic scale), of the absolute value of the difference between the individual circuit's output voltage and the ideal voltage for an ideal lowpass filter for that frequency (*i.e.*, the voltages shown in Fig. 19.1).

For example, one run of genetic programming synthesised the lowpass filter circuit of Fig. 19.2.

The evolved circuit of Fig. 19.2 is what is now called a cascade (ladder) of identical π sections [7, chapter 25]. The evolved circuit has the recognisable topology of the circuit for which George Campbell of American Telephone and Telegraph received US patent 1,227,113 in 1917 [27]. Claim 2 of Campbell's patent covered

> An electric wave filter consisting of a connecting line of negligible attenuation composed of a plurality of sections, each section including a capacity element and an inductance element, one of said elements of each section being in series with the line and the other in shunt across the line, said capacity and inductance elements having precomputed values dependent upon the upper limiting frequency and the lower limiting frequency of a range of frequencies it is desired to transmit without attenuation, the values of said capacity and inductance elements being so proportioned that the structure transmits with practically negligible attenuation sinusoidal currents of all frequencies lying between said two limiting frequencies, while attenuating and

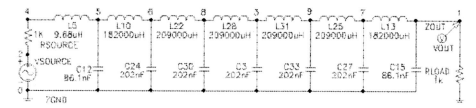

Fig. 19.2 Lowpass filter created by genetic programming that infringes on Campbell's patent.

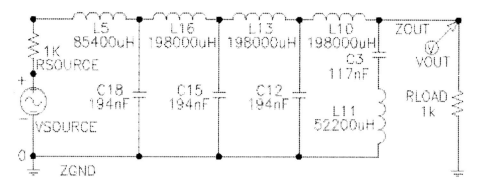

Fig. 19.3 Lowpass filter created by genetic programming that infringes on Zobel's patent.

approximately extinguishing currents of neighboring frequencies lying outside of said limiting frequencies.

In addition to possessing the topology of the Campbell filter, the numerical values of all the components in the evolved circuit closely approximate the numerical values taught in Campbell's 1917 patent.

Another run of genetic programming synthesised the lowpass filter circuit of Fig. 19.3. As before, this circuit was evolved using the previously described fitness measure based on frequency-domain behaviour.

This evolved circuit differs from the Campbell filter in that its final section consists of both a capacitor and inductor. This filter is an improvement over the Campbell filter because its final section confers certain performance advantages on the circuit. This circuit is equivalent to what is called a cascade of three symmetric T-sections and an M-derived half section [7, chapter 25]. Otto Zobel of American Telephone and Telegraph Company invented and received a patent for an "M-derived half section" used in conjunction with one or more "constant K" sections [28]. Again, the numerical values of all the components in this evolved circuit closely approximate the numerical values taught in Zobel's 1925 patent.

Seven circuits created using genetic programming infringe on previously issued patents [7]. Others duplicate the functionality of previously patented inventions in novel ways.

In both of the foregoing examples, genetic programming automatically created both the topology and sizing (component values) of the entire filter circuit by using a fitness measure expressed in terms of the signal observed at the single output point (the probe point labelled VOUT in the figures).

19.3.2 Squaring computational circuit

An analog electrical circuit whose output is a well-known mathematical function (*e.g.*, square, square root) is called a *computational circuit*.

We use a squaring computational circuit to illustrate the generality of the genetic programming process. Relatively few changes must be made in order to evolve automatically a squaring computational circuit.

First, computational circuits are typically constructed using capacitors, resistors, diodes, and transistors (instead of the inductors or capacitors used in the lowpass filter problem). Therefore, the function set must contain functions capable of inserting these components into a developing circuit.

Second, since the purpose of the circuit is different, the fitness measure must reflect the fact that the circuit's output is intended to be a voltage equal to the square of the circuit's input.

19.3.2.1 Preparatory steps for squaring computational circuit

For the problem of synthesising a computational circuit, capacitors, resistors, diodes, and transistors were used (instead of the inductors or capacitors used in the lowpass filter problem). The function set also included functions to provide connectivity to the positive and negative power supplies (in order to provide a source of energy for the transistors). Thus, the function set $\mathcal{F}_{\text{ccs-initial}}$ for each construction-continuing subtree was

$$\mathcal{F}_{\text{ccs-initial}} = \{R, C, SERIES, PARALLEL0, PARALLEL1, FLIP, NOOP, T\_GND\_0,$$
$$T\_GND\_1, T\_POS\_0, T\_POS\_1, T\_NEG\_0, T\_NEG\_1,$$
$$PAIR\_CONNECT\_0, PAIR\_CONNECT\_1, Q\_D\_NPN, Q\_D\_PNP,$$
$$Q\_3\_NPN0, K, Q\_3\_NPN11, Q\_3\_PNP0, K, Q\_3\_PNP11,$$
$$Q\_POS\_COLL\_NPN, Q\_GND\_EMIT\_NPN, Q\_NEG\_EMIT\_NPN,$$
$$Q\_GND\_EMIT\_PNP, Q\_POS\_EMIT\_PNP, Q\_NEG\_COLL\_PNP\}.$$

For the npn transistors, the Q2N3904 model was used. For pnp transistors, the Q2N3906 model was used.

Because filters discriminate on incoming signals based on frequency, the lowpass filter circuit was automatically synthesised using a fitness measure based on the behaviour of the circuit in the frequency domain. For the squaring computational circuit, SPICE is called to perform a DC sweep analysis at 21 equidistant voltages between -250 and $+250$ mV. Fitness is the sum, over these 21 fitness cases, of the absolute weighted deviation between the actual value of the voltage that is produced by the circuit and the target value for voltage (*i.e.*, the square of the input voltage).

19.3.2.2 Results for squaring computational circuit

Figure 19.4 shows a squaring circuit composed of transistors, capacitors, and resistors that was automatically synthesised using a fitness measure based on the behaviour of the circuit in the time domain [29].

This circuit was evolved using a fitness measure based on time-varying input signals. In particular, fitness was the sum, taken at certain sampled times for four dif-

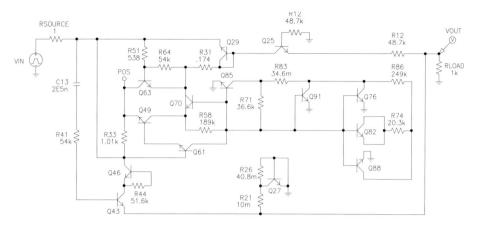

Fig. 19.4 Squaring circuit created by genetic programming.

ferent time-varying input signals, of the absolute value of the difference between the individual circuit's output voltage and the desired output voltage (*i.e.*, the square of the voltage of the input signal at the particular sampled time).

The four input signals were structured to provide a representative mixture of input values. All of the input signals produce outputs that are well within the range of voltages that can be handled by transistors (*i.e.*, below 4 V). For example, one of the input signals is a rising ramp whose value remains at 0 V up to 0.2 s and then rises to 2 V between 0.2 and 1.0 s. Figure 19.5 shows the output voltage produced by the evolved circuit for the rising ramp input superimposed on the (virtually indistinguishable) correct output voltage for the squaring function. As can be seen, as soon as the input signal becomes non-zero, the output is a parabolic-shaped curve representing the square of the incoming voltage.

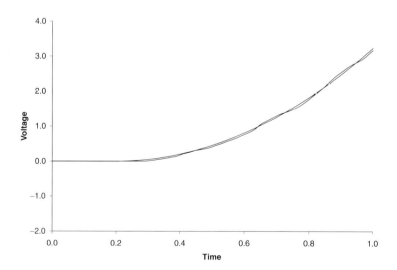

Fig. 19.5 Output for rising ramp input for squaring circuit.

19.4 Automatic synthesis of controllers

Genetic programming is capable of automatically creating both the topology and sizing (tuning) for controllers composed of time-domain blocks (such as integrators, differentiators, multipliers, adders, delays, gains, leads, and lags) merely by specifying the controller's effect on the to-be-controlled plant [30–37]. This automatic synthesis of controllers from data is performed by genetic programming even though there is no general mathematical method for creating both the topology and sizing for controllers from a high-level statement of the design goals for the controller.

In the PID type of controller, the controller's output is the sum of proportional (P), integrative (I), and derivative (D) terms based on the difference between the plant's output and the reference signal. Albert Callender and Allan Stevenson of Imperial Chemical Limited of Northwich, UK, received US Patent 2,175,985 in 1939 for the PI and PID controller.

Claim 1 of Callender and Stevenson [38] covers what is now called the PI controller,

> A system for the automatic control of a variable characteristic comprising means proportionally responsive to deviations of the characteristic from a desired value, compensating means for adjusting the value of the characteristic, and electrical means associated with and actuated by responsive variations in said responsive means, for operating the compensating means to correct such deviations in conformity with the sum of the extent of the deviation and the summation of the deviation.

Claim 3 of Callender and Stevenson [38] covers what is now called the PID controller,

> A system as set forth in claim 1 in which said operation is additionally controlled in conformity with the rate of such deviation.

The vast majority of automatic controllers used by industry are of the PI or PID type. However, it is generally recognised by leading practitioners in the field of control that PI and PID controllers are not ideal [39,40].

There is no pre-existing general-purpose analytic method (prior to genetic programming) for automatically creating both the topology and tuning of a controller for arbitrary linear and non-linear plants that can simultaneously optimise prespecified performance metrics. The performance metrics used in the field of control include, among others:

- minimising the time required to bring the plant output to the desired value (as measured by, say, the integral of the time-weighted absolute error);
- satisfying time-domain constraints (involving, say, overshoot and disturbance rejection);
- satisfying frequency-domain constraints (*e.g.*, bandwidth);
- satisfying additional constraints, such as limiting the magnitude of the control variable or the plant's internal state variables.

We employ a problem involving control of a two-lag plant (described by Dorf and Bishop [1, p. 707]) to illustrate the automatic synthesis of controllers by means of genetic programming. The problem entails synthesising the design of both the topology and parameter values for a controller for a two-lag plant such that plant output

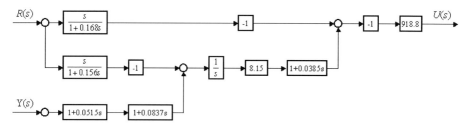

Fig. 19.6 Evolved controller that infringes on Jones' patent.

reaches the level of the reference signal so as to minimise the integral of the time-weighted absolute error, such that the overshoot in response to a step input is less than 2%, and such that the controller is robust in the face of significant variation in the plant's internal gain K and the plant's time constant τ.

Genetic programming routinely creates PI and PID controllers infringing on the 1942 of Callender and Stevenson patent during intermediate generations of runs of genetic programming on controller problems. However, the PID controller is not the best possible controller for this, and many other problems.

Figure 19.6 shows the block diagram for the best-of-run controller evolved during one run of this problem. In this figure, $R(s)$ is the reference signal; $Y(s)$ is the plant output; and $U(s)$ is the controller's output (control variable). This evolved controller is 2.42 times better than the Dorf and Bishop controller as measured by the criterion used by Dorf and Bishop [1]. In addition, this evolved controller has only 56% of the rise time in response to the reference input, has only 32% of the settling time, and is 8.97 times better in terms of suppressing the effects of disturbance at the plant input.

This genetically evolved controller differs from a conventional PID controller in that it employs a second-derivative processing block. Specifically, after applying standard manipulations to the block diagram of this evolved controller, the transfer function for the best-of-run controller can be expressed as a transfer function for a pre-filter and a transfer function for a compensator. The transfer function for the pre-filter, $G_{p32}(s)$, for the best-of-run individual from generation 32 is

$$G_{p32}(s) = \frac{1(1+0.1262s)(1+0.2029s)}{(1+0.03851s)(1+0.05146s)(1+0.08375s)(1+0.1561s)(1+0.1680s)}.$$

The transfer function for the compensator, $G_{c32}(s)$, is

$$G_{c32}(s) = \frac{7487(1+0.03851s)(1+0.05146s)(1+0.08375s)}{s}$$

$$= \frac{7487.05 + 1300.63s + 71.2511s^2 + 1.2426s^3}{s}.$$

The s^3 term (in conjunction with the s in the denominator) indicates a second derivative. Thus, the compensator consists of a second derivative in addition to proportional, integrative, and derivative functions. As it happens, Harry Jones of The Brown Instrument Company of Philadelphia received US Patent 2,282,726 for this kind of controller topology in 1942.

Claim 38 of the Jones patent [41] states

In a control system, an electrical network, means to adjust said network in response to changes in a variable condition to be controlled, control means responsive to network adjustments to control said condition, reset means including a reactance in said network adapted following an adjustment of said network by said first means to initiate an additional network adjustment in the same sense, and rate control means included in said network adapted to control the effect of the first mentioned adjustment in accordance with the second or higher derivative of the magnitude of the condition with respect to time.

Note that the human user of genetic programming did not preordain, prior to the run (*i.e.*, as part of the preparatory steps for genetic programming), that a second derivative should be used in the controller (or, for that matter, even that a P, I, or D block should be used). Genetic programming automatically discovered that the second-derivative element (along with the P, I, and D elements) was useful in producing a good controller for this particular problem. That is, necessity was the mother of invention.

Similarly, the human who initiated this run of genetic programming did not preordain any particular topological arrangement of proportional, integrative, derivative, second derivative, or other functions within the automatically created controller. Instead, genetic programming automatically created a controller for the given plant without the benefit of user-supplied information concerning the total number of processing blocks to be employed in the controller, the type of each processing block, the topological interconnections between the blocks, the values of parameters for the blocks, or the existence of internal feedback (none in this instance) within the controller.

19.5 Other examples

In the same vein as the foregoing examples, it should be mentioned that genetic programming is capable of discovering both the topological and numerical aspects of a satisfactory antenna design from a high-level specification of the antenna's behaviour. In one particular problem [42], genetic programming automatically discovered the design for a satisfactory antenna composed of wires for maximising gain in a preferred direction over a specified range of frequencies, having a reasonable value of voltage standing wave ratio when the antenna is fed by a transmission line with a specified characteristic impedance, and fitting into a specified bounding rectangle. The design that genetic programming discovered included:

1. the number of directors in the antenna;
2. the number of reflectors;
3. the fact that the driven element, the directors, and the reflector are all single straight wires;
4. the fact that the driven element, the directors, and the reflector are all arranged in parallel;
5. the fact that the energy source (via the transmission line) is connected only to the driven element, *i.e.*, the directors and reflectors are parasitically coupled.

The last three of the above characteristics discovered by genetic programming are the defining characteristics of an inventive design conceived in the early years of the field of antenna design [43–45].

Similarly, genetic programming is capable of discovering both the topology of the network of chemical reactions and the numerical parameters of all the reactions of the network starting with the observed time-domain concentrations of the final product substance(s). The approach used [46] involved:

1. establishing a representation for chemical networks involving symbolic expressions (S-expressions) and program trees that are composed of functions and terminals and that can be progressively bred (and improved) by genetic programming;
2. converting each individual program tree in the population into an analog electrical circuit representing a network of chemical reactions;
3. obtaining the behaviour of the individual network of chemical reactions by simulating the corresponding electrical circuit;
4. defining a fitness measure that measures how well the behaviour of an individual network matches the observed time-domain data concerning concentrations of product substances;
5. using the fitness measure to enable genetic programming to breed a population of improving program trees.

Figure 19.7 shows a network of chemical reactions that was evolved in this manner using genetic programming. This network closely matches the observed data for all data points, has the same topology as the correct metabolic pathway, and has almost exactly the same rate constants for the four reactions.

This network was evolved using only the time-domain concentration values of the final product C00165 (diacyl-glycerol). Genetic programming created the entire metabolic pathway, including

- topological features such as the internal feedback loop;
- topological features such as a bifurcation point, where one substance is distributed to two different reactions;

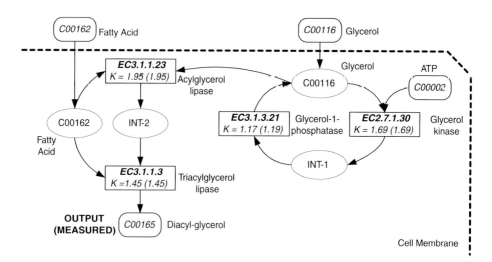

Fig. 19.7 Network of chemical reactions that was evolved using genetic programming.

- topological features such as an accumulation point, where one substance is accumulated from two sources;
- use of two intermediate substances (INT_1 and INT_2);
- numerical rates (sizing) for all reactions.

19.6 Conclusions

This chapter has demonstrated that a biologically motivated algorithm (genetic programming) is capable of automatically synthesising the design of both the topology of complex graphical structures and optimal or near-optimal numerical values for all elements of the structure possessing parameters.

References

[1] Dorf RC, Bishop RH. Modern control systems. 8th ed. Menlo Park (CA): Addison-Wesley, 1998.
[2] Voit EO. Computational analysis of biochemical systems. Cambridge: Cambridge University Press, 2000.
[3] Koza JR. Genetic programming: on the programming of computers by means of natural selection. Cambridge (MA): MIT Press, 1992.
[4] Koza JR, Rice JP. Genetic programming: the movie. Cambridge (MA): MIT Press, 1992.
[5] Koza JR. Genetic programming II: automatic discovery of reusable programs. Cambridge (MA): MIT Press, 1994.
[6] Koza JR. Genetic programming II videotape: the next generation. Cambridge (MA): MIT Press, 1994.
[7] Koza JR, Bennett FH III, Andre D, Keane MA. Genetic programming III: Darwinian invention and problem solving. San Francisco (CA): Morgan Kaufmann, 1999.
[8] Koza JR, Bennett FH III, Andre D, Keane MA, Brave S. Genetic programming III videotape: human-competitive machine intelligence. San Francisco (CA): Morgan Kaufmann, 1999.
[9] Holland JH. Adaptation in natural and artificial systems: an introductory analysis with applications to biology, control, and artificial intelligence. Ann Arbor (MI): University of Michigan Press, 1975 [2nd ed. Cambridge (MA): MIT Press, 1992].
[10] Banzhaf W, Nordin P, Keller RE, Francone FD. Genetic programming – an introduction. San Francisco (CA)/Heidelberg: Morgan Kaufmann/dpunkt, 1998.
[11] Langdon WB. Genetic programming and data structures: genetic programming + data structures = automatic programming! Amsterdam: Kluwer, 1998.
[12] Ryan C. Automatic re-engineering of software using genetic programming. Amsterdam: Kluwer, 1999.
[13] Wong ML, Leung KS. Data mining using grammar based genetic programming and applications. Amsterdam: Kluwer, 2000 [ISBN: 0-7923-7746-X].
[14] Kinnear KE Jr, editor. Advances in genetic programming. Cambridge (MA): MIT Press, 1994.
[15] Angeline PJ, Kinnear KE Jr, editors. Advances in genetic programming 2. Cambridge (MA): MIT Press, 1996.
[16] Spector L, Langdon WB, O'Reilly U-M, Angeline P, editors. Advances in genetic programming 3. Cambridge (MA): MIT Press, 1999.
[17] Koza JR, Goldberg DE, Fogel DB, Riolo RL, editors. Genetic Programming 1996: Proceedings of the First Annual Conference, 28–31 July, Stanford University. Cambridge (MA): MIT Press, 1996.
[18] Koza JR, Deb K, Dorigo M, Fogel DB, Garzon M, Iba H, et al., editors. Genetic Programming 1997: Proceedings of the Second Annual Conference, 13–16 July, Stanford University. San Francisco (CA): Morgan Kaufmann, 1997.
[19] Koza JR, Banzhaf W, Chellapilla K, Deb K, Dorigo M, Fogel DB, et al., editors. Genetic Programming 1998: Proceedings of the Third Annual Conference. San Francisco (CA): Morgan Kaufmann, 1998.

[20] Banzhaf W, Daida J, Eiben AE, Garzon MH, Honavar V, Jakiela M, et al., editors. GECCO-99: Proceedings of the Genetic and Evolutionary Computation Conference, 13–17 July, Orlando, FL, USA. San Francisco (CA): Morgan Kaufmann, 1999.
[21] Whitley D, Goldberg D, Cantu-Paz E, Spector L, Parmee I, Beyer H-G, editors. GECCO-2000: Proceedings of the Genetic and Evolutionary Computation Conference, 10–12 July, Las Vegas, NV. San Francisco (CA): Morgan Kaufmann, 2000.
[22] Banzhaf W, Poli R, Schoenauer M, Fogarty TC. In: Genetic Programming: First European Workshop. EuroGP'98. April 1998 Proceedings. Paris, France. Lecture notes in computer science, vol. 1391. Berlin (Germany): Springer, 1998.
[23] Poli R, Nordin P, Langdon WB, Fogarty TC. In: Genetic Programming: Second European Workshop. EuroGP'99. Proceedings. Lecture notes in computer science, vol. 1598. Berlin (Germany): Springer, 1999.
[24] Poli R, Banzhaf W, Langdon WB, Miller J, Nordin P, Fogarty TC. In: Genetic Programming: European Conference, EuroGP 2000, Edinburgh, Scotland, UK, April 2000 Proceedings. Lecture notes in computer science, vol. 1802. Berlin (Germany): Springer, 2000.
[25] Aaserud O, Nielsen IR. Trends in current analog design: a panel debate. Analog Integr Circuits Signal Process 1995;7(1):5–9.
[26] Quarles T, Newton AR, Pederson DO, Sangiovanni-Vincentelli A. SPICE 3 version 3F5 user's manual. Berkeley (CA): Department of Electrical Engineering and Computer Science, University of California, 1994.
[27] Campbell GA. Electric wave filter. US Patent 1,227,113, filed 15 July 1915, issued 22 May 1917.
[28] Zobel OJ. Wave filter. US Patent 1,538,964, filed 15 January 1921, issued 26 May 1925.
[29] Mydlowec W, Koza J. Use of time-domain simulations in automatic synthesis of computational circuits using genetic programming. In: Late Breaking Papers at the 2000 Genetic and Evolutionary Computation Conference, Las Vegas, NV. 2000; 187–97.
[30] Keane MA, Yu J, Koza JR. Automatic synthesis of both the topology and tuning of a common parameterized controller for two families of plants using genetic programming. In: Whitley D, Goldberg D, Cantu-Paz E, Spector L, Parmee I, Beyer H-G, editors. GECCO-2000: Proceedings of the Genetic and Evolutionary Computation Conference, 10–12 July, Las Vegas, NV. San Francisco (CA): Morgan Kaufmann, 2000; 496–504.
[31] Koza JR, Keane MA, Yu J, Bennett FH III, Mydlowec W, Stiffelman O. Automatic synthesis of both the topology and parameters for a robust controller for a non-minimal phase plant and a three-lag plant by means of genetic programming. In: Proceedings of 1999 IEEE Conference on Decision and Control, 1999; 5292–300.
[32] Koza JR, Keane, MA, Bennett FH III, Yu J, Mydlowec W, Stiffelman O. Automatic creation of both the topology and parameters for a robust controller by means of genetic programming. In: Proceedings of the 1999 IEEE International Symposium on Intelligent Control, Intelligent Systems, and Semiotics. Piscataway (NJ): IEEE, 1999; 344–52.
[33] Koza JR, Keane MA, Yu J, Bennett FH III, Mydlowec W. Automatic creation of human-competitive programs and controllers by means of genetic programming. Genet Prog Evolv Mach 2000;(1):121–64.
[34] Koza JR, Keane MA, Yu J, Mydlowec W, Bennett FH III. Automatic synthesis of both the topology and parameters for a controller for a three-lag plant with a five-second delay using genetic programming. In: Cagnoni S, Poli R, Smith GD, Corner D, Oates M, Hart E, editors. Real-World Applications of Evolutionary Computing. EvoWorkshops 2000. EvoIASP, Evo SCONDI, EvoTel, EvoSTIM, EvoRob, and EvoFlight, Edinburgh, Scotland, UK, April 2000, Proceedings. Lecture notes in computer science, vol. 1803. Berlin (Germany): Springer, 2000; 168–77 [ISBN 3-540-67353-9].
[35] Koza JR, Keane MA, Yu J, Mydlowec W, Bennett FH III. Automatic synthesis of both the control law and parameters for a controller for a three-lag plant with five-second delay using genetic programming and simulation techniques. In: Proceedings of the 2000 American Control Conference, 28–30 June, Chicago, IL. Evanston (IL): American Automatic Control Council, 2000; 453–9.
[36] Koza JR, Yu J, Keane MA, Mydlowec W. 2000. Evolution of a controller with a free variable using genetic programming. In: Poli R, Banzhaf W, Langdon WB, Miller J, Nordin P, Fogarty TC, editors. Genetic Programming: European Conference, EuroGP 2000, Edinburgh, Scotland, UK, April 2000, Pro-

ceedings. Lecture notes in computer science, vol. 1802. Berlin (Germany): Springer, 2000; 91–105 [ISBN 3-540-67339-3].

[37] Yu J, Keane, Martin A, Koza JR. Automatic design of both topology and tuning of a common parameterized controller for two families of plants using genetic programming. In: Proceedings of Eleventh IEEE International Symposium on Computer-Aided Control System Design (CACSD) Conference and Ninth IEEE International Conference on Control Applications (CCA) Conference, Anchorage, AK, 25–27 September, 2000; CACSD-234–42.

[38] Callender A, Stevenson AB. Automatic control of variable physical characteristics. US Patent 2,175,985, filed (in USA) 17 February 1936, (in UK) 13 February 1935, issued (USA) 10 October 1939.

[39] Astrom KJ, Hagglund T. PID controllers: theory, design, and tuning. 2nd ed. Research Triangle Park (NC): Instrument Society of America, 1995.

[40] Boyd SP, Barratt CH. Linear controller design: limits of performance. Englewood Cliffs (NJ): Prentice Hall, 1991.

[41] Jones HS. 1942. Control apparatus. US Patent 2,282,726, filed 25 October 1939, issued 12 May 1942.

[42] Comisky W, Yu J, Koza J. Automatic synthesis of a wire antenna using genetic programming. In: Late Breaking Papers at the 2000 Genetic and Evolutionary Computation Conference, Las Vegas, NV, 2000; 179–86.

[43] Uda S. Wireless beam of short electric waves. J Inst Electr Eng (Jpn) 1926;(March):273–82.

[44] Uda S. Wireless beam of short electric waves. J Inst Electr Eng (Jpn) 1927;(March):1209–19.

[45] Yagi H. Beam transmission of ultra short waves. Proc Inst Radio Eng 1928;26:714–41.

[46] Koza JR, Mydlowec W, Lanza G, Yu J, Keane MA. Reverse engineering and automatic synthesis of metabolic pathways from observed data using genetic programming. Stanford Medical Informatics technical report SMI-2000-0851, 2000.

Index

A

1-port 305
2-port 305
abduction   3, 9, 13, 16, 17, 67, 68, 89
  algorithm for   67
  creative   16
  model-based   67
  non-creative   16
absorption refrigerator   24
abstract
  entity concept   82
  view   24, 29
abstraction   188
  high-level   283
  level of   188
  multiple levels of   195
  qualitative   275
accelerometers   153
actualisation   13
adaptations – design   283; *see also* FAMING
adders   320
aggregation   219
aircraft   21
AIR-CYL   307
algorithm
  distributed   326
  genetic   321
  for inventive problem solving (ARIZ)   142
  see also abduction
Altshuller   131-2 *see also* TRIZ
amplifier   273
analogy   9, 245
  analogical elaboration   272
  analogical transfer   271
  between-domain   261
  derivational   245
  model-based (MBA)   271
  purpose-directed   283
  source analogue   272
  transformational   245
  within-domain   283

analogy-based design (ABD)   245
analogy-based retrieval   x, 245-68
  design support system using analogy (DESSUA)   263
  example   264
  knowledge representation   246
    behaviour   250
    causal knowledge   256
    design prototype   257
    FBS path   256
    function   253
    structure   246
  model   258
    evaluation   263
analysis   3, 285
  as deduction of consequences of conjecture   7
  as function   7
  functional   285
  as phase   7
  as problem analysis   7
  as reasoning from form to function   11
  vs. synthesis   37, 68
analysis-oriented thought process model   68, 89
analysis–synthesis evaluation model   5
approach   245
  assembly   8
  computational, to creative design   245
  diagrammatic   242
  function structures   ix
  grammar-based   xiv
  nine-screen (also multi-screen)   140
  pencil-and-paper   171
  trial-and-error   94
architecture   xi
arm and stopper   240
artefact   99, 271
artificial intelligence (AI)   201
attribute   248
  element   248
    value   248
      continuous   248

attribute *cont.*
　　discrete   248
attributes (artefact)   37
audit trail   209
auto-focus cameras   199
automobiles   286
axiomatic design   144
axioms   71
　GDT's   39
　　correspondence   39
　　operation   39
　　recognition   39

B

behaviour   3, 273
　actual   10
　additional   276
　causal   273
　　internal   273
　desired   308
　dynamic   205
　frequency-domain   326
　functional   15
　ideal device   157
　incidental   310
　intended   xi, 10
　kinematic   231
　multiple   303
　nominal   172
　qualitative   249
　quantitative   249
　relevant   246
　required (*see also* intended)   229
　types   250
　　aesthetic   250
　　spatial   250
　　temporal   250
　　unexpected   298
bicycles   20
blinds   262
block diagrams   111, 118
blocks   xi, 320
　functional building   ix
　signal-processing   320
bond graphs   158, 176, 204, 205, 305
boundary   53
　system   220
　of TS   53
brainstorming   viii
browsing   180

C

CADRE   298, 299
cam and follower   229
camcorders   199

can opener   248
candidate schematic descriptions   162, 163
　classifying behaviour   163
　generating   162
　modifying   163
capacitor   319
cascade (also ladder)   327
cascading   282
catalogues *see* design catalogues
catch-as-catch-can   20
causality
　chain
　　horizontal   96
　　vertical   96
　design   101
CFRL   292
chair   248
characteristics   93
　behavioural   230
　component   316
　design   109
　dynamic   258
　physical   303
　structural   93
　temporal   183
chemical reactions   320
circuit(s)
　computational   323
　　analog   319
　　electrical   322
　　mixed analog-digital   322
　electronic   xi
　　low-pass filter   323
circumscription   69
classifications   39
CLIPS   209
CLOS   204
clothes hanger   262
clustering   192
codified models   78
cognitive design process model   67
combination   288
Common Lisp   264
communication   5
compensator   332
completeness   53, 171, 310
complexity   53
　combinatorial   157
　degree of   53
compliance   236
components   273
computational
　linguistics   306
　support   vii
computer science   306

Index 341

concentration
  time-domain 334
  of catalysts 320
  of products 320
    intermediate 320
    of substrates 320
concept
  abstract 38
  allocation arithmetic 204
  alternative ix
  of continuity 38
  of entity (also entity concept) 38, 40
  functional 83
  of physical law 43
  variant ix
concept of functions 109–20
  advantages 110
  block diagrams 111, 118
  dynamic processes 110
  generally valid functions 112
  inappropriate use 118
  practical application 114, 118
    and fault finding 118
    and modular systems 118
    and planning 118
    steps 114
  principle solutions 114
  static processes 110
  task-specific functions 112
  use 110
  verb–noun pairs 110, 118
conceptual design synthesis x, 179–96
  approach 180
    program evaluation 188
    theory development 182
  FuncSION 188
  further developments and work 190, 194
  idea generation 180
  solution numbers management 191
  solutions exploration 180
  theory testing 186
  visualisation aids 193
concreteness 53
concurrent engineering 144
configuration space 230
configuration(s) 307
  device 307
  functional 306
  generic physical 179
  geometric 303
  of means 103
  space 230
  spatial 179
  structural 306
  topological 179

conjectures 5
constraint(s) 298
  compositional 298
  frequency-domain 331
  functional 306
  magnitude 192
  operational 110
  position 192
  satisfaction 296
  spatial 110
  structural 272
  time-domain 331
construction 9
context(s) 276
  causal 275
  defining 219
  functional 275
  problem-solving 276
  structural 275
contradiction 298
  notion of 134
  organisational 134
  physical 134
  psychological inertia 134
  technical 134
controller(s) 319
  electronic xi
  mechanical xi
  PI 331
  PID 331
convergence 38
cooking pot 249
coordinate measuring machines (CMM) 26
corkscrew 194
correspondences 89
cost reduction 229
creativity xii, 93
criteria 181
crossover 320
curtain 250, 264
customer wishes 65
cybernetics 94

D

data 5
  electronic 249
  geometrical 248
  topological 248
  collection 5
decision xii
decomposability 277
decomposition 77
  functional 77, 200
  hierarchical 200
  predetermined 307

decomposition *cont.*
  schematic-physical   157
  structural   307
  of an artefact   102
deduction   12, 17, 285
degree of freedom   290
delays   320
dependencies
  artefact   39
  causal
    cyclic   251
description
  description I   181
  description II   181
  design   xii
    concept   xii
    embodiment   xi
    problem statement   xii
    requirements   xii
  functional   38, 303
  physical   156
  schematic   153
descriptive studies   xiv
design(s)
  adaptive   115
  alternatives   245
  analog circuit   323
  analogy-based (ABD)   245
  by analysis   67
  antenna   320, 333
  architectural   245
  axioms   81
  behaviour-preserving   xi
  brief   61
  case-based   xiv
  catalogues   ix, 121–8
    appendix   121
    classifying criteria   121
    desirable forms   127
    main part   121
    object   122, 128
    one-dimensional   127
    operation   125, 128
    purpose   121
    requirements   127
    selection characteristics   121
    solution   123, 128
    three-dimensional   128
    two-dimensional   127
    types and structure   121
    use   128
  chemical engineering   204
  closed   220
  complete   309
  configuration   304

context   222
as converging process   46
as decision making   67
economical   311
embodiment   214
functional   214
function-based   200
gearbox   304
industrial   50
innovative   285
instrument   25
interior   264
of kinematic pairs   286
knowledge   35, 81
as knowledge-based activity   67
machine   229
for manufacture (DFM)   144
as mapping between function and
    attribute space   42
means   222
mechanical device   245
mechatronic   200
methodology   3
model-based   286
new   259
open   220
paradigms   286
parametric   306, 307
patterns   271–84
platforms   19
principles   viii, 19, 25, 32
problem(s)   9
  novel   49
as problem solving   67
procedure   62
process engineering   204
process, study of   3
  empirical   68
  methodological (also prescriptive)   68
  theoretical   68
proposal   16
research methodology   181
research   vii
science   51
scope   222
see also technical systems theory
shape   286
simulator   67
software engineering   245
source   258
specification(s)   37
  feasible   43
  numerical   173
  structural   245, 306
support system   68

synthesis   vii
    research   vii
        computational support   vii
        influencing factors   vii
        prescriptive approaches   vii
        understanding   vii
    as synthesis   67
    target   258
    tasks   200
    transmission   308
    working space(s)   213, 220
designing   viii
    engineering   50
    methodical   121
    phases
        clarifying problem   61
        conceptualising   61
        detailing   61
        embodying   61
    recursive nature of   103
DESSUA *see* analogy-based retrieval
diagram(s)   200
    metric   290
    data-flow   200
    electric circuit   305
    schematic   200
differentiators   320
DIICAD Entwurf system   x, 224
    architecture
        design modeller   224
        object model   220
        object-oriented database   224
        process control   220
        process model   220
        task solution   220
dimension–time–cost operators (DTC)   140
directors   333
discovery   285
disposable camera   233
distinct unlabelled simple directed path   184
domain(s)
    design   245
    functional   303
    physical   303
    retrieved   xi
    target   xi
domain(s)   ix
    design   245
    functional   303
    physical   303
    retrieved   xi
    target   xi
    theory   ix, 93–107
        artefact-oriented synthesis   105

design-process-oriented synthesis   101
    function–means tree   101
    basic design step   102
    decomposition   102
    configuration   103
    problem analysis   105
new designs   94, 99
    concretisation   101
    detailing   101
    steps   101
organ domain   95
part domain   95
past designs   94, 99
and systems theory   94
transformation domain   95
visualisation   98
    function–means tree   105
    product model   106
    product model development   106
door   249
door latches   182
drug infuser   204
dynamic systems   157
dynamic variability   322

E
earth-moving equipment   182
effect(s)   11
    output   53
    physical   11
effectors   53
eigenschaften (*see* behavioural properties)   94
element(s)   246
    abstract design   271
    basic types   183
    bond graph   308
    common   181
    design   246
    idealised   305
        compliances   153
        inertia   153
        resistance   153
    lumped-parameter   305
        generalised capacitances   158
        generalised gyrators   158
        generalised inertances   158
        generalised resistances   158
        generalised tramsformers   158
    machine   54
    primitive   248
    structural   248
    transformer   312
embedded control software   200
embodiment   31

empirical research   17
energy
   flow   114
   source   333
engineering
   civil   248
   design   viii
      research   67
   science   viii
      systems   6
   synthesis   vii
engines   20
   internal combustion (IC)   20
   Newcomen   23
   four-stroke   22
   steam   23
entity (and entities)   xii, 38, 82
   concepts   81
   logical   246
   of means   103; see also organs,
         wirkelements and relationships
   physical   246
   -set   38
environment   10
envisionment   290
enzymes   320
epicyclic gears   26
equations   242
   constraint   242
   differential   320
   Michelis–Menten   320
escapement   285
evaluation   xii, 246
events   247
evoloid gearwheels (also pinions)   122
exclusion set   202
explanation   250
exploration   xii
   resource   245
   solution space   179
extensibility   172

F
FABEL   282
FAMING   xi, 285–300
   case adaptation   295
   derivation   285
   FBS model inversion   294
   function   292
   innovation   288
   interaction   289
   kinematic pairs   286
   metric diagram   290
   model-based design   286

prototype-based design   287
S–B inversion   295
   combination   296
   constraints   298
   modification   297
SBF models   90, 287
fault tree   118
FBS model   88
FDL   306
feasibility   188
feature(s)   293
   functional   293
   shape   298
feedback   171, 172, 333
feedforward   282
filter   324
   attenuation of   324
   campbell   328
   electric wave   327
   ideal   326
   LC   324
finality   62
fitness measure   325
fixation   109
flow
   control   248
   data   248
   modulated source of (MSF)   205
flush tank   264
form (structure)   9
   and properties   9
formalism   247
fountain pen   262
four Cs   144
   collect   144
   construct   144
   create   144
   produce   144
frame cognition model   85
FuncSION   x, 188
   see also conceptual design synthesis
function(s)   viii, 3, 13
   analysis   77
   arithmetic-performing   324
   assisting (also termed partial functions)
         53
      auxiliary   53
      connecting   53
      controlling   53
      propelling   53
      regulating   53
      supporting   53
   automatically defined   324
   auxiliary   111

Index   345

carriers (also termed organs)   53
  as causal relation between environment
      and behaviour   292
  component-creating   323
  concept of   109
  derivative   332
  design   77
  desired   10
  development-controlling   323
  device   292
  disturbing   118
  elementary (*see* general)   7
  envisionment   292
  escapement   285
  fault   118
  general (*see* elementary)   7
  generally valid   114
  generic   271
  as input–output relationship   110
  integrative   332
  intended   x
  interesting   289
  kinematic   286
  logical   114
  main   111
  mechanical   x
  multiple   311
  of organ   97
  overall (as overall purpose of system)
      110
  power transformation   203
  primary   251
  primitive   251, 321
  as product behaviour at non-physical
      level   215
  proportional   332
  qualitative   300
  required   103
  as set of inputs and outputs   183
  as solution-neutral intended purpose
      109
  squaring   330
  as starting point for fault finding   118
  sub-   ix
  as subjective description of behaviours
      77
  subordinate (*see* subfunctions)   103
  as subset of behaviours   306
  superfluous   115
  task structure
    elementary subtasks   215
    main task   215
    subtasks   215
  task-specific   112
  temporal   242
  topology-modifying   323
  transfer   332
  as verb–noun environment   103
  working   53
functional differences   276
functional reasoning approach   xiv
functional representation   3, 16
function–means-based synthesis   x,
    199–211
  bond graphs   205
  example   205
  function–means trees   201
  hierarchical decomposition   200
  implementation   209
  key concepts   200
  scheme defined   200
  Schemebuilder   199–211
  simulation and evaluation   204
  software prototyping   201
function–means law   94, 99, 106
function-means trees (F/M-trees)   ix, 99,
    106, 201

G
gains   320
GDT-IDEAL   42
GDT-REAL   43
gear transmissions   xi
gearbox   217
gene deletion   320
gene duplication   320
general design language   107
general design theory (GDT)   viii, 35–47, 67
  axioms   39
  definitions   38–43
  design process   46
  extensional representation   45
  guidelines   36
  ideal knowledge   40
    summary   44
  insights of   47
  intensional representation   45
  knowledge structures   47
  models   43
  real knowledge   42
    summary   44
general theory of technical systems   93,
    106
generalisation, generalising   285, 288
generate and test   80
generation   321
  current   321
  idea   xii

generation *cont.*
    locus   234
    new   321
    old   321
    of wider range of ideas   180
genetic
    algorithms   xi
    programming   viii, 319–35
        examples   322–9, 331, 333, 334
        method   320
        two-step design process   319
geometrical form   9
geometry   303
goal(s)   246
    design   246
    as required behaviour   99
goals–means transitions   62; *see also* function–means trees
grammar(s)   307
    attribute   311
    bond graph   307
    formal   303
    graph   303
    shape   306
    transformational   xi
granularity   217
graph   312
    behaviour   251
    behaviourally equivalent   312
    behaviour-structure   262
    directed acyclic (DAG)   202
    of design requirements   306
    of functional elements   154
    functionally equivalent   307
    influence   251
    of kinematic states and transitions   290
    labelled cyclic   323
    maximal common sub-   283
    physical   304
    specifications   308
guidelines – VDI   214
gyroscope   277

H
hand dryer   251
heat exchangers   282
helicopter rotor blades   26
heuristic(s)   170
    based on domain knowledge   193
    elimination   192
hierarchical schematic diagrams   199
hierarchy (abstraction)   209
HI-RISE   307
Hubka's law   99

hydrogas suspensions   20
hydrolastic suspensions   20
hypermedia   209

I
ICED   287
ideal result (notion of)   134
IDeAL system   271–84
IFIP   220
indexes   259
    graph   259
    label   259
induction   9, 17
inference   11
influences   181
information   213
    contextual   213
        effective lines   216
        effective spaces   216
        effective surfaces   216
innoduction   3, 16, 17
innovative abduction   viii, 3
input vector   183
inputs   220
insight   viii, 19
    defined   23
    developing   24
    examples   23–29
insight-developing studies (IDSs)   24
    insufficient   25
inspiration   285
integration   285
integrators   320
interaction analysis   105
interaction(s)   298
    user   298
    synergistic   316
internal gear   229
interpretation   287
intuition   298
invention
    engineering   20
    mathematical   20
    machine   146
inventive steps   19, 22
    illumination   22
    incubation   22
    preparation   22
    verification   22
inversion   294
inverters   320
inverting amplifier   273
involute gearwheel   127
isolation (concept of)   166
I-TRIZ   146

Index

K
KEE 201
key issues 19
KIC 220
kinds 182
kinematic(s) 309
kinematic
  behaviour 233
    chains 286
  design (least constraint) 25
    motions 229
  pairs 286
  state 290
    motions 229
Kirchhoff's laws 158
knowledge viii
  actions on viii
  adaptation 287
  acquisition 287
  artefact 213
  artefactual xii
  -based system 106
  between-domain 263
  causal 257
  common-sense 263
  computational process 272
  deep 257
  derived from practical activities 51
  domain 178
  experimental 229
  functional-view 35
  generalised 246
  ideal 40
  ill-structured xii
  latent 144
  object 51
  operational 257
  operations
    conflict resolution 77
    information confirmation 77
    knowledge/information acquisition 77
    knowledge/information reorganisation 77
    knowledge/information revision 77
    object analysis 77
    solution synthesis 77
  about patterns of differences 275
  about patterns of modifications 275
  practice 51
  problem-oriented design 217
  process 213
  real 42
  relational 257
  shallow 257
  source 78
  structural-view 35
  synthesis 35
  temporal 261
  theoretical (also descriptive) 51
  theory (of) 35, 51
  well-structured xii
KRITIK 287

L
lags 320
language 300
  functional 300
  SBF 273
  spatial and functional representation 306
laws 159
  element constitutive 159
  power 320
  rate 320
layouts 306
  architectural 306
  dimensional 53
  sketch 53
leads 320
learning 277
length vector 183
levels (abstraction) 214
  embodiment 214
  functional 214
  physical principle 214
  requirement 214
life cycle engineering 50
light-emitting diodes 248
limiter 240
linguistics 306
links 203
LispWorks 188
literature, wider xii
lock 249
locker 264
locus types 236
  along range limit 236
  along region boundary 235
  through free region 237
logic 5
  inductive 5
  of systems methodology 6
  prepositional 11
  logical working space 78

M
magnitudes 182
manipulator (robotic) 205
mapping 246
  analogical 256

mapping *cont.*
  data   261
  rules   261
  unit (MU)   261
muscular dystrophy (MD)   188
matching   21, 258
  behaviour   258
  function   260
  pattern   238
matrix   312
means   ix
mechanical clocks   285
mechanism(s)   x, 288
  fixed-axis   290
  forward-reverse   288
  generic teleological (GTM)   271
  library   233
  operating modes   242
  scotch-yoke   183
  shutter release   233
  two-dimensional elementary   285
Merkmale   94
metabolic pathways   320
method(s)   140
  miniature-people   140
  iterative refinement   298
  for solving physical contradictions   141
  the standard   141
  for using and/or operating subject   50
  Yourdon structured   200
metric space approach   xiv
metrology   25
minimal characterisation (also compact representation)   165
mnemonic icons   176
mobile arm support (MAS)   188
mode of action   14
model(s)   80
  activity   213
  -based analogy (MBA)   271–84
    design patterns   271, 275
    evaluation   282
    GTM aquisition   276
    related research   282
    SBF   273
    transfer   271, 277
  behaviour–function (BF)   275
  behaviour-specific   276
  calculation   80
  chromosome   53
  cognitive design process   67
  computable design process   67
  design object   68
  dynamic   160
  dynamic simulation   199

frame recognition   85
  framing   85
  moving   85
  naming   85
  reflecting   85
geometric   303
knowledge operation   67
meta   43
qualitative   287
quantitative   287
reflection in action   xiii
situated   xiii
solid   307
structural   294
structure-behaviour-function (SBF)   271
thought-process
  analysis-oriented   67
  synthesis-oriented   67
topological   282
TRIZ   131
modelling   viii
  adaptive   283
  bond-graph   204
  design-object   77
  design-process   77
  engineering design process   67
  function   76
  function-behaviour-state (FBS)   76
  in media   78
  operations   68
  qualitative   286
  of synthesis   viii
modification
  of cases   288
  design   283
modularity   303
modulated sources of effort (MSE)   205
modules (production)   118
morphological
  charts   viii
  matrix   ix
  method   7
morphology   100
morphology of design   7
motion
  input   233
  leaping   237
  output   233
  part   290
  rotational   271
  speed dependence levels   234
  translational   271
  rotational   271
  types   232
multiple model-based reasoning   68, 78, 89

Index 349

multipliers   320
mutation   320

N
needs   215
nitric acid cooler   287
non-manifold   307
N-port   305
nuclear industry   205

O
object-dependent models   78
object-independent model   78
objective(s)   307
   design   307
   system   6
object-oriented programming   282
ontology   75
op-amp   276
open loop gain   276
operand   95
   biological objects   95
   data   95
   energy   95
   material   95
operations   321
   genetic   321
   logical   67
     derivation of solution from requirements and axioms   79
     derivation of theorems from axioms   79
     describing requirements   79
     extraction of facts   79
     formation of hypotheses (also selection of axioms)   79
     observation of phenomena   79
     selection of axioms   79
     verification of theorems against facts   79
     verification of theorems against other known axioms   79
     verification of theorems against requirements   79
   modelling   67
     building model   78
     introducing new modelling system   78
     maintenance of models in different modelling systems   78
     modification of model   78
     modification of knowledge base of modelling system   78
     reasoning about model   78
     selection of modelling system   78
operators   178

opportunism   19
optimal   311
ordering   246
organ(s)   99
   domain   94
   structures   105
     assisting   53
       auxiliary   53
       connecting   53
       controlling   53
       propelling   53
       regulating   53
     control   53
     propulsion   53
     working   53
     behaviour   97
     function   97
     properties   97
originality   188
oscillation   285
output(s)   220
   vector   183

P
p-aesthetic   21
pairs (kinematic)   285
paper punches   182
parameters   273
   engineering   311
   motion   230
   numerical   319
   orientation   230
part(s)   62
   behaviour   97
   domain   94
   shapes   285
passband ripple   324
patent analysis   133
pattern   271
   causal   271
   functional   271
   locus   234
   complete   217
   design   271
     causal   271
     functional   271
     topological   283
   incomplete   217
   object   217
   process   217
   solution   216
pawl   229
pendulum   285
peristaltic pump   203
perspiration   285

phase (of design process)   16
phenomena (physical)   7
philosophy of science   4
PHINEAS   283
phototropic spectacles   249
physical
    law   43
    representation   3, 16
    system   9
physico-chemical form   9
pilotless aircraft fin actuator   209
place vocabulary   242
plan (design)   245
planning (spatial)   230
plausible inference   9
pluggable metamodel mechanism   76
pneumatic cylinders   153
Poincaré's sieve (mechanical sieve)   22
population   321
    of compositions   321
    initial   321
    of offspring   321
    of programs   320
ports   203
position(s)   182, 230
power spine   162, 167, 172
predicates   287
"prefer pivots to slides and flexures to either"   27
preprocessing   191
prescription   181
pressure gauges   153
presumption of fact   9
primitives   207
    abstract   308
    behavioural   304, 308
    graph   307
        aggregrate   307
        redistribute   307
        decompose   307
principle(s)   321
    Darwinian   321
    of developmental biology   320
    incidental behaviour   310
    of natural selection   320
    no-function-in-structure   286
    physical   214
    solution   114
    technical   7
    working   7, 214
    working   7
Pro/Engineer   224
PROBER   257
problem(s)   222
    abductive   285

analysis   3
    combinatorial   308
    complex   222
    design   179
        target   272
    known design   181
    nature of   19
    real-world design   300
    solution-neutral   109
    unfamiliar   245
problem-oriented designknowledge   217
problem-solving   3, 6, 16
    cycle   3
    functions   16
        communicating results   16
        problem definition   16
        selecting objectives   16
        selecting best system   16
        systems analysis   16
        systems synthesis   16
    process   viii
process(es)   viii
    analogical reasoning   245
    dynamic structure mapping (DSMP)   262
    evolutionary   325
    extensional   47
    external   246
    intensional   47
    internal   246
    intuitive   49
    pattern-matching   238
    process structure mapping (PSMP)   263
    solution-finding   132, 220
    systematic   49
    technical   53
    transformation   56
ProDEVELOP   224
product   9
    compendia   x
    design   v
    family   107
    model   107
    planning   49
    programming   5
    propagation   204
        of information   204
        parallel   251
        series   251
    properties (and property)   10
        behavioural   273
        design   218
        external   56
        functional   38
        intensive   10

internal   56
object   43
observable   41
structural   94
to-be   215
protocol data   viii
prototypes   218
  design   218
  preprogrammed   287
prototyping   201
  rapid software   201
  virtual   199
  approach   xiv
pseudo-vector   183
punch-bowl   286
purpose   247

Q

quality   50
quality function deployment (QFD)   62
quantity   160
  input   160
  output   160
  physical   242
quasi-static   249

R

rachet wheel   229
rachets   286
range limit segments   231
reasoning   287
  case-based   287
  deductive   4
  ends to means   17
  form to function   11
  function to form   3, 16
  inductive   4
  levels
    action-level (also meta-level)   69
    object-level   69
  logical   68
  operations   89
  about physical principles   100
  plausible   9
  reductive (abductive and deductive)   71
  structure to behaviour   11
  symbolic   204
receptors   53
recombination   320
reconfiguration   306
redesign, redesigning   25, 65
reduction of fact   9
redundancy   188
reference model   viii
refinement process   37

reflectors   333
reformulation   131
region   231
  blocked   231
  boundary   230
  free   230
region diagram   242
reinterpretation   288
relation(s)   249
  assembly   97
  causal   245
  changeable   249
  contact   286
  dependency   250
  detachable   247
  fixed   249
  generic   271
  generic functional   271
  higher-order   245
  includes   273
  parallel-connected   273
  part-of   273
  structure–behaviour   94
relationship(s)   220
  attribute   76
  behaviour   276
  among concepts   76
  contact   290
  hierarchical (between (sub-)problems, or
    between (sub-)solutions)   103
  logical   115
  physical   115
  spatial   184
  symmetrical (between problem and
    solution)   103
  between values   230
representation   250
  behaviour (also behavioural)   250
  extensional   45
  geometrical   304
  intensional   45
  knowledge   246
  object-oriented   201
  physical   303
  qualitative configuration space   238
  shape   304
  spatial   306
  two-dimensional boundary   239
requirements   16, 303
  behavioural   303
  customers'   61
  functional   179
  initial design   245
  physical   303
  product   214

requirements *cont.*
  overall   111
  user-defined   322
resistor   319
retrieval   x, 229–43
  analogy   246
  behaviour-based   238
  design case   242
  catalogue   80
  compliance   236
  configuration spaces   230
    matching   234
  examples   238
  further behavioural information   232
  index   229
  key   233
  kinematic behaviour   231, 233
  locus patterns   234–8
  mechanism libraries   229
  motions   233–8
  related studies   242
  required behaviour   233
  timing charts   233
  see also analogy-based retrieval
retriever   259
  analogy   259
  concept   259
  design   259
reuse   209
robot gripper   214
robust design   144
rotating cylinder valve engine   21
rule(s)   209
  to apply   209
  bond-graph-based   199
  debugging   173
  design   306
  of envisionment   xi
  of replacement   xi

S

scales   264
scape wheel   285
schema   273
schematic
  descriptions   153
    synthesis problem   x, 153–77
    bond graphs   176
    completeness   171
    computer implementation   173
    domain description   157
    example   167
    example specification   161
    extensibility   172
    importance and utility   156, 171

power spine   162, 167, 172
problem domain   153
related work   174
schematic synthesis defined   154
SISO systems   155, 157
solution technique   153, 162
type number   160, 175
Schemebuilder *see* function–means-based synthesis
schemes   200
search   xiv
  adaptive   xiv
  bi-directional   191
  unidirectional   191
selection   xii
self-locking   310
semantic information   81
sensors   x
servo controller   211
sewing machine   234
sharing   171
  function   171
  of parts   229
  of subassemblies   229
signal(s) (*see also* data, information)   109
  flow of   114
  reference   320
  sinusoidal   326
  time-varying input   329
similarity   192
  concepts of   192
  grouping solutions using   192
  literal   245
  measures of   179
  morphological   192
  structural   307
simplification   242
simulated annealing   81
simulation   204
  composable   204
  qualitative   273
simulators   204
simulink   204
SIT   146
situation (design)   271
sizing   319
slider and wheel   240
small, fast principle   26
solid model   88
solution(s)   215
  alternative   180
  basic technical function   121
  known design   181
  optimum design   121
  partial design   199

Index 353

(physical) as product behaviour based on
    physical laws 215
  principles 109
  redundant 188
  search 109
  similar 190
  subsystem 200
  synthesis 79
solution-neutral formulation 109–10
space 229
  attribute 38
  closure 46
  configuration 229
  design searching 245
  function 38
  mathematical 242
  physical 234
  quantity dimension 242
  search 286
specification 277
  functional 277
  numerical 173
  necessary 294
  parameter 238
  sufficient 294
  system 307
speedometers 153
SPICE 325
spur gear 229
standard terms 89
state transitions 77
state(s) 273
  behavioural 273
  kinematic 291
    no contacts 291
    one contact 291
    two contacts 291
  of mechanism 230
statements
  design and debug 306
  hypothetical 10
  normative 10
  statistical generalisation 9
  statistical specification 9
  stopband ripple 324
  strategy 306
  transformational 303
string copy program 248
structure(s) 271
  abstract 271
  causal 175
  complex 319
  constructional 53
  data 247
  design 247

design conjecture 259
  dynamic 249
  function ix
  junction 304
  organ 53
  physical 271
  process 247
modular product 107
  relational 245
  static 249
  topological, of objects 38
structure–function relation 37
structure-mapping engine (SME) 283
subfunctions 111
subproblem 9
subspaces 220
substances 273
subsumptions 298
subsystem 3
subtracters 320
support of design synthesis 213–27
  design process 213
  design working spaces 220
  DIICAD Entwurf system 224
  object patterns 217
  process patterns 218
  solution patterns 217
symbolic
  expressions (also s-expressions) 334
  geometry x
SYN 282
synthesis 93
  activities 93
  as activity 35
  vs. analysis 37, 68
  approach xii
    artefact-oriented 94
    design-process-oriented 94
  assembly as 7
  automatic 319
  bond graphs for 305
  compositional 179, 191
  computational x
    as design problem and solution
      generation xi
    as design solution generation xi
    as designing xi
    as exploration xi
    as problem solving xi
  as converging process 46
  defined viii, xi, 16, 35, 37
  defining 3–17
  design 245
  in design process 4, 16
  etymology 4

synthesis *cont.*
  example   11, 13
  as function of problem-solving   3
  as function   7
  functional   196, 205
  general sense   3
  human aspect of   93
  hydraulic circuit   209
  influencing factors   xii
  as integration   8
  logic of   viii
  model of   ix
  modelling   viii, 67–89
    formal   69
    function   76
    implementation strategy   80
    multiple model-based reasoning   74
    reasoning framework   77
    reference model   68, 85, 89
    related work   68, 69
    thought-process models   73
    verification and testing   84, 87, 89
  nature of   xii
  objective of   6
  pattern of reasoning   3
  as phase of design process   3
  in problem solving   3, 6, 16
  as reasoning   9, 11
  schematic   154
  schemes   202 of
  scientific model of   67
  steps
    concretisation   101
    detailing   101
    from one domain to another   101
  synthesised artefact   93
  topological   199
synthesis-oriented thought-process model   68, 89
system(s)   202
  air circulation   282
  associative   xiv
  assumption truth maintenance (ATMS)   202
  boundary   110
  CAD   221
  chemical control   249
  configuration   107
  context management   199
  control   271
  deductive   9
  design   245
  dynamic   153
  Edinburgh Designer (EDS)   201
  energy-transforming   204
  flow   8
  hydraulic   x
  hypothetical   6
  IDeAL   271
  intelligent design support   67
  interactive   259
  KBDS   204
  KEEworlds   202
  life cycle   6
  lumped-parameter   157
  machine   54
  man–machine   100
  MAX   204
  mechanical   x
  mechanism retrieval   229
  mechatronic   x, 199
  modelling and simulation   204
  modular   118
  motor   23
  multidisciplinary   199
  natural   213
  non-linear   300
  Parsytec computer   326
  physical   305
  proprioceptive (self-sensing or feedback)   23
  single-input–single-output (SISO)   x
  software   x
  structural   x
  technical   4, 50
  technological   131
  theory   94
  three degrees of freedom   205
  transformation   53
    actions   53
    operand   53
    operations   53
    operators   53
    secondary inputs and outputs   53
  two degrees of freedom   205
  visual   23
systematic
  design processes, checking   32
  invention   19
  synthesis   28

T
table of options   30
task (domain theory)   97
technical systems   49–66
  causality   62
  clarifying   63
  classes   53
  conceptualising   63
  degree of complexity   53

design specification 61
detailing 64
development 56
embodying 64
engineering design 50
finality 62
functions 53
general 56
goal–means transitions 62
knowledge 51
life cycle 56
machine elements 54
models 51
operators 60
organs 53
output effects 53
phases 61
procedure 60
properties 56
science 53
structures 53
technical processes 53
theory viii, 49–66
technology 96
techoptimizer 146
telechiric manipulator 199
tentative solutions 9, 16
terminals 321
theodolite 26
theorems 72
theory (and theories) 283
 adaptive modelling 283
 of artefact being designed 93
 of axiomatic design viii
 axiomatic set 67
 case-based 283
 describing design process 93
 design 3
  methodology 95
 domain 93
 elements 106
 general design (GDT) viii
 of inventive problem solving (TRIZ) 131
 knowledge level, of designing xiii
 MBA 271
 normative 271
 normative computational 283
 qualitative process 204
 structure mapping 245
 systems x
 of technical systems viii
 universal design (UDT) 69
thermodynamics 300
thinking 107

creative 107
lateral 268
thought-process models 68–89
time constant 332
time-to-market 131
timing charts 233
tool(s) 194
 browsing and clustering 194
 problem-solving 141
topology 332
 controller 332
 of constraint network 296
 kinematic 292
 power path 312
torque converters 305
transference 246
transformation(s) x, xi, 95, 303–16
 approach 303
 behaviour 95
 behaviour-preserving 311
 bond graphs 305
 component-directed 311
 design 303
 domain 94
 example 308
 graph 307
 matrix 312
 physical descriptions 310
 related work 304
 representation 305, 309
transistor 319
transmissions 286
TRIZ ix, 131–48
 advantages 132
 after Altshuller 145
 ARIZ algorithm 141
 associated fields 145
 basic notions 134
 case study 137
 classification 133
 creativity 147
 derivation 131
 as design asset 147
 development 145
 genesis 131
 integration 142, 144, 146
 key points 144
 laws of development 134, 137
 literature 148
 potential 144
 problem-solving 141
 psychological inertia 140
tuning 320
turbine
 steam 23

turbine *cont.*
  gas   26
  wind   26
two-lag plant   331
type number   160
type number   160, 175

U
understanding   285
unit design cycle   69
  steps
    awareness of problem   69
    decision   69
    development   69
    evaluation   69
    suggestion   69
universal design theory (UDT)   216
universal virtues   96
use   10
  conditions of   10
  mode of   10
user interface (UI)194
utilising basic problem solving principles   100

V
vacuum cleaners   305
value engineering (VE)   143
variable(s)   248
  attribute   248
  behaviour   250
    exogenous (indirect)   250
    structural (direct)   250
  effort   158
  element   247
  flow   158
  relation   247
variants   188
variation
  methods   100
  operations   121
VDI2221   7
vector   290
  transformer   242
vepoles   141
verb–noun pairs   118
vibration absorber   307
visual thinking   22, 23
visual variation methods   103
visualisation   180
vocabulary   86
voltage standing wave ratio   333
voltmeter   160
VT   307

W
wall light fitting   262
water faucet   264
wave energy collector (WEC)   19
wave energy collector   19, 24
"where possible, transfer complexity to the software"   28
wirk elements   96
  wirk field   96
  wirk surface   96
  wirk volume   96
wirkung   96
working space   x
worlds (also contexts)   202
worm and wheel   232